Graduate Texts in Mathematics 31

Graduate Texts in Mathematics

1 TAKEUTI/ZARING. Introduction to Axiomatic Set Theory. 2nd ed.
2 OXTOBY. Measure and Category. 2nd ed.
3 SCHAEFFER. Topological Vector Spaces.
4 HILTON/STAMMBACH. A Course in Homological Algebra.
5 MACLANE. Categories for the Working Mathematician.
6 HUGHES/PIPER. Projective Planes.
7 SERRE. A Course in Arithmetic.
8 TAKEUTI/ZARING. Axiometic Set Theory.
9 HUMPHREYS. Introduction to Lie Algebras and Representation Theory.
10 COHEN. A Course in Simple Homotopy Theory.
11 CONWAY. Functions of One Complex Variable. 2nd ed.
12 BEALS. Advanced Mathematical Analysis.
13 ANDERSON/FULLER. Rings and Categories of Modules.
14 GOLUBITSKY/GUILLEMIN. Stable Mappings and Their Singularities.
15 BERBERIAN. Lectures in Functional Analysis and Operator Theory.
16 WINTER. The Structure of Fields.
17 ROSENBLATT. Random Processes. 2nd ed.
18 HALMOS. Measure Theory.
19 HALMOS. A Hilbert Space Problem Book. 2nd ed., revised.
20 HUSEMOLLER. Fibre Bundles. 2nd ed.
21 HUMPHREYS. Linear Algebraic Groups.
22 BARNES/MACK. An Algebraic Introduction to Mathematical Logic.
23 GREUB. Linear Algebra. 4th ed.
24 HOLMES. Geometric Functional Analysis and its Applications.
25 HEWITT/STROMBERG. Real and Abstract Analysis.
26 MANES. Algebraic Theories.
27 KELLEY. General Topology.
28 ZARISKI/SAMUEL. Commutative Algebra. Vol. I.
29 ZARISKI/SAMUEL. Commutative Algebra. Vol. II.
30 JACOBSON. Lectures in Abstract Algebra I: Basic Concepts.
31 JACOBSON. Lectures in Abstract Algebra II: Linear Algebra.
32 JACOBSON. Lectures in Abstract Algebra III: Theory of Fields and Galois Theory.
33 HIRSCH. Differential Topology.
34 SPITZER. Principles of Random Walk. 2nd ed.
35 WERMER. Banach Algebras and Several Complex Variables. 2nd ed.
36 KELLEY/NAMIOKA et al. Linear Topological Spaces.
37 MONK. Mathematical Logic.
38 GRAUERT/FRITZSCHE. Several Complex Variables.
39 ARVESON. An Invitation to C^*-Algebras.
40 KEMENY/SNELL/KNAPP. Denumerable Markov Chains. 2nd ed.
41 APOSTOL. Modular Functions and Dirichlet Series in Number Theory.
42 SERRE. Linear Representations of Finite Groups.
43 GILLMAN/JERISON. Rings of Continuous Functions.
44 KENDIG. Elementary Algebraic Geometry.
45 LOÈVE. Probability Theory I. 4th ed.
46 LOÈVE. Probability Theory II. 4th ed.
47 MOISE. Geometric Topology in Dimensions 2 and 3.

continued after Index

Nathan Jacobson

Lectures in Abstract Algebra

II. Linear Algebra

Springer-Verlag
New York Berlin Heidelberg Tokyo

Nathan Jacobson

Yale University
Department of Mathematics
New Haven, Connecticut 06520

AMS Subject Classification
15-01

Library of Congress Cataloging in Publication Data

Jacobson, Nathan, 1910–
Lectures in abstract algebra.
(Graduate texts in mathematics; 31)
Reprint of the 1951-1964 ed. published by Van Nostrand,
New York in The University series in higher mathematics,
M. A. Stone, L. Nirenberg, and S. S. Chern, eds.
Includes index.
CONTENTS: ---2. Linear
algebra.
1. Algebra, Abstract. I. Title. II. Series.
QA162.J3 1975 512'.02 75-15564

Printed and bound by Halliday Lithograph, West Hanover, Massachusetts.
Printed in the United States of America.

9 8 7 6 5 4 3 2 (Second printing, 1984)

ISBN 0-387-90123-X Springer-Verlag New York Berlin Heidelberg Tokyo
ISBN 3-540-90123-X Springer-Verlag Berlin Heidelberg New York Tokyo

TO

MICHAEL

PREFACE

The present volume is the second in the author's series of three dealing with abstract algebra. For an understanding of this volume a certain familiarity with the basic concepts treated in Volume I: groups, rings, fields, homomorphisms, is presupposed. However, we have tried to make this account of linear algebra independent of a detailed knowledge of our first volume. References to specific results are given occasionally but some of the fundamental concepts needed have been treated again. In short, it is hoped that this volume can be read with complete understanding by any student who is mathematically sufficiently mature and who has a familiarity with the standard notions of modern algebra.

Our point of view in the present volume is basically the abstract conceptual one. However, from time to time we have deviated somewhat from this. Occasionally formal calculational methods yield sharper results. Moreover, the results of linear algebra are not an end in themselves but are essential tools for use in other branches of mathematics and its applications. It is therefore useful to have at hand methods which are constructive and which can be applied in numerical problems. These methods sometimes necessitate a somewhat lengthier discussion but we have felt that their presentation is justified on the grounds indicated. A student well versed in abstract algebra will undoubtedly observe short cuts. Some of these have been indicated in footnotes.

We have included a large number of exercises in the text. Many of these are simple numerical illustrations of the theory. Others should be difficult enough to test the better students. At any rate a diligent study of these is essential for a thorough understanding of the text.

At various stages in the writing of this book I have benefited from the advice and criticism of many friends. Thanks are particularly due to A. H. Clifford, to G. Hochschild, and to I. Kaplansky for suggestions on earlier versions of the text. Also I am greatly indebted to W. H. Mills, Jr. for painstaking help with the proofs and for last minute suggestions for improvements of the text.

N. J.

New Haven, Conn.
September, 1952

CONTENTS

CHAPTER IV: SETS OF LINEAR TRANSFORMATIONS

CHAPTER V: BILINEAR FORMS

CHAPTER VI: EUCLIDEAN AND UNITARY SPACES

CHAPTER VII: PRODUCTS OF VECTOR SPAÇES

Chapter I

FINITE DIMENSIONAL VECTOR SPACES

In three-dimensional analytic geometry, vectors are defined geometrically. The definition need not be recalled here. The important fact from the algebraic point of view is that a vector v is completely determined by its three coördinates (ξ, η, ζ) (relative to a definite coordinate system). It is customary to indicate this by writing $v = (\xi, \eta, \zeta)$, meaning thereby that v is the vector whose x-, y-, and z-coordinates are, respectively, ξ, η, and ζ. Conversely, any ordered triple of real numbers (ξ, η, ζ) determines a definite vector. Thus there is a 1–1 correspondence between vectors in 3-space and ordered triples of real numbers.

There are three fundamental operations on vectors in geometry: addition of vectors, multiplication of vectors by scalars (numbers) and the scalar product of vectors. Again, we need not recall the geometric definitions of these compositions. It will suffice for our purposes to describe the algebraic processes on the triples that correspond to these geometric operations. If $v = (\xi, \eta, \zeta)$ and $v' = (\xi', \eta', \zeta')$, then the *sum*

$$v + v' = (\xi + \xi', \eta + \eta', \zeta + \zeta').$$

The *product* ρv of the vector v by the real number ρ is the vector

$$\rho v = (\rho\xi, \rho\eta, \rho\zeta)$$

and the *scalar product* (v, v') of v and v' is the real number

$$(v, v') = \xi\xi' + \eta\eta' + \zeta\zeta'.$$

A substantial part of analytic geometry—the theory of linear dependence and of linear transformations—depends only on the first two of these concepts. It is this part (in a generalized form) which constitutes the main topic of discussion in these Lectures. The concept of scalar product is a metric one, and this will be relegated to a relatively minor role in our discussion.

The study of vectors relative to addition and multiplication by numbers can be generalized in two directions. First, it is not necessary to restrict oneself to the consideration of triples; instead, one may consider n-tuples for any positive integer n. Second, it is not necessary to assume that the coordinates ξ, η, $\cdots$ are real numbers. To insure the validity of the theory of linear dependence we need suppose only that it is possible to perform rational operations. Thus any field can be used in place of the field of real numbers. It is fairly easy to go one step further, namely, to drop the assumption of commutativity of the basic number system.

We therefore begin our discussion with a given division ring Δ. For example, Δ may be taken to be any one of the following systems: 1) the field of real numbers, 2) the field of complex numbers, 3) the field of rational numbers, 4) the field of residues modulo p, or 5) the division ring of real quaternions.

Let n be a fixed positive integer and let $\Delta^{(n)}$ denote the totality of n-tuples $(\xi_1, \xi_2, \cdots, \xi_n)$ with the ξ_i in Δ. We call these n-tuples *vectors*, and we call $\Delta^{(n)}$ *the vector space of n-tuples over* Δ. If $y = (\eta_1, \eta_2, \cdots, \eta_n)$, we regard $x = y$ if and only if $\xi_i = \eta_i$ for $i = 1, 2, \cdots, n$. Following the pattern of the three-dimensional real case, we introduce two compositions in $\Delta^{(n)}$: addition of vectors and multiplication of vectors by elements of Δ. First, if x and y are arbitrary vectors, we define their *sum* $x + y$ to be the vector

$$x + y = (\xi_1 + \eta_1, \xi_2 + \eta_2, \cdots, \xi_n + \eta_n).$$

As regards to multiplication by elements of Δ there are two possibilities: *left multiplication* defined by

$$\rho x = (\rho\xi_1, \rho\xi_2, \cdots, \rho\xi_n)$$

and right multiplication defined by

$$x\rho = (\xi_1\rho, \xi_2\rho, \cdots, \xi_n\rho).$$

Either of these can be used. Parallel theories will result from the two choices. In the sequel we give preference to left multiplication. It goes without saying that all of our results may be transferred to results on right multiplication.

The first eight chapters of this volume will be devoted to the study of the systems $\Delta^{(n)}$ relative to the compositions we have just defined. The treatment which we shall give will be an axiomatic one in the sense that our results will all be derived from a list of simple properties of the systems $\Delta^{(n)}$ that will serve as axioms. These axioms define the concept of a *finite dimensional (abstract) vector space* and the systems $\Delta^{(n)}$ are instances of such spaces. Moreover, as we shall see, any other instance of a finite dimensional vector space is essentially equivalent to one of the systems $\Delta^{(n)}$.

Thus the shift to the axiomatic point of view is not motivated by the desire to gain generality. Its purposes are rather to clarify the discussion by focusing attention on the essential properties of our systems, and to make it easier to apply the results to other concrete instances. Finally, the broadening of the point of view leads naturally to the consideration of other, more general, concepts which will be useful in studying vector spaces. The most important of these is the concept of a module which will be our main tool in the theory of a single linear transformation (Chapter III). In order to prepare the ground for this application we shall consider this concept from the beginning of our discussion.

The present chapter will be devoted to laying the foundations of the theory of vector spaces. The principal concepts that we shall consider are those of basis, linear dependence, subspace, factor space and the lattice of subspaces.

1. Abstract vector spaces. We now list the properties of the compositions in $\Delta^{(n)}$ from which the whole theory of these systems will be derived. These are as follows:

A1 $$(x + y) + z = x + (y + z).$$

A2 $$x + y = y + x.$$

A3 There exists an element 0 such that $x + 0 = x$ for all x.

A4 For any vector x there exists a vector $-x$ such that $x + (-x) = 0$.

S1 $$\alpha(x + y) = \alpha x + \alpha y.$$

S2 $$(\alpha + \beta)x = \alpha x + \beta x.$$

S3 $$(\alpha\beta)x = \alpha(\beta x).$$

S4 $$1x = x.$$

F There exist a finite number of vectors $e_1, e_2, \cdots, e_n$ such that every vector can be written in one and only one way in the form $\xi_1 e_1 + \xi_2 e_2 + \cdots + \xi_n e_n$.

The verifications of A1, A2, S1–S4 are immediate. We can prove A3 by observing that $(0, 0, \cdots, 0)$ has the required property and A4 by noting that, if $x = (\xi_1, \cdots, \xi_n)$, then we can take $-x = (-\xi_1, \cdots, -\xi_n)$. To prove F we choose for e_i,

$$e_i = (0, 0, \cdots, 0, \overset{i}{1}, 0, \cdots, 0), \quad i = 1, 2, \cdots, n. \tag{1}$$

Then $\xi_i e_i$ has ξ_i in its ith place, 0's elsewhere. Hence $\sum_1^n \xi_i e_i = (\xi_1, \xi_2, \cdots, \xi_n)$. Hence if $x = (\xi_1, \xi_2, \cdots, \xi_n)$, then x can be written as the "linear combination" $\Sigma \xi_i e_i$ of the vectors e_i. Also our relation shows that, if $\Sigma \xi_i e_i = \Sigma \eta_i e_i$, then $(\xi_1, \xi_2, \cdots, \xi_n) = (\eta_1, \eta_2, \cdots, \eta_n)$ so that $\xi_i = \eta_i$ for $i = 1, 2, \cdots, n$. This is what is meant by the uniqueness assertion in F.

The properties A1–A4 state that $\Delta^{(n)}$ is a commutative group under the composition of addition. The properties S1–S4 are properties of the multiplication by elements of Δ and relations between this composition and the addition composition. Property F is the fundamental finiteness condition.

We shall now use these properties to define an *abstract vector space*. By this we mean a system consisting of 1) a commutative group $\mathfrak{R}$ (composition written as $+$), 2) a division ring Δ, 3) a function defined for all the pairs (ρ, x), ρ in Δ, x in $\mathfrak{R}$, having values ρx in $\mathfrak{R}$ such that S1–S4 hold. In analogy with the geometric case of n-tuples we call the elements of $\mathfrak{R}$ *vectors* and the elements of Δ *scalars*. In our discussion the emphasis will usually be placed on $\mathfrak{R}$. For this reason we shall also refer to $\mathfrak{R}$ somewhat inexactly as a "vector space over the division ring Δ." (Strictly speaking $\mathfrak{R}$ is only the group part of the vector space.) If F holds in addition to the other assumptions, then we say that $\mathfrak{R}$ is *finite dimensional*, or that $\mathfrak{R}$ *possesses a finite basis over* Δ.

The system consisting of $\Delta^{(n)}$, Δ, and the multiplication ρx defined above is an example of a finite dimensional vector space. We shall describe next a situation in the theory of rings which gives rise to vector spaces. Let $\mathfrak{R}$ be an arbitrary ring with an identity element 1 and suppose that $\mathfrak{R}$ contains a division subring Δ that contains 1. For the product ρx, ρ in Δ, and x in $\mathfrak{R}$ we take the ring product ρx. Then S1–S3 are consequences of the distributive and associative laws of multiplication, and S4 holds since the identity element of Δ is the identity of $\mathfrak{R}$. Hence the additive group $\mathfrak{R}$, the division ring Δ and the multiplication ρx constitute a vector space. This space may or may not be finite dimensional. For example, if $\mathfrak{R}$ is the field of complex numbers and Δ is the subfield of real numbers, then $\mathfrak{R}$ is finite dimensional; for any complex number can be written in one and only one way as $\xi + \eta\sqrt{-1}$ in terms of the "vectors" $1, \sqrt{-1}$. Another example of this type is $\mathfrak{R} = \Delta[\lambda]$, the polynomial ring in the transcendental element (indeterminate) λ with coefficients in the division ring Δ. We shall see that this vector space is not finite dimensional (see Exercise 1, p. 13). Similarly we can regard the polynomial ring $\Delta[\lambda_1, \lambda_2, \cdots, \lambda_r]$ where the λ_i are algebraically independent (independent indeterminates) as a vector space over Δ.

Other examples of vector spaces can be obtained as subspaces of the spaces defined thus far. Let $\mathfrak{R}$ be any vector space over Δ and let $\mathfrak{S}$ be a subset of $\mathfrak{R}$ that is a subgroup and that is closed under multiplication by elements of Δ. By this we mean that if $y \in \mathfrak{S}$ and ρ is arbitrary in Δ then $\rho y \in \mathfrak{S}$. Then it is clear that the trio consisting of $\mathfrak{S}$, Δ and the multiplication ρy is a vector space; for, since S1–S4 hold in $\mathfrak{R}$, it is obvious that they hold also in the subset $\mathfrak{S}$. We call this a *subspace* of the given vector space, and also we shall call $\mathfrak{S}$ a subspace of $\mathfrak{R}$. As an example, let $\mathfrak{R} = \Delta[\lambda]$ and let $\mathfrak{S}$ be the subset of polynomials of degree $< n$. It is immediate that $\mathfrak{S}$ is a subspace. Moreover, it is finite dimensional since any polynomial of degree $<n$ can be expressed in one and only one way as a linear combination of the polynomials $1, \lambda, \cdots, \lambda^{n-1}$.

EXERCISE

1. Show that the totality $\mathfrak{S}$ of homogeneous quadratic polynomials $\sum_{i \leq j} \alpha_{ij}\lambda_i\lambda_j$, α_{ij} in Δ, is a finite dimensional subspace of $\Delta[\lambda_1, \lambda_2]$.

2. Right vector spaces. As we have pointed out at the beginning the system $\Delta^{(n)}$ of n-tuples can also be studied relative to addition and to right multiplication by scalars. This leads us to define the concept of a *right vector space*. By this we mean a system consisting of a commutative group $\mathfrak{R}'$, a division ring Δ and a function of pairs (ρ, x'), ρ in Δ, x' in $\mathfrak{R}'$, having values $x'\rho$ in $\mathfrak{R}'$ and satisfying:

S′1 $$(x' + y')\alpha = x'\alpha + y'\alpha.$$

S′2 $$x'(\alpha + \beta) = x'\alpha + x'\beta.$$

S′3 $$x'(\alpha\beta) = (x'\alpha)\beta.$$

S′4 $$x'1 = x' \text{ for all } x' \text{ in } \mathfrak{R}'.$$

Obviously the theory based on this definition will parallel that of left vector spaces. It should be noted, however, that a right space over Δ cannot be regarded as a left space over Δ if this division ring is not commutative. For if we write $\alpha x'$ for $x'\alpha$, then we have by S′3

$$(\alpha\beta)x' = x'(\alpha\beta) = (x'\alpha)\beta = \beta(\alpha x').$$

Hence S3: $(\beta\alpha)x' = \beta(\alpha x')$ holds only if

$$[(\alpha\beta) - (\beta\alpha)]x' = 0$$

for all x'. This together with S4 implies that $\alpha\beta = \beta\alpha$ for all α, β.

On the other hand, let Δ' be a division ring anti-isomorphic to Δ and let $\alpha \rightarrow \alpha'$ be any anti-isomorphism of Δ onto Δ'. Then if $\mathfrak{R}'$ is a right vector space over Δ, $\mathfrak{R}'$ may be considered a left vector space over Δ'. This can be done by defining $\alpha' x'$ to be $x'\alpha$. Then

$$(\alpha'\beta')x' = (\beta\alpha)'x' = x'(\beta\alpha) = (x'\beta)\alpha = (\beta' x')\alpha = \alpha'(\beta' x'),$$

so that S3 is now satisfied. The verification of the other rules is also immediate.

3. $\mathfrak{o}$-modules. Before embarking on the systematic study of finite dimensional vector spaces we shall consider briefly the generalization to modules which will be very useful later on. This generalization is obtained by replacing in our definition the division ring Δ by any ring $\mathfrak{o}$ that has an identity. Thus we define a (left) *$\mathfrak{o}$-module* to be a system consisting of a commutative group $\mathfrak{R}$, a ring $\mathfrak{o}$ with an identity and a function of pairs (ρ, x), ρ in $\mathfrak{o}$ and x in $\mathfrak{R}$ with values ρx in $\mathfrak{R}$ satisfying S1–S4.* It is evident from our definitions that a vector space is simply a Δ-module where Δ is a division ring.

Besides the special case of a vector space we note the following important instance of an $\mathfrak{o}$-module: Let $\mathfrak{R}$ be any commutative group written additively and let $\mathfrak{o}$ be the ring of integers. If $x \in \mathfrak{R}$ and $\alpha \in \mathfrak{o}$, we define

$$\alpha x = \begin{cases} x + x + \cdots + x, & \alpha \text{ times if } \alpha > 0 \\ 0 \text{ if } \alpha = 0 & \\ -(x + x + \cdots + x), & -\alpha \text{ times if } \alpha < 0. \end{cases}$$

Then S1–S4 are the well-known laws of multiples in $\mathfrak{R}$.

We note also that any ring with an identity $\mathfrak{o}$ can be regarded as an $\mathfrak{o}$-module. As the group part $\mathfrak{R}$ we take the additive group of $\mathfrak{o}$ and we define ax for a in $\mathfrak{o}$ and x in $\mathfrak{R}$ to be the ring product. Properties S1–S4 are immediate consequences of the associative, distributive and identity laws for multiplication.

As in the case of vector spaces a subset $\mathfrak{S}$ of a module $\mathfrak{R}$ determines a *submodule* if $\mathfrak{S}$ is a subgroup of $\mathfrak{R}$ that is closed relative to the multiplication by arbitrary elements of $\mathfrak{o}$. Now let

* This definition is a slight departure from the usual one in which $\mathfrak{o}$ need not have an identity and only S1–S3 are assumed. We make this change here since we shall be interested only in rings with identities in this volume. *Right $\mathfrak{o}$-modules* are obtained in the obvious way by replacing S1–S4 by S′1–S′4.

$S = (x_\alpha)$ be an arbitrary subset of $\mathfrak{R}$ and let $[S]$ denote the totality of sums of the form

$$\xi_1 x_{\alpha_1} + \xi_2 x_{\alpha_2} + \cdots + \xi_m x_{\alpha_m} \tag{2}$$

where the ξ_i are arbitrary in $\mathfrak{o}$ and the x_{α_i} are arbitrary in S. We assert that $[S]$ is a submodule. Clearly $[S]$ is closed under addition and under multiplication by elements of $\mathfrak{o}$. Also it is easy to see (Exercise 1, below) that $0x = 0$ and $(-\xi)x = -\xi x$ hold in any module, and these imply that $[S]$ contains 0 and the negative of any element in $[S]$. Hence $[S]$ is a submodule of $\mathfrak{R}$. We note also that $[S]$ contains the elements $x_\alpha = 1x_\alpha$ of S and that $[S]$ is contained in every submodule of $\mathfrak{R}$ that contains S. Because of these properties we shall say that $[S]$ is the submodule *generated by the set* S.

If $[S] = \mathfrak{R}$, then the set S is said to be a *set of generators* for $\mathfrak{R}$. If $\mathfrak{R} = [e_1, e_2, \cdots, e_n]$ for some finite set $S = (e_1, e_2, \cdots, e_n)$, then we say that $\mathfrak{R}$ is *finitely generated.* If there exists a set of generators S such that every x can be written in one and only one way in the form $\Sigma \xi_i e_{\alpha_i}$, e_{α_i} in S, then $\mathfrak{R}$ is called a *free module* and the set S is called a *basis.* Thus condition F states that a finite dimensional vector space is a free Δ-module with a finite basis.

It is easy to construct, for any n, a free $\mathfrak{o}$-module with n base elements. The construction is the same as that of $\Delta^{(n)}$. We let $\mathfrak{o}^{(n)}$ denote the totality of n-tuples $(\xi_1, \xi_2, \cdots, \xi_n)$ with components ξ_i in $\mathfrak{o}$. Addition and multiplication by elements of $\mathfrak{o}$ are defined as before. If the e_i are defined by (1), it can be seen as in the case of $\Delta^{(n)}$ that these elements serve as a basis for $\mathfrak{o}^{(n)}$.

We consider now the fundamental concept of equivalence for $\mathfrak{o}$-modules. Let $\mathfrak{R}$ and $\bar{\mathfrak{R}}$ be two $\mathfrak{o}$-modules defined with respect to the same ring $\mathfrak{o}$. We shall say that $\mathfrak{R}$ and $\bar{\mathfrak{R}}$ are $\mathfrak{o}$-*isomorphic* or simply *equivalent* if there is a 1–1 correspondence, $x \to \bar{x}$ of $\mathfrak{R}$ onto $\bar{\mathfrak{R}}$ such that

$$\overline{x + y} = \bar{x} + \bar{y}, \quad \overline{\alpha x} = \alpha \bar{x}. \tag{3}$$

Thus $x \to \bar{x}$ is an isomorphism between the groups $\mathfrak{R}$ and $\bar{\mathfrak{R}}$ satisfying $\overline{\alpha x} = \alpha \bar{x}$ for all α and x. Such a mapping will be called an $\mathfrak{o}$-*isomorphism* or an *equivalence.*

If $x = \Sigma\alpha_i e_i$, then by (3) $\bar{x} = \overline{\Sigma\alpha_i e_i} = \Sigma\overline{\alpha_i e_i} = \Sigma\alpha_i\bar{e}_i$. Hence if the elements e_i are generators for $\mathfrak{R}$, then the corresponding elements $\bar{e}_i$ are generators for $\bar{\mathfrak{R}}$. If $\Sigma\alpha_i\bar{e}_i = \Sigma\beta_i\bar{e}_i$, then $\Sigma\alpha_i e_i = \Sigma\beta_i e_i$. It follows from this that, if $\mathfrak{R}$ is a free module with basis e_i, then $\bar{\mathfrak{R}}$ is free with basis $\bar{e}_i$. These remarks illustrate the general principle that equivalent modules have the same properties, and need not be distinguished in our discussion.

Suppose now that $\mathfrak{R}$ and $\bar{\mathfrak{R}}$ are two free $\mathfrak{o}$-modules and suppose that both of these modules have bases of n elements. Let the basis for $\mathfrak{R}$ be $e_1, e_2, \cdots, e_n$ and that for $\bar{\mathfrak{R}}$ be $\bar{e}_1, \bar{e}_2, \cdots, \bar{e}_n$. Then if x is any element of $\mathfrak{R}$, we write $x = \Sigma\xi_i e_i$, and we associate with this element the element $\bar{x} = \Sigma\xi_i\bar{e}_i$ of $\bar{\mathfrak{R}}$. Since the e_i and the $\bar{e}_i$ are bases, this correspondence is 1–1 of $\mathfrak{R}$ onto $\bar{\mathfrak{R}}$. Moreover, if $y = \Sigma\eta_i e_i$, then $\bar{y} = \Sigma\eta_i\bar{e}_i$ while $x + y = \Sigma(\xi_i + \eta_i)e_i$ and

$$\overline{x + y} = \Sigma(\xi_i + \eta_i)\bar{e}_i = \Sigma\xi_i\bar{e}_i + \Sigma\eta_i\bar{e}_i = \bar{x} + \bar{y}.$$

Also

$$\overline{\alpha x} = \Sigma(\alpha\xi_i)\bar{e}_i = \alpha\Sigma\xi_i\bar{e}_i = \alpha\bar{x}.$$

Hence $\mathfrak{R}$ and $\bar{\mathfrak{R}}$ are equivalent. This proves the following

Theorem 1. *Any two free* $\mathfrak{o}$*-modules which have bases of* n *elements are equivalent.*

In particular we see that any finite dimensional vector space with a basis of n elements is equivalent to the space $\Delta^{(n)}$ of n-tuples. This substantiates the assertion made before that the study of finite dimensional vector spaces is equivalent to the study of the concrete systems $\Delta^{(n)}$.*

EXERCISES

1. Prove the following rules for any $\mathfrak{o}$-module: 1) $\alpha 0 = 0$, 2) $\alpha(-x) = -\alpha x$, 3) $0x = 0$, 4) $(-\alpha)x = -\alpha x$.
2. Show that any subset of an $\mathfrak{o}$-module which is closed relative to addition and to multiplication by elements of $\mathfrak{o}$ is a submodule.
3. If $\mathfrak{R}$ is a vector space, then $\alpha x = 0$ only if $\alpha = 0$ or $x = 0$.

4. Linear dependence. From now on, unless otherwise stated, $\mathfrak{R}$ will be a finite dimensional vector space over Δ with basis e_1,

* A fuller account of the theory of modules can be found in Chapter VI of Volume I of these Lectures. However, the present discussion should be adequate for our purposes.

$e_2, \cdots, e_n$. It is easy to see that this basis is not uniquely determined. For example, the set $e_1 + e_2, e_2, e_3, \cdots, e_n$ is a second basis and, if $\alpha \neq 0$, the set $\alpha e_1, e_2, \cdots, e_n$ is also a basis. A fundamental theorem we shall prove in the next section is that the number of vectors in any basis is the same. Hence the number n, which we shall call the *dimensionality* of $\mathfrak{R}$ over Δ, is an invariant. As a necessary preliminary to the proof of this theorem we investigate now the fundamental concept of linear dependence of vectors.

We say that a vector x is *linearly dependent* on a set of vectors S if $x \varepsilon [S]$. This is equivalent to saying that $x = \sum_1^m \xi_i x_i$ for suitable ξ_i in Δ and suitable x_i in S. This proves the first of the following obvious properties of linear dependence: 1) If x is linearly dependent on a set S, then x is linearly dependent on a finite subset of S; 2) x is linearly dependent on the set $S = (x)$; 3) if x is linearly dependent on S and T is a set containing S, then x is linearly dependent on T; 4) if x is linearly dependent on S and if every $x_\alpha \varepsilon S$ is linearly dependent on the set T, then x is linearly dependent on T.

The vectors $x_1, x_2, \cdots, x_m$ are *linearly dependent* if there exist β_i not all 0 in Δ such that $\beta_1 x_1 + \beta_2 x_2 + \cdots + \beta_m x_m = 0$. Since $\beta x = 0$ if and only if either $\beta = 0$ or $x = 0$, a set consisting of a single vector x is linearly dependent if and only if $x = 0$. If $m > 0$ and the x_i are linearly dependent, then we can suppose that, say, $\beta_m \neq 0$. Then

$$x_m = -\beta_m{}^{-1} \sum_1^{m-1} \beta_j x_j = \Sigma(-\beta_m{}^{-1})\beta_j x_j$$

so that x_m is linearly dependent on $(x_1, x_2, \cdots, x_{m-1})$. Conversely if x_m is linearly dependent on $(x_1, x_2, \cdots, x_{m-1})$, then the vectors $x_1, x_2, \cdots, x_m$ are linearly dependent. Thus a set of more than one vector is a linearly dependent set if and only if one of the vectors in the set is linearly dependent on the remaining ones. If $r \leq m$ and $x_1, \cdots, x_r$ is a dependent set, then so is $x_1, \cdots, x_m$; for if $\sum_1^r \beta_i x_i = 0$, then $\sum_1^m \beta_j x_j = 0$ if we take $\beta_{r+1} = \cdots = \beta_m = 0$.

If $x_1, \cdots, x_m$ are not linearly dependent, then these vectors are said to be *linearly independent*. The last property noted for dependent sets may also be stated in the following way: Any non-vacuous subset of a linearly independent set is a linearly independent set. In particular, every vector in a linearly independent set must be $\neq 0$.

The following property will be used a number of times. Hence we state it formally as

Lemma 1. *If $x_1, x_2, \cdots, x_m$ are linearly independent and $x_1, x_2, \cdots, x_m, x_{m+1}$ are linearly dependent, then x_{m+1} is linearly dependent on $x_1, \cdots, x_m$.*

Proof. We have $\beta_1 x_1 + \beta_2 x_2 + \cdots + \beta_m x_m + \beta_{m+1} x_{m+1} = 0$ where some $\beta_k \neq 0$. If $\beta_{m+1} = 0$, this implies that $x_1, \cdots, x_m$ are linearly dependent contrary to assumption. Hence $\beta_{m+1} \neq 0$. We may therefore solve for x_{m+1} obtaining an expression for it in terms of $x_1, \cdots, x_m$.

We shall also require the following

Lemma 2. *Let $x_1, x_2, \cdots, x_m$ be a set of $m > 1$ vectors and define $x_i' = x_i$ for $i = 1, 2, \cdots, m - 1$ and $x_m' = x_m + \rho x_1$. Then the x_i are linearly independent if and only if the x_i' are linearly independent.*

Proof. Suppose that the x_i are linearly independent and let β_i be elements of Δ such that $\Sigma \beta_i x_i' = 0$. Then

$$\beta_1 x_1 + \beta_2 x_2 + \cdots + \beta_{m-1} x_{m-1} + \beta_m (x_m + \rho x_1) = 0$$

so that

$$(\beta_1 + \beta_m \rho) x_1 + \beta_2 x_2 + \cdots + \beta_m x_m = 0.$$

Hence $\beta_1 + \beta_m \rho = \beta_2 = \cdots = \beta_m = 0$, and this implies that all the $\beta_i = 0$. This proves that the x_i' are linearly independent. Now $x_i = x_i'$ for $i = 1, 2, \cdots, m - 1$ and $x_m = x_m' - \rho x_1'$; hence the relation between the two sets of vectors is a symmetric one. We can conclude therefore that if the x_i' are linearly independent, then so are the x_i.

Evidently we can generalize this lemma to prove that the two sets $x_1, x_2, \cdots, x_m$ and $x_1', x_2', \cdots, x_m'$ where $x_1' = x_1$ and $x_j' = x_j + \rho_j x_1, j = 2, \cdots, m$ are either both dependent or both independent; for we can obtain the second set from the first by a sequence of replacements of the type given in Lemma 2.

We come now to one of the fundamental results of the theory of vector spaces.

Theorem 2. *If $\mathfrak{R}$ has a basis of n vectors, then any $n + 1$ vectors in $\mathfrak{R}$ are linearly dependent.*

Proof. We prove the theorem by induction on n. Let $e_1, e_2, \cdots, e_n$ be a basis and let $x_1, x_2, \cdots, x_{n+1}$ be vectors in $\mathfrak{R}$. The theorem is clear for $n = 1$; for, in this case, $x_1 = \alpha_1 e_1$, $x_2 = \alpha_2 e_1$ and either $x_1 = 0$ or $x_2 = \alpha_2 \alpha_1^{-1} x_1$. We assume now that the result has already been established for spaces that have bases of $n - 1$ vectors. Suppose that the vectors $x_1, x_2, \cdots, x_{n+1}$ are linearly independent and let

$$(4) \qquad \begin{aligned} x_1 &= \alpha_{11} e_1 + \alpha_{12} e_2 + \cdots + \alpha_{1n} e_n \\ x_2 &= \alpha_{21} e_1 + \alpha_{22} e_2 + \cdots + \alpha_{2n} e_n \\ &\cdots\cdots\cdots\cdots\cdots\cdots \\ x_{n+1} &= \alpha_{n+1,1} e_1 + \alpha_{n+1,2} e_2 + \cdots + \alpha_{n+1,n} e_n \end{aligned}$$

be the expressions for the x's in terms of the basis. Now we may assume $x_1 \neq 0$. Hence we can suppose that one of the α_{1i}, say α_{1n}, is $\neq 0$. Then the set $x_1', x_2', \cdots, x_{n+1}'$ where $x_1' = x_1$ and $x_j' = x_j - \alpha_{jn} \alpha_{1n}^{-1} x_1$, $j > 1$, is a linearly independent set. It follows that the vectors $x_2', x_3', \cdots, x_{n+1}'$ are linearly independent. But by (4) these x_j' do not involve e_n, that is, $x_j' \in \mathfrak{S} \equiv [e_1, e_2, \cdots, e_{n-1}]$. Since the e_i, $i \leq n - 1$, form a basis for $\mathfrak{S}$, this contradicts the fact that the theorem holds for $n - 1$, and the proof is complete.

Remarks. 1) Since any non-vacuous subset of a linearly independent set of vectors is a linearly independent set, Theorem 2 evidently implies that any $r > n$ vectors in a space with a basis of n vectors are linearly dependent.

2) Let S be a set of vectors and let $x_1, x_2, \cdots, x_r$ be linearly independent vectors in S. Either every set $(x_1, x_2, \cdots, x_r, x)$, x in S, is linearly dependent or there exists an $x_{r+1} \in S$ such that $(x_1, x_2, \cdots, x_{r+1})$ is independent. Similarly either every set $(x_1, x_2, \cdots, x_{r+1}, x)$, x in S, is dependent or there is an x_{r+2} in S

such that $(x_1, x_2, \cdots, x_{r+2})$ is independent. After a finite number of steps we obtain $(x_1, x_2, \cdots, x_m)$, x_i in S, a linearly independent subset of S such that any larger subset of S is linearly dependent. Thus any linearly independent subset of a set of vectors S can be imbedded in a maximal linearly independent subset of S.

3) The method of proof of Theorem 2 can be used to test a given finite set $x_1, x_2, \cdots, x_m$ for linear dependence. If $x_1 = 0$ the set is certainly dependent. Otherwise, we can replace this set by $x_1, x_2', \cdots, x_m'$ where x_1 involves, say e_n, but the x_i' do not, and such that the second set is linearly independent if and only if the original set is linearly independent. Now it is easy to see that since x_1 involves e_n while the x_i' do not, then $x_1, x_2', \cdots, x_m'$ is linearly independent if and only if $x_2', x_3', \cdots, x_m'$ is linearly independent. This reduces the problem to one of testing $m - 1$ vectors in a space with a basis of $n - 1$ vectors.

EXERCISES

1. Prove that the vector space $\Delta[\lambda]$ of polynomials in λ is infinite dimensional.
2. Test for linear dependence:

   (a) $(2, -5, 2, -3), (-1, -3, 3, -1), (1, 1, -1, 0), (-1, 1, 0, 1)$
   (b) $(2, -3, 0, 4), (6, -7, -4, 10), (0, -1, 2, 1)$
   (c) $(1, 1, 1, 1), (1, 2, 3, 4), (1, 4, 9, 16), (1, 8, 27, 64)$.

3. Show that the vectors $x_i = \sum_{j=1}^{n} \alpha_{ij} e_j$, $i = 1, 2, \cdots, m$, are linearly dependent if and only if the system of equations

$$
\begin{aligned}
\xi_1\alpha_{11} + \xi_2\alpha_{21} + \cdots + \xi_m\alpha_{m1} &= 0 \\
\xi_1\alpha_{12} + \xi_2\alpha_{22} + \cdots + \xi_m\alpha_{m2} &= 0 \\
\cdot \quad\quad \cdot \quad\quad \cdots \quad\quad \cdot \quad\quad & \cdot \\
\xi_1\alpha_{1n} + \xi_2\alpha_{2n} + \cdots + \xi_m\alpha_{mn} &= 0
\end{aligned} \tag{5}
$$

has a non-trivial solution $(\xi_1, \xi_2, \cdots, \xi_m) = (\beta_1, \beta_2, \cdots, \beta_m) \neq (0, 0, \cdots, 0)$. Use this to prove that any system (5) whose coefficients α_{ij} are in a division ring Δ has a non-trivial solution in Δ, provided the number m of unknowns exceeds the number n of equations.

(A similar result can be proved for "right-handed" systems $\Sigma\alpha_{ij}\xi_j = 0$ by using right vector spaces.)

5. Invariance of dimensionality. A set of vectors (f) has been called a set of generators for $\mathfrak{R}$ if every x can be expressed in the form $\Sigma\xi_i f_i$ for suitable f_i in (f) and suitable ξ_i in Δ. If $e_1, e_2, \cdots, e_n$ is a basis, these elements are, of course, generators. Moreover, they are linearly independent; for if $\Sigma\beta_i e_i = 0$, then

$$\beta_1 e_1 + \beta_2 e_2 + \cdots + \beta_n e_n = 0e_1 + 0e_2 + \cdots + 0e_n.$$

Hence by the uniqueness of the representation, each $\beta_i = 0$. Conversely, any finite set of generators $f_1, f_2, \cdots, f_m$ which are linearly independent form a basis. Thus if $\Sigma\xi_i f_i = \Sigma\eta_i f_i$, then $\Sigma(\xi_i - \eta_i)f_i = 0$. Hence $\xi_i - \eta_i = 0$ and $\xi_i = \eta_i$ for $i = 1, 2, \cdots, m$. It follows from Theorem 2 that the number m of vectors in any basis $f_1, f_2, \cdots, f_m$ does not exceed n. By reversing the roles of the e's and the f's, we obtain $n \leq m$. Hence $m = n$. This proves the following fundamental

Theorem 3. *Any basis of $\mathfrak{R}$ contains n vectors.*

The number n of elements in a basis is therefore uniquely determined. We shall call this number the *dimensionality* of $\mathfrak{R}$ over Δ.

We have seen that if $\mathfrak{R}$ and $\bar{\mathfrak{R}}$ are equivalent free $\mathfrak{o}$-modules, then any basis $e_1, e_2, \cdots, e_n$ for $\mathfrak{R}$ yields a basis $\bar{e}_1, \bar{e}_2, \cdots, \bar{e}_n$ for $\bar{\mathfrak{R}}$. It follows that equivalent vector spaces have the same dimensionality. In particular we see that the spaces $\Delta^{(m)}$ and $\Delta^{(n)}$ are not equivalent if $m \neq n$.

We prove next the following

Theorem 4. *If $f_1, f_2, \cdots, f_r$ are linearly independent, then we can supplement these vectors with $n - r$ vectors chosen from a basis $e_1, e_2, \cdots, e_n$ to obtain a basis.*

Proof. We consider the set $(f_1, f_2, \cdots, f_r; e_1, e_2, \cdots, e_n)$, and we choose in this set a maximum linearly independent set $(f_1, f_2, \cdots, f_r; e_{i_1}, e_{i_2}, \cdots, e_{i_h})$ including the f_i. If we add any of the e's to this set, we obtain a dependent set. Hence by Lemma 1 of § 4 every e_i is linearly dependent on the set $(f_1, \cdots, f_r; e_{i_1}, \cdots, e_{i_h})$. Hence any x is dependent on this set, and the set is a basis.

The number h of e's that are added is, of course, $n - r$. In particular we see that, *if $r = n$, then the f_i constitute a basis.*

Suppose next that the vectors $f_1, f_2, \cdots, f_m$ are generators. We select from this set a maximal linearly independent subset, and we assume that the notation has been chosen so that $f_1, f_2, \cdots, f_r$ is such a subset. Then for any i, $(f_1, f_2, \cdots, f_r, f_i)$ is a linearly dependent set. Hence f_i and consequently every x is linearly dependent on $f_1, f_2, \cdots, f_r$. The latter set is therefore a basis, and, by Theorem 3, $r = n$. Thus we see that *any set of generators contains at least n elements and contains a subset of n elements that forms a basis.*

EXERCISES

1. If $f_1 = (1, -1, 2, 3)$ and $f_2 = (3, 0, 4, -2)$, find vectors f_3 and f_4 so that f_1, f_2, f_3, f_4 is a basis.
2. Find a maximum linearly independent subset in the following set of vectors: $(2, -3, 0, 4)$, $(-1, \frac{3}{2}, 0, -2)$, $(1, -1, 2, 1)$, $(6, -7, 8, 8)$.
3. Prove that any finitely generated Δ-module, Δ a division ring, is a finite dimensional vector space.

6. Bases and matrices. In considering finite sets of vectors, we shall now regard the order of these vectors as material. Thus we consider ordered sets. In particular we distinguish between the basis $e_1, e_2, \cdots, e_n$ and the basis $e_{i_1}, e_{i_2}, \cdots, e_{i_n}$ where the i's form a permutation of $1, 2, \cdots, n$. Let $(e_1, e_2, \cdots, e_n)$ be a particular ordered set which forms a basis and let $(x_1, x_2, \cdots, x_r)$ be an ordered set of arbitrary vectors. We write $x_i = \sum_1^n \alpha_{ij} e_j$, $i = 1, 2, \cdots, r$. The elements α_{ij} are uniquely determined. Hence the matrix

$$(\alpha) = \begin{bmatrix} \alpha_{11} & \alpha_{12} & \cdots & \alpha_{1n} \\ \alpha_{21} & \alpha_{22} & \cdots & \alpha_{2n} \\ \cdot & \cdot & \cdots & \cdot \\ \alpha_{r1} & \alpha_{r2} & \cdots & \alpha_{rn} \end{bmatrix} \tag{6}$$

is uniquely determined by the ordered set $(x_1, x_2, \cdots, x_r)$ and the ordered basis $(e_1, e_2, \cdots, e_n)$. We call this matrix *the matrix of $(x_1, x_2, \cdots, x_r)$ relative to $(e_1, e_2, \cdots, e_n)$.*

It will be well to recall at this point the basic facts concerning matrix multiplication.* Let (α) be an $r \times n$ matrix (r rows, n

* Cf. § 4, Chapter II of Volume I of these Lectures.

columns) with elements in Δ. As above we denote the element in the (i, j)-position, that is, in the intersection of the ith row and jth column by α_{ij}. Similarly let (β) be an $n \times m$ matrix with elements β_{jk} in Δ. We define the *product* $(\alpha)(\beta)$ to be the $r \times m$ matrix whose element in the (i, k) position is

$$\sigma_{ik} = \alpha_{i1}\beta_{1k} + \alpha_{i2}\beta_{2k} + \cdots + \alpha_{in}\beta_{nk}. \tag{7}$$

If (γ) is an $m \times q$ matrix with elements in Δ, then the products $[(\alpha)(\beta)](\gamma)$ and $(\alpha)[(\beta)(\gamma)]$ are defined as $r \times q$ matrices. The (i, l) elements of these products are respectively

$$\sum_{j,k} (\alpha_{ij}\beta_{jk})\gamma_{kl}, \quad \sum_{j,k} \alpha_{ij}(\beta_{jk}\gamma_{kl}).$$

Thus we have the associative law: $[(\alpha)(\beta)](\gamma) = (\alpha)[(\beta)(\gamma)]$.

If we stick to square matrices of a definite size, say $n \times n$, then the product is again a matrix of the same type. Since the associative law holds, we can say that the totality Δ_n of these matrices is a semi-group. Also it is immediate that the matrix

$$1 = \begin{bmatrix} 1 & & & 0 \\ & 1 & & \\ & & \ddots & \\ 0 & & & 1 \end{bmatrix}$$

is the identity in Δ_n in the sense that $(\alpha)1 = (\alpha) = 1(\alpha)$ for all $(\alpha) \in \Delta_n$. As usual for semi-groups we call a matrix (α) a *unit* if there exists a (β) such that $(\alpha)(\beta) = 1 = (\beta)(\alpha)$. These matrices are also called *non-singular* or *regular* matrices in Δ_n. It is easy to verify that the totality of units of any semi-group with an identity constitutes a group.* In particular the totality $L(\Delta, n)$ of non-singular matrices is a group relative to multiplication. As in any group the inverse (β) of (α) is uniquely determined. As usual we write $(\beta) = (\alpha)^{-1}$.

We return now to the consideration of finite dimensional vector spaces. Let $(e_1, e_2, \cdots, e_n)$ and $(f_1, f_2, \cdots, f_n)$ be ordered bases for the vector space $\mathfrak{R}$ over Δ and, as before, let (α) be the

* See, for example, these Lectures, Volume I, p. 24.

matrix of (f_i) relative to (e_i). Next let $(g_1, g_2, \cdots, g_n)$ be a third ordered basis and let

$$(\beta) = \begin{bmatrix} \beta_{11} & \beta_{12} & \cdots & \beta_{1n} \\ \beta_{21} & \beta_{22} & \cdots & \beta_{2n} \\ \cdot & \cdot & \cdots & \cdot \\ \beta_{n1} & \beta_{n2} & \cdots & \beta_{nn} \end{bmatrix}$$

be the matrix of $(g_1, g_2, \cdots, g_n)$ relative to $(f_1, f_2, \cdots, f_n)$. Then $g_j = \Sigma \beta_{jk} f_k$. Since $f_k = \Sigma \alpha_{ki} e_i$,

$$g_j = \Sigma \beta_{jk} f_k = \Sigma \beta_{jk} \alpha_{ki} e_i = \Sigma \gamma_{ji} e_i$$

where $\gamma_{ji} = \sum_k \beta_{jk} \alpha_{ki}$. This shows that the matrix of $(g_1, g_2, \cdots, g_n)$ relative to $(e_1, e_2, \cdots, e_n)$ is the product $(\beta)(\alpha)$ of the matrices (β) and (α). If, in particular, $g_i = e_i$, then $(\beta)(\alpha)$ is the matrix of the $(e_1, e_2, \cdots, e_n)$ relative to $(e_1, e_2, \cdots, e_n)$. Since $e_i = e_i$, it is evident that this matrix must be the identity matrix 1. Hence $(\beta)(\alpha) = 1$. By reversing the roles of $(e_1, e_2, \cdots, e_n)$ and $(f_1, f_2, \cdots, f_n)$, we obtain also $(\alpha)(\beta) = 1$. Thus we have proved

Theorem 5. *The matrix of any ordered basis* $(f_1, f_2, \cdots, f_n)$ *relative to the ordered basis* $(e_1, e_2, \cdots, e_n)$ *is non-singular.*

Conversely, let (α) be an element of $L(\Delta, n)$. Let $(\beta) = (\alpha)^{-1}$. Define f_i by $f_i = \Sigma \alpha_{ij} e_j$. Then we assert that $(f_1, f_2, \cdots, f_n)$ is a basis for $\mathfrak{R}$ over Δ. Thus the elements

$$\sum_i \beta_{ki} f_i = \sum_{i,j} \beta_{ki} \alpha_{ij} e_j = \sum_j \delta_{kj} e_j \tag{8}$$

where δ_{kj} is the *Kronecker "delta,"* that is, $\delta_{kj} = 0$ if $k \neq j$ and $= 1$ if $k = j$. Thus $\Sigma \beta_{ki} f_i = e_k$ and the e_k are dependent on the f's. Hence every x is dependent on the f's. Thus the f's are generators. Since their number is n, they form a basis.

We have therefore established a 1–1 correspondence between the different ordered bases and the units in Δ_n: If $(e_1, e_2, \cdots, e_n)$ is a particular ordered basis, then every ordered basis is obtained by taking a unit (α) in Δ_n and defining $f_i = \Sigma \alpha_{ij} e_j$.

There is no difficulty, of course, in duplicating the above results for right vector spaces. We need only to settle on the definition of the matrix of $(x_1', x_2', \cdots, x_r')$ relative to the basis

$(e_1', e_2', \cdots, e_n')$ for the right space $\mathfrak{R}'$. We do this by writing $x_j' = \Sigma e_i'\alpha_{ij}$ and by defining the matrix of $(x_1', x_2', \cdots, x_r')$ relative to $(e_1', e_2', \cdots, e_n')$ to be (α). Thus in this case the matrix is the transposed * of the matrix which appears in the equations $x_j' = \Sigma e_i'\alpha_{ij}$. As before we obtain a 1–1 correspondence between the various ordered bases and the elements of $L(\Delta, n)$.

EXERCISES

1. Prove that, if $(\alpha_{11}, \alpha_{12}, \cdots, \alpha_{1n}), (\alpha_{21}, \alpha_{22}, \cdots, \alpha_{2n}), \cdots, (\alpha_{r1}, \alpha_{r2}, \cdots, \alpha_{rn})$ are (left) linearly independent, then there exist α_{ij}; $i = r + 1, \cdots, n, j = 1, 2, \cdots, n$, such that $(\alpha) = (\alpha_{ij})$ is a unit.

2. Let Δ be a finite division ring containing q elements. Show that the number of units in Δ_n is

$$N = (q^n - 1)(q^n - q) \cdots (q^n - q^{n-1}).$$

7. Applications to matrix theory. The correspondence between bases and units in Δ_n enables us to apply our results on bases to obtain some simple but non-trivial theorems on matrices with elements in a division ring. We prove first the following

Theorem 6. *If* (α) *and* $(\beta) \varepsilon \Delta_n$ *and* $(\beta)(\alpha) = 1$, *then also* $(\alpha)(\beta) = 1$ *so that* (α) *and* $(\beta) \varepsilon L(\Delta, n)$.

Proof. If $(\beta)(\alpha) = 1$, the equation (8) shows that if $f_i = \Sigma\alpha_{ij}e_j$, then the e_k are dependent on the f's. The argument given above then shows that the f's form a basis. Hence the matrix (α) of $(f_1, f_2, \cdots, f_n)$ relative to $(e_1, e_2, \cdots, e_n)$ is a unit. Since the inverse is unique, it follows that $(\beta) = (\alpha)^{-1}$.

Theorem 7. *If* (α) *is not a right (left) zero divisor in* Δ_n, *then* $(\alpha) \varepsilon L(\Delta, n)$.

Proof. We have to show that the vectors $f_i = \Sigma\alpha_{ij}e_j$ form a basis. By Theorem 4 it suffices to show that the f's are linearly independent. Suppose therefore that $\Sigma\beta_i f_i = 0$. Then $\Sigma\beta_i\alpha_{ij}e_j = 0$ and hence $\sum_i \beta_i\alpha_{ij} = 0$ for $j = 1, 2, \cdots, n$. Thus if

$$(\beta) = \begin{bmatrix} \beta_1 & \beta_2 & \cdots & \beta_n \\ 0 & 0 & \cdots & 0 \\ \cdot & \cdot & \cdots & \cdot \\ 0 & 0 & \cdots & 0 \end{bmatrix}$$

* The transposed of the matrix (α_{ij}) is the matrix with element α_{ij} in its (j, i)-position.

then $(\beta)(\alpha) = 0$. Since (α) is not a right zero divisor, this implies that $(\beta) = 0$. Hence each $\beta_i = 0$. This proves that the f_i are linearly independent and completes the proof for the case in which (α) is not a right zero divisor. The proof for the case (α) not a left zero divisor can be obtained in the same way by using right vector spaces. The details are left to the reader.

We shall obtain next a set of generators for the group $L(\Delta, n)$. Consider the matrices of the form

$$T_{pq}(\beta) = \begin{bmatrix} 1 & & & & \vdots & & \\ & \ddots & & & \vdots & & \\ & & 1 & \cdots & \beta & \cdots & \cdot \\ & & & \ddots & \vdots & & \\ & & & & 1 & & \\ & & & & & \ddots & \\ & & & & & & 1 \end{bmatrix} \begin{matrix} \\ \\ p \\ \\ \\ \\ \\ \end{matrix}$$

(with β in column q)

$$D_p(\gamma) = \begin{bmatrix} 1 & & & \vdots & & & \\ & \ddots & & \vdots & & & \\ & & 1 & \cdot & & & \\ & & & \gamma & & & \\ & & & & 1 & & \\ & & & & & \ddots & \\ & & & & & & 1 \end{bmatrix} \qquad \gamma \neq 0$$

(with γ in column p)

$$P_{pq} = \begin{array}{c} \\ \end{array}\begin{array}{ccccccccc} & & p & & & q & & & \\ \end{array}$$

$$P_{pq} = \left[\begin{array}{cccccccc} 1 & & & & & & & \\ & \ddots & \vdots & & & \vdots & & \\ & 1 & & & & & & \\ & & 0 & \cdots & & 1 & \cdots & \\ & & & 1 & & & & \\ & & \vdots & \ddots & & \vdots & & \\ & & & & 1 & & & \\ & & 1 & \cdots & & 0 & \cdots & \\ & & & & & & 1 & \\ & & & & & & \ddots & \\ & & & & & & & 1 \end{array}\right]\begin{array}{c} \\ \\ \\ p \\ \\ \\ \\ q \\ \\ \\ \\ \end{array}$$

in which the elements not indicated are 0. We call these matrices *elementary matrices* of respective *types* I, II and III. These matrices belong to $L(\Delta, n)$; for $T_{pq}(\beta)^{-1} = T_{pq}(-\beta)$, $D_p(\gamma)^{-1} = D_p(\gamma^{-1})$ and $P_{pq}^{-1} = P_{pq}$. We shall now prove the following

Theorem 8. *Any matrix (α) in $L(\Delta, n)$ is a product of elementary matrices.*

Proof. We note first that, if $(f_1, f_2, \cdots, f_n)$ is an ordered basis, then so are the following sets:

$$(f_1, f_2, \cdots, f_{p-1}, f_p', f_{p+1}, \cdots, f_n), \quad f_p' = f_p + \beta f_q, \quad q \neq p$$

$$(f_1, f_2, \cdots, f_{p-1}, f_p', f_{p+1}, \cdots, f_n), \quad f_p' = \gamma f_p, \quad \gamma \neq 0$$

$$(f_1, \cdots, f_{p-1}, f_p', f_{p+1}, \cdots, f_{q-1}, f_q', f_{q+1}, \cdots, f_n),$$
$$f_p' = f_q, \quad f_q' = f_p.$$

Moreover, the matrices of these bases relative to $(f_1, f_2, \cdots, f_n)$ are elementary matrices of types I, II, or III.

Now let (α) be any matrix in $L(\Delta, n)$ and define $f_i = \Sigma\alpha_{ij}e_j$ where the e's constitute a basis for an n dimensional vector space.

Then $(f_1, f_2, \cdots, f_n)$ is an ordered basis. We wish to show that we can go from this basis to the basis $(e_1, e_2, \cdots, e_n)$ by a sequence of "elementary replacements" of the types indicated above. This is trivial if $n = 1$, and we can suppose that it has already been proved for $(n - 1)$ dimensional vector spaces. Now the f_i cannot all belong to $[e_2, e_3, \cdots, e_n]$. Hence one of the α_{i1}, say α_{p1}, is $\neq 0$. We interchange f_1, f_p to obtain the basis $(f_1', f_2, \cdots, f_{p-1}, f_p', f_p, \cdots, f_n)$ in which f_1' has a non-zero coefficient for e_1 in its expression in terms of the e_i. Next we replace f_2 by $f_2^* = f_2 + \beta f_1'$ where β is chosen so that $f_2^* \varepsilon [e_2, e_3, \cdots, e_n]$. A sequence of such elementary replacements yields the basis $(f_1', f_2^*, f_3^*, \cdots, f_n^*)$ where the $f_i^* \varepsilon [e_2, e_3, \cdots, e_n]$. The vectors $f_2^*, f_3^*, \cdots, f_n^*$ are linearly independent so that they constitute a basis for $[e_2, e_3, \cdots, e_n]$. Hence by the induction assumption we can pass by a finite sequence of elementary replacements to the basis $(f_1', e_2, e_3, \cdots, e_n)$. Next we obtain $(f_1'', e_2, e_3, \cdots, e_n)$ in which $f_1'' = f_1' + \mu e_2$ does not involve e_2. A finite sequence of such replacements yields $(\gamma e_1, e_2, \cdots, e_n)$ and then $(e_1, e_2, \cdots, e_n)$. We can now conclude the proof; for the matrix (α) of the basis $(f_1, f_2, \cdots, f_n)$ relative to the basis $(e_1, e_2, \cdots, e_n)$ is the product of the matrices of successive bases in our sequence, and these are elementary matrices.

EXERCISES

1. Express the following matrix as a product of elementary matrices

$$\begin{bmatrix} 2 & -1 & 1 & -1 \\ -5 & -3 & 1 & 1 \\ 2 & 3 & -1 & 0 \\ -3 & -1 & 0 & 1 \end{bmatrix}.$$

2. Verify that

$$\begin{bmatrix} 0 & 1 \\ 1 & 0 \end{bmatrix} = \begin{bmatrix} 1 & 1 \\ 0 & 1 \end{bmatrix} \begin{bmatrix} 1 & 0 \\ -1 & 1 \end{bmatrix} \begin{bmatrix} 1 & 1 \\ 0 & 1 \end{bmatrix} \begin{bmatrix} -1 & 0 \\ 0 & 1 \end{bmatrix}.$$

Generalize this result and use the generalization to prove that the elementary matrices of types I and II suffice to generate $L(\Delta, n)$.

3. Prove that, if $\delta \neq 0$, $\begin{bmatrix} \delta & 0 \\ 0 & \delta^{-1} \end{bmatrix}$ is a product of elementary matrices of type 1. Hence prove that any matrix in $L(\Delta, n)$ has the form $(\beta)D_n(\gamma)$ where (β) is a product of elementary matrices of type I and $D_n(\gamma)$ is defined above.

8. Rank of a set of vectors. Determinantal rank. Let $S = (x_\alpha)$ be an arbitrary subset of the vector space $\mathfrak{R}$ and as before let $[S]$ denote the subspace spanned by S. If $(x_1, x_2, \cdots, x_r)$ is a maximal linearly independent set of vectors chosen from the set S, then every vector in S and hence in $[S]$ is linearly dependent on the x_i. Hence $(x_1, x_2, \cdots, x_r)$ is a basis for $[S]$. The theorem on invariance of dimensionality now shows that r is uniquely determined by S, that is, any two maximal linearly independent subsets of a set S have the same cardinal number. We call this number the *rank* of the set S. Of course, the rank r is $\leq n$ and $r = n$ if and only if $[S] = \mathfrak{R}$. These remarks show in particular that, if $S = \mathfrak{S}$ is a subspace, then $\mathfrak{S} = [S]$ is finite dimensional with dimensionality $\leq n$. Moreover dim $\mathfrak{S} = n$ only if $\mathfrak{S} = \mathfrak{R}$.

We shall now apply the concept of rank of a set of vectors to the study of matrices with elements in a division ring Δ. Let (α) be an arbitrary $r \times n$ matrix with elements in Δ and let $(e_1, e_2, \cdots, e_n)$ be an arbitrary ordered basis for $\mathfrak{R}$. We introduce the *row vectors* $x_i = \sum_{j=1}^{n} \alpha_{ij}e_j$, $i = 1, 2, \cdots, r$, of $\mathfrak{R}$ and we define the *row rank* of (α) to be the rank of the set $(x_1, x_2, \cdots, x_r)$. A different choice of basis yields the same result. For, if $(f_1, f_2, \cdots, f_n)$ is a second basis for $\mathfrak{R}$ (or for another n-dimensional space), then the mapping $\Sigma\xi_i e_i \rightarrow \Sigma\xi_i f_i$ is an equivalence which maps x_i into $y_i = \Sigma\alpha_{ij}f_j$. Hence dim $[x_1, x_2, \cdots, x_r] =$ dim $[y_1, y_2, \cdots, y_r]$.

In a similar fashion we define the column rank of (α). Here we introduce a right vector space $\mathfrak{R}'$ of r dimensions with basis $(e_1', e_2', \cdots, e_r')$. Then we define the *column rank* of (α) to be the rank of the set $(x_1', x_2', \cdots, x_n')$ where $x_i' = \Sigma e_j'\alpha_{ji}$. The x_i' are called *column vectors* of (α). We shall prove in the next chapter that the two ranks of a matrix are always equal. In the special case $\Delta = \Phi$ a field (commutative) this equality can be established by showing that these ranks coincide with still another rank which can be defined in terms of determinants.

We recall first that a *minor* of the matrix (α), α_{ij} in Φ, is a determinant of a square matrix that is obtained by striking out a certain number of rows and columns from the matrix (α). For example, the minors of second order have the form $\begin{vmatrix} \alpha_{pr} & \alpha_{ps} \\ \alpha_{qr} & \alpha_{qs} \end{vmatrix}$.

We say that (α) has *determinantal rank* ρ if every $(\rho+1)$-rowed minor has the value 0, but there exists a ρ-rowed minor $\neq 0$ in (α). The following theorem will enable us to prove the equality of row rank and determinantal rank. The proof will make use of well-known theorems on determinants.

Theorem 9. *The vectors* $x_i = \Sigma\alpha_{ij}e_j$, $i = 1, 2, \cdots, r$, *are linearly independent if and only if* (α) *is of determinantal rank* r.

Proof. Evidently the determinantal rank $\rho \leq n$. Also the x's are linearly independent only if $r \leq n$. Hence we may assume that $r \leq n$. Suppose first that the x's are dependent, so that, say, $x_1 = \beta_2 x_2 + \cdots + \beta_r x_r$. Then $\alpha_{1j} = \beta_2\alpha_{2j} + \beta_3\alpha_{3j} + \cdots + \beta_r\alpha_{rj}$ for $j = 1, 2, \cdots, n$. Hence

$$(\alpha) = \begin{bmatrix} \sum_2^r \beta_k\alpha_{k1} & \sum_2^r \beta_k\alpha_{k2} & \cdots & \sum_2^r \beta_k\alpha_{kn} \\ \alpha_{21} & \alpha_{22} & \cdots & \alpha_{2n} \\ \cdot & \cdot & \cdots & \cdot \\ \alpha_{r1} & \alpha_{r2} & \cdots & \alpha_{rn} \end{bmatrix}.$$

Since the first row of any r-rowed minor is a linear combination of the other rows, each r-rowed minor vanishes. Hence $\rho < r$. Conversely, suppose that $\rho < r$. It is clear that the determinantal rank is unaltered when the rows or the columns of (α) are permuted. Such permutations give matrices of the x's in some other order relative to the e's in some other order. Hence there is no loss in generality in assuming that

$$\beta = \begin{vmatrix} \alpha_{11} & \alpha_{12} & \cdots & \alpha_{1\rho} \\ \alpha_{21} & \alpha_{22} & \cdots & \alpha_{2\rho} \\ \cdot & \cdot & \cdots & \cdot \\ \alpha_{\rho 1} & \alpha_{\rho 2} & \cdots & \alpha_{\rho\rho} \end{vmatrix} \neq 0.$$

Now let β_i, $i = 1, 2, \cdots, \rho + 1$, be the cofactor of $\alpha_{i,\rho+1}$ in

$$\begin{vmatrix} \alpha_{11} & \alpha_{12} & \cdots & \alpha_{1,\rho+1} \\ \alpha_{21} & \alpha_{22} & \cdots & \alpha_{2,\rho+1} \\ \cdot & \cdot & \cdots & \cdot \\ \alpha_{\rho+1,1} & \alpha_{\rho+1,2} & \cdots & \alpha_{\rho+1,\rho+1} \end{vmatrix}.$$

Then $\beta_{\rho+1} = \beta \neq 0$ and $\beta_1\alpha_{1j} + \beta_2\alpha_{2j} + \cdots + \beta_{\rho+1}\alpha_{\rho+1,j} = 0$ for $j = 1, 2, \cdots, n$. Hence $\beta_1 x_1 + \beta_2 x_2 + \cdots + \beta_{\rho+1}x_{\rho+1} = 0$ where $\beta_{\rho+1} \neq 0$. Thus the x's are dependent. This completes the proof.

Again let r be arbitrary and assume that the vectors $x_1, x_2, \cdots, x_\rho$ form a basis for the set of x's. Then by the above theorem there exists a non-vanishing ρ-rowed minor in the first ρ rows of (α). Moreover, since any $\rho + 1$ x's are linearly dependent, every $\rho + 1$-rowed minor in (α) vanishes. Hence the determinantal rank equals the row rank ρ. If we apply the same arguments to right vector spaces, we can show that the column rank and the determinantal rank are equal. As a consequence, we see that in the commutative case the two ranks (row and column) of a matrix are equal.

We have seen that the matrix $(\alpha) \varepsilon L(\Phi, n)$ if and only if the row vectors $(x_1, x_2, \cdots, x_n)$, $x_i = \Sigma\alpha_{ij}e_j$, form a basis for $\mathfrak{R}$. The latter condition is equivalent to the statement that the row rank of (α) is n. Hence the above result shows that $(\alpha) \varepsilon L(\Phi, n)$ if and only if the determinant of this matrix is not zero in Φ. This result can also be proved directly (cf. these Lectures, Volume I, p. 59). As a matter of fact, the inverse of (α) can be expressed in a simple fashion by means of determinants in the following way. Let A_{ij} be the cofactor of the element α_{ji} in (α) and set $\beta_{ij} = A_{ij}[\det(\alpha)]^{-1}$. Then $(\beta_{ij}) = (\alpha)^{-1}$. This follows easily from the expansion theorems for determinants. A proof is given in Volume I, p. 59.

EXERCISES

1. Prove that if $\Delta = \Phi$ is commutative and the elements α_i are all different, then

$$\begin{bmatrix} 1 & \alpha_1 & \alpha_1^2 & \cdots & \alpha_1^{n-1} \\ 1 & \alpha_2 & \alpha_2^2 & \cdots & \alpha_2^{n-1} \\ \cdot & \cdot & \cdot & \cdots & \cdot \\ 1 & \alpha_n & \alpha_n^2 & \cdots & \alpha_n^{n-1} \end{bmatrix}$$

is in $L(\Phi, n)$. (Hint: The determinant of this matrix is a so-called Vandermonde determinant. Prove that its value is $\prod_{i>j}(\alpha_i - \alpha_j)$.)

2. Prove that if $\Delta = \Phi$ is a field and $(\alpha) \varepsilon L(\Phi, n)$, then the transposed matrix $(\alpha)' \varepsilon L(\Phi, n)$.

3. Prove the following converse of Ex. 2: If $(\alpha)' \varepsilon L(\Delta, 2)$ for every $(\alpha) \varepsilon L(\Delta, 2)$, then Δ is a field.

4. Calculate the inverse of

$$\begin{bmatrix} 1 & -1 & 2 & 3 \\ 0 & 1 & -1 & 1 \\ 2 & 1 & 1 & 0 \\ 3 & -5 & 1 & 7 \end{bmatrix}.$$

9. Factor spaces. Any subspace $\mathfrak{S}$ of $\mathfrak{R}$ is, of course, a subgroup of the additive group $\mathfrak{R}$. Since $\mathfrak{R}$ is commutative, we can define the factor group $\bar{\mathfrak{R}} = \mathfrak{R}/\mathfrak{S}$. The elements of this group are the cosets $\bar{x} = x + \mathfrak{S}$, and the composition in $\bar{\mathfrak{R}}$ is given by

$$\bar{x} + \bar{y} = \overline{x + y}.$$

Now let α be any element of Δ. Then if $x \equiv y \pmod{\mathfrak{S}}$, that is, $x - y = z \in \mathfrak{S}$, also $\alpha z \in \mathfrak{S}$; hence $\alpha x \equiv \alpha y \pmod{\mathfrak{S}}$. Thus the coset $\overline{\alpha x}$ is uniquely determined by the coset $\bar{x}$ and by the element $\alpha \in \Delta$. We now *define* this coset to be the product $\alpha\bar{x}$, and we can verify without difficulty that $\bar{\mathfrak{R}}$, Δ and the composition $(\alpha, \bar{x}) \to \alpha\bar{x}$ constitute a vector space. We shall call this vector space *the factor space* of $\mathfrak{R}$ relative to the subspace $\mathfrak{S}$.

Now let $(f_1, f_2, \cdots, f_r)$ be a basis for $\mathfrak{S}$. We extend this to a basis $(f_1, f_2, \cdots, f_r, f_{r+1}, \cdots, f_n)$ for $\mathfrak{R}$, and we shall now show that the cosets $\bar{f}_{r+1}, \cdots, \bar{f}_n$ form a basis for $\bar{\mathfrak{R}} = \mathfrak{R}/\mathfrak{S}$. Let $\bar{x}$ be any coset and write $x = \sum_1^n \alpha_i f_i$; then $\bar{x} = \overline{\sum_1^n \alpha_i f_i} = \sum_1^n \overline{\alpha_i f_i}$ $= \sum_1^n \alpha_i \bar{f}_i = \sum_{r+1}^n \alpha_j \bar{f}_j$ since $\bar{f}_i = 0$ for $i \leq r$. Thus $(\bar{f}_{r+1}, \cdots, \bar{f}_n)$ is a set of generators for $\bar{\mathfrak{R}}$. On the other hand, if $\sum_{r+1}^n \beta_j \bar{f}_j = 0$, then $\sum_{r+1}^n \beta_j f_j \in \mathfrak{S}$ and so $\sum_{r+1}^n \beta_j f_j = \sum_1^r \gamma_k f_k$. This implies that all the $\beta_j = 0$. Thus $(\bar{f}_{r+1}, \cdots, \bar{f}_n)$ is a basis. We have therefore proved that the dimensionality of $\bar{\mathfrak{R}}$ is the difference of the dimensionalities of $\mathfrak{R}$and of $\mathfrak{S}$.

10. Algebra of subspaces. The totality L of subspaces of a vector space $\mathfrak{R}$ over a division ring Δ constitutes an interesting type of algebraic system with respect to two compositions which we proceed to define. We consider first the system L relative to the relation of set inclusion. With respect to this relation L is a *partially ordered set.** By this we mean that the relation $\mathfrak{S}_1 \supseteq \mathfrak{S}_2$ is defined for some pairs in L and that

1. $\mathfrak{S} \supseteq \mathfrak{S}$,
2. if $\mathfrak{S}_1 \supseteq \mathfrak{S}_2$ and $\mathfrak{S}_2 \supseteq \mathfrak{S}_1$, then $\mathfrak{S}_1 = \mathfrak{S}_2$,
3. if $\mathfrak{S}_1 \supseteq \mathfrak{S}_2$ and $\mathfrak{S}_2 \supseteq \mathfrak{S}_3$, then $\mathfrak{S}_1 \supseteq \mathfrak{S}_3$.

Thus the relation is reflexive, asymmetric and transitive.

Consider now any two subspaces $\mathfrak{S}_1$ and $\mathfrak{S}_2$. The logical intersection $\mathfrak{S}_1 \cap \mathfrak{S}_2$ is also a subspace, and this space acts as a greatest lower bound relative to the inclusion relation. By this we mean that $\mathfrak{S}_1 \cap \mathfrak{S}_2$ is contained in $\mathfrak{S}_1$ and $\mathfrak{S}_2$, and $\mathfrak{S}_1 \cap \mathfrak{S}_2$ contains every $\mathfrak{S}'$ which is contained in $\mathfrak{S}_1$ and $\mathfrak{S}_2$. The set theoretic sum $\mathfrak{S}_1 \cup \mathfrak{S}_2$ of two spaces need not be a subspace. As a substitute for this set we therefore take the space $[\mathfrak{S}_1 \cup \mathfrak{S}_2]$ spanned by the set $\mathfrak{S}_1 \cup \mathfrak{S}_2$. We denote this space by $\mathfrak{S}_1 + \mathfrak{S}_2$ and we call it *the join* of $\mathfrak{S}_1$ and $\mathfrak{S}_2$. It has the properties of a least upper bound: $\mathfrak{S}_1 + \mathfrak{S}_2 \supseteq \mathfrak{S}_1$ and $\mathfrak{S}_2$, and $\mathfrak{S}_1 + \mathfrak{S}_2$ is contained in every subspace $\mathfrak{S}$ which contains $\mathfrak{S}_1$ and $\mathfrak{S}_2$. It is immediate that these properties characterize $\mathfrak{S}_1 + \mathfrak{S}_2$, that is, any subspace that has these properties coincides with $\mathfrak{S}_1 + \mathfrak{S}_2$. Also it is immediate from this characterization or from the definition of $\mathfrak{S}_1 + \mathfrak{S}_2$ as $[\mathfrak{S}_1 \cup \mathfrak{S}_2]$ that this space is the set of vectors of the form $y_1 + y_2$ where the $y_i \in \mathfrak{S}_i$.

A partially ordered set in which any two elements have a greatest lower bound and a least upper bound is called a *lattice.* Hence we call L *the lattice of subspaces* of the space $\mathfrak{R}$. In this section we derive the basic properties of this lattice. First we note the following properties that hold in any lattice.

1. The associative and commutative laws hold for the compositions $\cap$ and $+$.

These follow easily from the definitions. The rules for $\cap$ are, of course, familiar to the reader.

We note next some special properties of the lattice L.

2. There exists a zero element in L, that is, an element 0 such that

$$\mathfrak{S} \cap 0 = 0 \quad \text{and} \quad \mathfrak{S} + 0 = \mathfrak{S}$$

for all $\mathfrak{S}$.

* Cf. Volume I, Chapter VII, for the concepts considered in this section.

The subspace consisting of the 0 vector only has these properties. Dually the whole space $\mathfrak{R}$ acts as an "all" element in the sense that

$$\mathfrak{S} + \mathfrak{R} = \mathfrak{R} \quad \text{and} \quad \mathfrak{S} \cap \mathfrak{R} = \mathfrak{S}$$

for all $\mathfrak{S}$.

The distributive law $\mathfrak{S}_1 \cap (\mathfrak{S}_2 + \mathfrak{S}_3) = \mathfrak{S}_1 \cap \mathfrak{S}_2 + \mathfrak{S}_1 \cap \mathfrak{S}_3$ does not hold without restriction in L. For example, let x_1 and x_2 be independent vectors and set $\mathfrak{S}_1 = [x_1]$, $\mathfrak{S}_2 = [x_2]$ and $\mathfrak{S}_3 = [x_1 + x_2]$. Then $\mathfrak{S}_2 + \mathfrak{S}_3 = [x_1, x_2]$ so that $\mathfrak{S}_1 \cap (\mathfrak{S}_2 + \mathfrak{S}_3) = \mathfrak{S}_1$. On the other hand, $\mathfrak{S}_1 \cap \mathfrak{S}_2$ and $\mathfrak{S}_1 \cap \mathfrak{S}_3 = 0$ so that $\mathfrak{S}_1 \cap \mathfrak{S}_2 + \mathfrak{S}_1 \cap \mathfrak{S}_3 = 0$. We shall show that a certain weakening of the distributive law does hold in L. This is the following rule:

3. If $\mathfrak{S}_1 \supseteq \mathfrak{S}_2$, then $\mathfrak{S}_1 \cap (\mathfrak{S}_2 + \mathfrak{S}_3) = \mathfrak{S}_1 \cap \mathfrak{S}_2 + \mathfrak{S}_1 \cap \mathfrak{S}_3 = \mathfrak{S}_2 + \mathfrak{S}_1 \cap \mathfrak{S}_3$.

Proof. We note first that $\mathfrak{S}_1 \cap \mathfrak{S}_2 \subseteq \mathfrak{S}_1 \cap (\mathfrak{S}_2 + \mathfrak{S}_3)$ and $\mathfrak{S}_1 \cap \mathfrak{S}_3 \subseteq \mathfrak{S}_1 \cap (\mathfrak{S}_2 + \mathfrak{S}_3)$. Hence

$$\mathfrak{S}_1 \cap \mathfrak{S}_2 + \mathfrak{S}_1 \cap \mathfrak{S}_3 \subseteq \mathfrak{S}_1 \cap (\mathfrak{S}_2 + \mathfrak{S}_3).$$

Next let $z \in \mathfrak{S}_1 \cap (\mathfrak{S}_2 + \mathfrak{S}_3)$. Then $z = y_1$ in $\mathfrak{S}_1$ and $z = y_2 + y_3$ where y_2 and y_3 are in $\mathfrak{S}_2$ and $\mathfrak{S}_3$ respectively. Hence $y_3 = y_1 - y_2 \in \mathfrak{S}_1 + \mathfrak{S}_2 = \mathfrak{S}_1$. Thus $y_3 \in \mathfrak{S}_1 \cap \mathfrak{S}_3$ and $z = y_2 + y_3 \in \mathfrak{S}_2 + \mathfrak{S}_1 \cap \mathfrak{S}_3$. This proves that $\mathfrak{S}_1 \cap (\mathfrak{S}_2 + \mathfrak{S}_3) \subseteq \mathfrak{S}_2 + \mathfrak{S}_1 \cap \mathfrak{S}_3$. Hence **3.** holds.

A lattice in which **3.** holds is called a *modular* (or *Dedekind*) lattice. We shall show next that L is a *complemented* lattice in the sense that the following property holds:

4. For any $\mathfrak{S}$ in L there exists an $\mathfrak{S}^*$ in L such that

$$\mathfrak{S} + \mathfrak{S}^* = \mathfrak{R}, \quad \mathfrak{S} \cap \mathfrak{S}^* = 0.$$

Proof. If $(f_1, f_2, \cdots, f_r)$ is a basis for $\mathfrak{S}$, these vectors are linearly independent and can therefore be supplemented by vectors $f_{r+1}, \cdots, f_n$ to give a basis $(f_1, f_2, \cdots, f_n)$ for $\mathfrak{R}$. We set $\mathfrak{S}^* = [f_{r+1}, f_{r+2}, \cdots, f_n]$. Then $\mathfrak{S} + \mathfrak{S}^* = [f_1, f_2, \cdots, f_n] = \mathfrak{R}$. Moreover, any vector y in $\mathfrak{S} \cap \mathfrak{S}^*$ is linearly dependent on $f_1, f_2, \cdots, f_r$ and on $f_{r+1}, f_{r+2}, \cdots, f_n$. Since $f_1, f_2, \cdots, f_n$ are linearly independent, this implies that $y = 0$. Hence $\mathfrak{S} \cap \mathfrak{S}^* = 0$.

A subspace $\mathfrak{S}^*$ satisfying the above condition is called a *complement* of the subspace $\mathfrak{S}$ in $\mathfrak{R}$. We note finally that the following *chain conditions* hold in L:

5. If $\mathfrak{S}_1 \supseteq \mathfrak{S}_2 \supseteq \cdots$ is an infinite descending chain of subspaces, then there exists an integer r such that $\mathfrak{S}_r = \mathfrak{S}_{r+1} = \cdots$.

If $\mathfrak{S}_1 \subseteq \mathfrak{S}_2 \subseteq \cdots$ is an infinite ascending chain of subspaces then there exists an integer r such that $\mathfrak{S}_r = \mathfrak{S}_{r+1} = \cdots$.

Both of these are clear since the dimensionality of a subspace is a non-negative integer * and since $\mathfrak{S} \supset \mathfrak{S}'$ implies that $\dim \mathfrak{S} > \dim \mathfrak{S}'$.

EXERCISES

1. Prove that, if $\mathfrak{S}_1 \cup \mathfrak{S}_2 = \mathfrak{S}_1 + \mathfrak{S}_2$, then either $\mathfrak{S}_1 \supseteq \mathfrak{S}_2$ or $\mathfrak{S}_2 \supseteq \mathfrak{S}_1$.
2. Prove that, if $\dim \mathfrak{S} = r$, then the dimensionality of any complement is $n - r$.
3. Prove the general dimensionality relation:

$$\dim (\mathfrak{S}_1 + \mathfrak{S}_2) = \dim \mathfrak{S}_1 + \dim \mathfrak{S}_2 - \dim (\mathfrak{S}_1 \cap \mathfrak{S}_2).$$

4. Show that if $\mathfrak{S}$ is any subspace $\neq 0$ and $\neq \mathfrak{R}$, then $\mathfrak{S}$ has more than one complement.

11. Independent subspaces, direct sums. We consider next a concept which we shall see is a generalization of the notion of linear independence of vectors. Let $\mathfrak{S}_1, \mathfrak{S}_2, \cdots, \mathfrak{S}_r$ be a finite set of subspaces of $\mathfrak{R}$. Then we say that these subspaces are *independent* if

$$\mathfrak{S}_i \cap (\mathfrak{S}_1 + \cdots + \mathfrak{S}_{i-1} + \mathfrak{S}_{i+1} + \cdots + \mathfrak{S}_r) = 0 \tag{9}$$

for $i = 1, 2, \cdots, r$. If $x_1, x_2, \cdots, x_r$ are vectors in $\mathfrak{R}$, then necessary and sufficient conditions that linear independence holds for these are: 1) $x_i \neq 0$ for $i = 1, 2, \cdots, r$; 2) the spaces $[x_i]$ are independent. Thus suppose that 1) and 2) hold and let $\Sigma \beta_i x_i = 0$. Then $-\beta_i x_i = \sum_{j \neq i} \beta_j x_j \; \varepsilon \; [x_i] \cap ([x_1] + \cdots + [x_{i-1}] + [x_{i+1}] + \cdots + [x_r])$. Hence by 2), $-\beta_i x_i = 0$. Since $x_i \neq 0$, this implies that each $\beta_i = 0$. Next assume that the x_i are linearly independent. Then certainly each $x_i \neq 0$. Furthermore, if $x \; \varepsilon \; [x_i] \cap ([x_1] + \cdots + [x_{i-1}] + [x_{i+1}] + \cdots + [x_r])$, then $x = \beta_i x_i =$

* We assign to the space 0 the dimensionality 0.

$\sum_{j \neq i} \beta_j x_j$. Hence by the linear independence of the x's, $\beta_i = 0$ and so $x = 0$.

Let $\mathfrak{S}_1, \mathfrak{S}_2, \cdots, \mathfrak{S}_r$ be arbitrary independent subspaces and set $\mathfrak{S} = \mathfrak{S}_1 + \mathfrak{S}_2 + \cdots + \mathfrak{S}_r$. If $y \varepsilon \mathfrak{S}$, $y = y_1 + y_2 + \cdots + y_r$ where $y_i \varepsilon \mathfrak{S}_i$. We assert that this representation is unique, that is, if $y = y_1' + y_2' + \cdots + y_r'$ where $y_i' \varepsilon \mathfrak{S}_i$, then $y_i = y_i'$, $i = 1, 2, \cdots, r$. Thus if $\Sigma y_i = \Sigma y_i'$, then $\Sigma z_i = 0$ for $z_i = y_i - y_i'$ in $\mathfrak{S}_i$. Then

$$-z_i = \sum_{j \neq i} z_j \varepsilon \mathfrak{S}_i \cap (\mathfrak{S}_1 + \cdots + \mathfrak{S}_{i-1} + \mathfrak{S}_{i+1} + \cdots + \mathfrak{S}_r).$$

Hence $z_i = 0$ and $y_i = y_i'$. The converse of this result holds also; for if (9) fails for some i, then there is a vector $z_i \neq 0$ in this intersection. Thus $z_i = \sum_{j \neq i} z_j$, and we have two distinct representations of this element as a sum of elements out of the spaces $\mathfrak{S}_k$. We have therefore proved

Theorem 10. *A necessary and sufficient condition that the spaces $\mathfrak{S}_1, \mathfrak{S}_2, \cdots, \mathfrak{S}_r$ be independent is that every vector in $\mathfrak{S} = \mathfrak{S}_1 + \mathfrak{S}_2 + \cdots + \mathfrak{S}_r$ have a unique representation in the form Σy_i, y_i in $\mathfrak{S}_i$.*

A second important characterization of independence of subspaces is furnished by

Theorem 11. *The spaces $\mathfrak{S}_i$ are independent if and only if dim $(\mathfrak{S}_1 + \mathfrak{S}_2 + \cdots + \mathfrak{S}_r) = \Sigma$ dim $\mathfrak{S}_i$.*

Proof. Suppose first that the $\mathfrak{S}_i$ are independent and let $(f_{1i}, f_{2i}, \cdots, f_{n_i i})$ be a basis for $\mathfrak{S}_i$. Then if $\Sigma \beta_{ji} f_{ji} = 0$, $\Sigma y_i = 0$ where $y_i = \sum_j \beta_{ji} f_{ji} \varepsilon \mathfrak{S}_i$. Hence for each i, $0 = y_i = \Sigma \beta_{ji} f_{ji}$. Then $\beta_{ji} = 0$ since the f_{ji} for a fixed i are linearly independent. This proves that all the f's are linearly independent. Hence the f's form a basis for $\mathfrak{S} = \mathfrak{S}_1 + \mathfrak{S}_2 + \cdots + \mathfrak{S}_r$. Their number Σn_i, where $n_i = \dim \mathfrak{S}_i$, is the dimensionality of $\mathfrak{S}$. Thus $\dim \mathfrak{S} = \Sigma \dim \mathfrak{S}_i$. Conversely suppose that this dimensionality relation holds and, as before, let the f_{ji} form a basis for $\mathfrak{S}_i$. The number of these f's is $\Sigma \dim \mathfrak{S}_i = \dim \mathfrak{S}$. On the other hand, these f_{ji} are generators for $\mathfrak{S}$. It follows that they form a basis, and consequently they are linearly independent. It follows directly from this that, if $\Sigma y_i = \Sigma y_i'$, y_i, y_i' in $\mathfrak{S}_i$, then $y_i = y_i'$. Hence the $\mathfrak{S}_i$ are independent.

If $\mathfrak{R}_1, \mathfrak{R}_2, \cdots, \mathfrak{R}_r$ are independent subspaces and $\mathfrak{R} = \mathfrak{R}_1 + \mathfrak{R}_2 + \cdots + \mathfrak{R}_r$, then we say that $\mathfrak{R}$ is a *direct sum* of the subspaces $\mathfrak{R}_i$. We indicate this by writing $\mathfrak{R} = \mathfrak{R}_1 \oplus \mathfrak{R}_2 \oplus \cdots \oplus \mathfrak{R}_r$. If this is the case, every vector of $\mathfrak{R}$ can be written in one and only one way as a sum of vectors in the subspace $\mathfrak{R}_i$.

EXERCISES

1. Prove that the following are necessary and sufficient conditions that the subspaces $\mathfrak{S}_i$ be independent.

$$\mathfrak{S}_1 \cap \mathfrak{S}_2 = 0, \quad (\mathfrak{S}_1 + \mathfrak{S}_2) \cap \mathfrak{S}_3 = 0,$$

$$(\mathfrak{S}_1 + \mathfrak{S}_2 + \mathfrak{S}_3) \cap \mathfrak{S}_4 = 0, \quad \cdots \quad .$$

2. Prove that if $\mathfrak{R} = \mathfrak{R}_1 \oplus \mathfrak{R}_2 \oplus \cdots \oplus \mathfrak{R}_r$ and each $\mathfrak{R}_i = \mathfrak{R}_{i1} \oplus \cdots \oplus \mathfrak{R}_{in_i}$, then $\mathfrak{R} = \mathfrak{R}_{11} \oplus \cdots \oplus \mathfrak{R}_{1n_1} \oplus \mathfrak{R}_{21} \oplus \cdots \oplus \mathfrak{R}_{2n_2} \oplus \cdots \oplus \mathfrak{R}_{rn_r}$.

Chapter II

LINEAR TRANSFORMATIONS

In this chapter we discuss the simplest properties of linear transformations and of certain algebraic systems determined by these mappings. Two particular types of linear transformations are of special interest, namely, the linear transformations of a vector space into itself and the linear transformations of a space into the one-dimensional space Δ. The former type constitute a ring while the latter, called linear functions, form a right vector space. There is a natural way of associating with a linear transformation of the vector space $\mathfrak{R}_1$ into the vector space $\mathfrak{R}_2$ a transposed linear transformation of the conjugate space of linear functions on $\mathfrak{R}_2$ into the conjugate space of $\mathfrak{R}_1$. We consider the properties of the transposition mapping. The relation between linear transformations and matrices is discussed. Also we define rank and nullity for arbitrary linear transformations. Finally we study a special type of linear transformation called a projection, and we establish a connection between transformations of this type and direct decompositions of the vector space.

1. Definition and examples. The differentiation mapping $\phi(\lambda) \rightarrow \phi'(\lambda)$ in the vector space $\Phi[\lambda]$ of polynomials with real coefficients has the properties

$$[\phi(\lambda) + \psi(\lambda)]' = \phi'(\lambda) + \psi'(\lambda), \quad [\alpha\phi(\lambda)]' = \alpha\phi'(\lambda).$$

This is an example of a linear transformation. Another example is the mapping of $\Delta^{(3)}$ defined by

$$(\xi, \eta, \zeta) \rightarrow \xi\lambda + \eta\mu + \zeta\nu$$

where λ, μ, ν are fixed elements of the basic division ring Δ. In general, let $\mathfrak{R}_1$ and $\mathfrak{R}_2$ be vector spaces over the same division ring Δ. Then we call a mapping A of $\mathfrak{R}_1$ into $\mathfrak{R}_2$ a *linear transformation* if

$$(1) \qquad (x + y)A = xA + yA, \quad (\alpha x)A = \alpha(xA)$$

for all x, y in $\mathfrak{R}_1$ and all α in Δ. As usual xA denotes the image in $\mathfrak{R}_2$ of the element x. We shall also denote the image set, that is, the set of images xA, by $\mathfrak{R}_1 A$. The statement that A is a mapping *into* $\mathfrak{R}_2$ allows the possibility that $\mathfrak{R}_1 A \subset \mathfrak{R}_2$ (i.e. $\mathfrak{R}_1 A$ is a proper subset of $\mathfrak{R}_2$).

The concept of linear transformation is a special case of that of $\mathfrak{o}$-homomorphism of one $\mathfrak{o}$-module $\mathfrak{R}_1$ into a second one. The generalization is obtained by replacing "α in Δ" by "α in $\mathfrak{o}$" in the above definition. It may be recalled that $\mathfrak{o}$-homomorphisms which are 1–1 have been introduced in Chapter I. Such mappings have been called equivalences or $\mathfrak{o}$-isomorphisms. The existence of an $\mathfrak{o}$-isomorphism of $\mathfrak{R}_1$ onto $\mathfrak{R}_2$ is our criterion for equivalence of the modules $\mathfrak{R}_i$.

The first condition in (1) states that A is a homomorphism of the additive group $\mathfrak{R}_1$ into the additive group $\mathfrak{R}_2$ while the second can be interpreted as a type of commutativity of A with α. This is strictly the case when $\mathfrak{R}_1 = \mathfrak{R}_2 = \mathfrak{R}$; for we can introduce for each α the mapping α_l which sends x into αx. We call α_l the *scalar multiplication* determined by α. It is clear that α_l is an endomorphism of $\mathfrak{R}$, that is, a homomorphism of $\mathfrak{R}$, regarded as a group, into itself. Now $x(\alpha_l A) = (\alpha x)A$ and $x(A\alpha_l) = \alpha(xA)$; hence A is a linear transformation in the vector space $\mathfrak{R}$ if and only if A is an endomorphism of $\mathfrak{R}$ which commutes with all the endomorphisms α_l.

Besides the linear transformations of a vector space into itself a second noteworthy type of linear transformation is a *linear function*. This is defined to be a mapping $x \rightarrow f(x)$ of a vector space $\mathfrak{R}$ into the division ring Δ such that

$$(2) \qquad f(x + y) = f(x) + f(y), \quad f(\alpha x) = \alpha f(x).$$

It is clear that a mapping of this type can be regarded as a linear transformation of $\mathfrak{R}$ into the one-dimensional vector space Δ.

The latter is obtained by using the additive group of Δ as group, Δ as division ring and left multiplication $\alpha\xi$ as multiplication by scalars. The element 1 (or any non-zero element) is a basis for Δ over Δ. The second example considered above is an instance of a linear function on $\Delta^{(3)}$.

EXERCISES

1. Show that the differentiation mapping $\phi(\lambda) \rightarrow \phi'(\lambda)$ is a linear transformation in the vector space $\mathfrak{R}$ of polynomials of degree $<n$.
2. Show that the difference operator $\phi(\lambda) \rightarrow \phi(\lambda + 1) - \phi(\lambda)$ is a linear transformation in $\Phi[\lambda]$.

2. Compositions of linear transformations. We consider now ways of combining linear transformations. It should be remarked that all the results of this section apply equally well to the more general case of $\mathfrak{o}$-homomorphisms of modules. However, for the sake of simplicity we shall state the results only for the special case that is of primary interest in the sequel.

Suppose first that A and B are linear transformations of a vector space $\mathfrak{R}_1$ into the same space $\mathfrak{R}_2$. We define a mapping $A + B$ of $\mathfrak{R}_1$ into $\mathfrak{R}_2$ by the equation

$$x(A + B) = xA + xB \tag{3}$$

for any x in $\mathfrak{R}_1$. Thus to obtain the effect of $A + B$ on x we add the images xA and xB. Clearly $A + B$ is a (single-valued) transformation of $\mathfrak{R}_1$ into $\mathfrak{R}_2$. Since

$$\begin{aligned}(x + y)(A + B) &= (x + y)A + (x + y)B = xA + yA + xB + yB\\ &= xA + xB + yA + yB = x(A + B) + y(A + B)\end{aligned}$$

and

$$\begin{aligned}(\alpha x)(A + B) &= (\alpha x)A + (\alpha x)B = \alpha(xA) + \alpha(xB)\\ &= \alpha(xA + xB) = \alpha(x(A + B)),\end{aligned}$$

$A + B$ is a linear transformation of $\mathfrak{R}_1$ into $\mathfrak{R}_2$.

We now denote the totality of linear transformations of $\mathfrak{R}_1$ into $\mathfrak{R}_2$ by $\mathfrak{L}(\mathfrak{R}_1, \mathfrak{R}_2)$, and we shall show that this set, together with the addition composition just introduced, is a commutative group. We note first that the associative and commutative laws hold; for we have the following relations:

$$x[(A + B) + C] = x(A + B) + xC = xA + xB + xC,$$

$$x[A + (B + C)] = xA + x(B + C) = xA + xB + xC,$$

$$x(A + B) = xA + xB, \quad x(B + A) = xB + xA.$$

Thus $(A + B) + C$ and $A + (B + C)$ have the same effect on any x in $\mathfrak{R}_1$, and this is what is meant by saying that the transformations $A + (B + C)$ and $(A + B) + C$ are equal. Similarly $A + B = B + A$. Next we define the mapping 0 by the condition $x0 = 0$, the zero vector in $\mathfrak{R}_2$. It is immediate that this mapping is in $\mathfrak{L}(\mathfrak{R}_1, \mathfrak{R}_2)$ and that $A + 0 = A = 0 + A$ for all A in $\mathfrak{L}(\mathfrak{R}_1, \mathfrak{R}_2)$. Hence 0 acts as identity element for the additive composition. Finally, if A is any member of $\mathfrak{L}(\mathfrak{R}_1, \mathfrak{R}_2)$ we define $-A$ to be the mapping such that $x(-A) = -xA$. It is easy to verify that $-A \varepsilon \mathfrak{L}(\mathfrak{R}_1, \mathfrak{R}_2)$. Moreover, $-A$ acts as the inverse of A since $x(A + (-A)) = xA - xA = 0$ for all x. This completes the verification that $\mathfrak{L}(\mathfrak{R}_1, \mathfrak{R}_2)$, $+$ is a commutative group.

We introduce next a second composition for linear transformations. This is defined for any A in $\mathfrak{L}(\mathfrak{R}_1, \mathfrak{R}_2)$ and any B in $\mathfrak{L}(\mathfrak{R}_2, \mathfrak{R}_3)$, and it is taken to be the resultant of A followed by B. As usual, we denote the resultant as AB. Hence by definition $x(AB) = (xA)B$. Consequently

$$(x + y)(AB) = ((x + y)A)B = (xA + yA)B = (xA)B + (yA)B$$
$$= x(AB) + y(AB)$$

and

$$(\alpha x)(AB) = ((\alpha x)A)B = (\alpha(xA))B = \alpha((xA)B) = \alpha(x(AB)).$$

This shows that $AB \varepsilon \mathfrak{L}(\mathfrak{R}_1, \mathfrak{R}_3)$.

As is well known, the product AB is an associative one, that is, if $A \varepsilon \mathfrak{L}(\mathfrak{R}_1, \mathfrak{R}_2)$, $B \varepsilon \mathfrak{L}(\mathfrak{R}_2, \mathfrak{R}_3)$ and $C \varepsilon \mathfrak{L}(\mathfrak{R}_3, \mathfrak{R}_4)$, then

$$(AB)C = A(BC); \tag{4}$$

for

$$x((AB)C) = (x(AB))C = ((xA)B)C$$

and

$$x(A(BC)) = (xA)(BC) = ((xA)B)C.$$

We prove next the important distributive laws: If $A \varepsilon \mathfrak{L}(\mathfrak{R}_1, \mathfrak{R}_2)$, $B, C \varepsilon \mathfrak{L}(\mathfrak{R}_2, \mathfrak{R}_3)$ and $D \varepsilon \mathfrak{L}(\mathfrak{R}_3, \mathfrak{R}_4)$, then

$$(5) \qquad A(B + C) = AB + AC, \quad (B + C)D = BD + CD.$$

These follow from the following equations:

$$\begin{aligned} x(A(B + C)) &= (xA)(B + C) = (xA)B + (xA)C \\ &= x(AB) + x(AC) = x(AB + AC) \\ x((B + C)D) &= (xB + xC)D = (xB)D + (xC)D \\ &= x(BD) + x(CD) = x(BD + CD). \end{aligned}$$

We now specialize the foregoing results to the case of the linear transformations in a single vector space $\mathfrak{R}$. It is clear that $\mathfrak{L} = \mathfrak{L}(\mathfrak{R}, \mathfrak{R}), +, \cdot$ is a ring; for $\mathfrak{L}(\mathfrak{R}, \mathfrak{R}), +$ is a commutative group, $\mathfrak{L}(\mathfrak{R}, \mathfrak{R})$ is closed under $\cdot$, and this composition is associative and distributive with respect to addition. It is evident also that $\mathfrak{L}$ contains the identity mapping $x \rightarrow x$ and that this mapping, denoted as 1, is the identity in the ring $\mathfrak{L}$ (i.e. $A1 = A = 1A$ for all A).*

Suppose next that Φ is the center of the ring Δ. Of course, Φ is a subfield of Δ. We observe that the scalar multiplications γ_l determined by the elements $\gamma \varepsilon \Phi$ are linear transformations; for, γ_l is an endomorphism and $(\alpha x)\gamma_l = \gamma(\alpha x) = (\gamma\alpha)x = (\alpha\gamma)x = \alpha(\gamma x) = \alpha(x\gamma_l)$. Thus if Φ_l denotes the set of multiplications by the elements γ of Φ, then $\mathfrak{L} \supseteq \Phi_l$. In particular if $\Delta = \Phi$ is commutative, then $\mathfrak{L}$ contains all the scalar multiplications. We now show that any of the groups $\mathfrak{L}(\mathfrak{R}_1, \mathfrak{R}_2)$ can be regarded as a vector space over the field Φ. For this purpose we define γA for γ in Φ and A in $\mathfrak{L}(\mathfrak{R}_1, \mathfrak{R}_2)$ to be the mapping $x \rightarrow \gamma(xA) = (\gamma x)A$. Since this is the resultant of A and γ_l (in $\mathfrak{R}_2$) or of γ_l (in $\mathfrak{R}_1$) with A and each of these is linear, γA is in $\mathfrak{L}(\mathfrak{R}_1, \mathfrak{R}_2)$. It

* A reader familiar with the theory of endomorphisms of a commutative group such as is given in Volume I, pp. 78–82, will note that these results can also be obtained by the following reasoning: The set $\mathfrak{E}$ of endomorphisms of $\mathfrak{R}, +$ is a ring relative to the addition composition $x(A + B) = xA + xB$ and the multiplication composition as resultant. The set $\mathfrak{L} = \mathfrak{L}(\mathfrak{R}, \mathfrak{R})$ is the subset of $\mathfrak{E}$ of elements commuting with the scalar multiplication α_l. Since the totality of elements of a ring commuting with the elements of a given subset form a subring, it is clear that $\mathfrak{L}$ is a subring of $\mathfrak{E}$.

is easy to verify that the function γA satisfies the rules for multiplication by scalars in a vector space. In this way we can regard $\mathfrak{L}(\mathfrak{R}_1, \mathfrak{R}_2)$ as a vector space over Φ.

If we combine the results of the last two paragraphs, we see that the set $\mathfrak{L} = \mathfrak{L}(\mathfrak{R}, \mathfrak{R})$ is at the same time a ring and a vector space over a field Φ. The ring addition is the same as the vector space addition. Moreover, we have the following relations connecting multiplication and scalar multiplication:

$$\gamma(AB) = (\gamma A)B = A(\gamma B). \tag{6}$$

A system having these properties is called an *algebra* (or *hypercomplex number system*) *over the field* Φ. Hence when we wish to study $\mathfrak{L}$ relative to all three operations at the same time, we shall refer to this system as *the algebra of linear transformations* in $\mathfrak{R}$.

EXERCISES

1. Show that $\mathfrak{L}(\mathfrak{R}_1, \mathfrak{R}_2)$ is an $\mathfrak{L}(\mathfrak{R}_1, \mathfrak{R}_1)$-module relative to the composition AX, A in $\mathfrak{L}(\mathfrak{R}_1, \mathfrak{R}_1)$, X in $\mathfrak{L}(\mathfrak{R}_1, \mathfrak{R}_2)$ as the resultant linear transformation. Similarly show that $\mathfrak{L}(\mathfrak{R}_1, \mathfrak{R}_2)$ can be regarded as a right $\mathfrak{L}(\mathfrak{R}_2, \mathfrak{R}_2)$ module.
2. Prove that if α_l is a linear transformation, then α is in the center Φ of Δ.
3. Verify that, if $C \in \mathfrak{L}(\mathfrak{R}_2, \mathfrak{R}_2)$ and $X \in \mathfrak{L}(\mathfrak{R}_1, \mathfrak{R}_2)$, then the mapping $X \rightarrow XC$ is an $\mathfrak{L}(\mathfrak{R}_1, \mathfrak{R}_1)$ endomorphism of $\mathfrak{L}(\mathfrak{R}_1, \mathfrak{R}_2)$.

3. The matrix of a linear transformation. We shall now show that a linear transformation of one finite dimensional vector space $\mathfrak{R}_1$ into a second finite dimensional space $\mathfrak{R}_2$ can be completely described by means of a finite matrix with elements in the underlying division ring Δ.

Let $\mathfrak{R}_i$, $i = 1, 2$, be n_i dimensional, let $(e_1, e_2, \cdots, e_{n_1})$ be an ordered basis for $\mathfrak{R}_1$, $(f_1, f_2, \cdots, f_{n_2})$ an ordered basis for $\mathfrak{R}_2$ and let $A \in \mathfrak{L}(\mathfrak{R}_1, \mathfrak{R}_2)$. We note first that the action of A on any x is determined by the images e_iA, $i = 1, 2, \cdots, n_1$. Thus x can be written as $\sum_1^{n_1} \xi_i e_i$. Hence $xA = (\Sigma\xi_i e_i)A = \Sigma(\xi_i e_i)A = \Sigma\xi_i(e_iA)$. Thus xA is determined by the expression for x and by the images e_iA. Now write

$$e_iA = \alpha_{i1}f_1 + \alpha_{i2}f_2 + \cdots + \alpha_{in_2}f_{n_2}, \quad i = 1, 2, \cdots, n_1. \tag{7}$$

Then we obtain the matrix

$$(8)\qquad \begin{bmatrix} \alpha_{11} & \alpha_{12} & \cdots & \alpha_{1n_2} \\ \alpha_{21} & \alpha_{22} & \cdots & \alpha_{2n_2} \\ \cdot & \cdot & \cdots & \cdot \\ \alpha_{n_11} & \alpha_{n_22} & \cdots & \alpha_{n_1n_2} \end{bmatrix}$$

as the matrix of $(e_1A, \cdots, e_{n_1}A)$ relative to $(f_1, \cdots, f_{n_2})$. Clearly if the two ordered bases and the matrix (8) are known, then the effect of A on any x can be deduced; for (7) holds and as we have seen $xA = \sum_{i,j} \xi_i\alpha_{ij}f_j$.

This connection can be expressed in terms of matrix multiplication as follows. Write $x = \Sigma\xi_ie_i$ as a row $(\xi_1, \xi_2, \cdots, \xi_{n_1})$ and similarly $y = \Sigma\eta_jf_j$ as $(\eta_1, \eta_2, \cdots, \eta_{n_2})$. Then the "row vector" associated with $y = xA$ is obtained by performing the matrix multiplication:

$$(9)\qquad (\xi_1, \xi_2, \cdots, \xi_{n_1}) \begin{bmatrix} \alpha_{11} & \alpha_{12} & \cdots & \alpha_{1n_2} \\ \alpha_{21} & \alpha_{22} & \cdots & \alpha_{2n_2} \\ \cdot & \cdot & \cdots & \cdot \\ \alpha_{n_11} & \alpha_{n_12} & \cdots & \alpha_{n_1n_2} \end{bmatrix}.$$

Thus $xA = y = \Sigma\eta_jf_j$ where $\eta_j = \Sigma\xi_i\alpha_{ij}$, and this is what we obtain from (9).

It should be emphasized that the matrix (α) depends on the choice of bases in the two spaces. For this reason we call (α) *the matrix of A relative to the ordered bases* $(e_1, e_2, \cdots, e_{n_1})$ *and* $(f_1, f_2, \cdots, f_{n_2})$. If $\mathfrak{R}_1 = \mathfrak{R}_2 = \mathfrak{R}$, then it is natural to use just one ordered basis, that is, to take the $f_i = e_i$. In this case we refer to (α) as *the matrix of A relative to* $(e_1, e_2, \cdots, e_n)$.

The result that we have obtained is that any $A \,\varepsilon\, \mathfrak{L}(\mathfrak{R}_1, \mathfrak{R}_2)$ determines an $n_1 \times n_2$ matrix with elements in Δ. We now note the converse: that any $n_1 \times n_2$ matrix defines a linear transformation of $\mathfrak{R}_1$ into $\mathfrak{R}_2$. We note first that, if $(e_1, e_2, \cdots, e_{n_1})$ is a basis in $\mathfrak{R}_1$ and $(u_1, u_2, \cdots, u_{n_1})$ is an arbitrary ordered set of n_1 vectors in $\mathfrak{R}_2$, then there exists a linear transformation A mapping e_i into u_i for $i = 1, 2, \cdots, n_1$. Thus consider the mapping $\Sigma\xi_ie_i \rightarrow \Sigma\xi_iu_i$. Since there is only one way of writing a vector $x \,\varepsilon\, \mathfrak{R}_1$ as $\Sigma\xi_ie_i$, this mapping is single-valued and, since any

$x \varepsilon \mathfrak{R}_1$ can be written in the form $\Sigma\xi_i e_i$, the mapping is defined on the whole of $\mathfrak{R}_1$. If $y = \Sigma\eta_i e_i$ is a second vector in $\mathfrak{R}_1$, $y \to \Sigma\eta_i u_i$ and $x + y = \Sigma(\xi_i + \eta_i)e_i \to \Sigma(\xi_i + \eta_i)u_i = \Sigma\xi_i u_i + \Sigma\eta_i u_i$. Hence the transformation is a homomorphism. Moreover, $\alpha x = \Sigma(\alpha\xi_i)e_i \to \Sigma(\alpha\xi_i)u_i = \alpha(\Sigma\xi_i u_i)$ so that the mapping is linear. Clearly $e_i = 1e_i \to 1u_i = u_i$ as required. Now let (α) be any $n_1 \times n_2$ matrix and let $(f_1, f_2, \cdots, f_{n_2})$ be a basis in $\mathfrak{R}_2$. Then we define $u_i = \Sigma\alpha_{ij}f_j$, $i = 1, 2, \cdots, n_1$, and we can determine a linear transformation A such that $e_iA = u_i$. Evidently the matrix of A relative to $(e_1, e_2, \cdots, e_{n_1})$, $(f_1, f_2, \cdots, f_{n_2})$ is the given matrix (α). We have, therefore, shown that the correspondence $A \to (\alpha)$ is 1–1 between $\mathfrak{L}(\mathfrak{R}_1, \mathfrak{R}_2)$ and the set of $n_1 \times n_2$ matrices with elements in Δ.

EXERCISES

1. Let $\mathfrak{R}$ be the vector space of polynomials of degree $< n$ with real coefficients and let D denote the differentiation operator. Show that D is *nilpotent* in the sense that $D^n = 0$. Determine the matrix of D relative to $(1, \lambda, \cdots, \lambda^{n-1})$ and also relative to $(1, \lambda/1!, \cdots, \lambda^{n-1}/(n-1)!)$.

2. Let $\mathfrak{R}$ be as in 1. and let U be the linear operator $f(\lambda) \to f(\lambda + 1)$. Prove that

$$U = 1 + \frac{D}{1!} + \frac{D^2}{2!} + \cdots + \frac{D^{n-1}}{(n-1)!}.$$

3. Determine the matrix of $\delta = U - 1$ relative to the basis $(e_0, e_1, \cdots, e_{n-1})$ where

$$e_0 = 1, \quad e_i = \frac{\lambda(\lambda - 1) \cdots (\lambda - i + 1)}{i!}.$$

4. Let $\mathfrak{R}$ be the set of complex numbers regarded as a vector space over the subfield of real numbers. Show that the mapping $x \to \bar{x}$ (complex conjugate) is linear and determine its matrix relative to the basis $(1, i)$.

4. Compositions of matrices. As before, let $\mathfrak{R}_i$, $i = 1, 2$, be n_i dimensional vector spaces over Δ and let $(e_1, e_2, \cdots, e_{n_1})$, $(f_1, f_2, \cdots, f_{n_2})$ be bases for these two spaces. Let A and B be linear transformations of $\mathfrak{R}_1$ into $\mathfrak{R}_2$, (α) and (β), respectively, their matrices relative to the given bases. Then

$$e_iA = \Sigma\alpha_{ij}f_j, \quad e_iB = \Sigma\beta_{ij}f_j. \tag{10}$$

Hence

$$e_i(A + B) = e_iA + e_iB = \Sigma(\alpha_{ij} + \beta_{ij})f_j. \tag{11}$$

This shows that the matrix of $A + B$ is obtained from the matrices (α) and (β) by adding elements in the same position. Accordingly we define the *sum* of two $n_1 \times n_2$ matrices (α) and (β) as the matrix whose (i, j)-element is $\alpha_{ij} + \beta_{ij}$. It is easy to verify that the set of $n_1 \times n_2$ matrices is a commutative group relative to this addition. As a matter of fact, except for a difference of notation, this is a special case of the result noted in Chapter I that the set of n-tuples form a group under addition as addition of components. The 0 matrix is the matrix that has 0 in every place, and $-(\alpha)$ has the element $-\alpha_{ij}$ in its (i, j) position. Now the result that we established in (11)—namely, that if $A \to (\alpha)$ and $B \to (\beta)$ in the correspondence between linear transformations and matrices, then $A + B \to (\alpha) + (\beta)$—is equivalent to the statement that $A \to (\alpha)$ is a group isomorphism.

We consider next a third vector space $\mathfrak{R}_3$ with basis $(g_1, g_2, \cdots, g_{n_3})$. Let C be a linear transformation of $\mathfrak{R}_2$ into $\mathfrak{R}_3$ and let (γ) be the matrix of C relative to $(f_1, f_2, \cdots, f_{n_2})$, $(g_1, g_2, \cdots, g_{n_3})$. Then

$$f_jC = \Sigma \gamma_{jk} g_k \tag{12}$$

so that

$$e_i(AC) = \Big(\sum_j \alpha_{ij} f_j\Big)C = \sum_j \alpha_{ij}(f_jC) = \sum_{j,k} \alpha_{ij}\gamma_{jk} g_k. \tag{13}$$

This shows that the matrix of AC has the element $\sum_j \alpha_{ij}\gamma_{jk}$ in its (i, k)-position; hence this matrix is the product $(\alpha)(\gamma)$ as defined in Chapter I.

The associative law for matrix multiplication has been established before (p. 16). We now prove distributivity. The (i, k) element of $[(\alpha) + (\beta)](\gamma)$ is

$$\sum_j (\alpha_{ij} + \beta_{ij})\gamma_{jk}$$

while the (i, k) element of $(\alpha)(\gamma) + (\beta)(\gamma)$ is

$$\sum_j \alpha_{ij}\gamma_{jk} + \sum_j \beta_{ij}\gamma_{jk}.$$

Hence by the distributive law in Δ,

$$[(\alpha) + (\beta)](\gamma) = (\alpha)(\gamma) + (\beta)(\gamma).$$

Similarly we can verify that

$$(\alpha)[(\beta) + (\gamma)] = (\alpha)(\beta) + (\alpha)(\gamma).$$

We remark also that it is easy to deduce the associative and distributive laws for matrices from the corresponding laws for linear transformations (cf. Exercise 1 below).

We note next that the set of $n_1 \times n_2$ matrices with elements in Δ can be regarded as a vector space (or a right vector space) over Δ. This is clear since the set of matrices is essentially the same as the set of n_1n_2-tuples over Δ. As before, we define $\rho(\alpha)$ to be the $n_1 \times n_2$ matrix whose elements are ρ times the corresponding elements of Δ. Clearly the vector space that we obtain in this way is n_1n_2 dimensional. If we use our correspondence between matrices and linear transformations, we can carry over this discussion to the set $\mathfrak{L}(\mathfrak{R}_1, \mathfrak{R}_2)$ of linear transformations of $\mathfrak{R}_1$ into $\mathfrak{R}_2$. However, unless $\Delta = \Phi$ is commutative, the scalar multiplication in $\mathfrak{L}(\mathfrak{R}_1, \mathfrak{R}_2)$ obtained in this way depends on the choice of the bases in $\mathfrak{R}_1$ and $\mathfrak{R}_2$.

On the other hand, let $\Delta = \Phi$. Then, as we saw in § 2, there is a way of defining a multiplication of linear transformations by elements of Φ which is independent of the choice of bases. The product γA, γ in Φ, is taken to be the resultant $\gamma_l A$. Relative to this composition $\mathfrak{L}(\mathfrak{R}_1, \mathfrak{R}_2)$ is a vector space over Φ. We now note that, if $(e_1, e_2, \cdots, e_{n_1})$ is a basis for $\mathfrak{R}_1$, then $e_i\gamma_l = \gamma e_i$. Hence the matrix of γ_l relative to this basis is the diagonal matrix

$$\text{(14)} \qquad \operatorname{diag}\{\gamma, \gamma, \cdots, \gamma\} \equiv \begin{bmatrix} \gamma & & & 0 \\ & \gamma & & \\ & & \ddots & \\ 0 & & & \gamma \end{bmatrix}.$$

Consequently if (α) is the matrix of A relative to $(e_1, e_2, \cdots, e_{n_1})$, $(f_1, f_2, \cdots, f_{n_2})$, then $\gamma(\alpha)$ is the matrix of γA relative to this pair of bases. This means that the scalar multiplication $\gamma(\alpha)$ corresponds to γA.

Another way of stating this result is the following. Let $A \rightarrow (\alpha)$ be the correspondence which associates with a linear transformation $A \in \mathfrak{L}(\mathfrak{R}_1, \mathfrak{R}_2)$ its matrix (α) relative to the bases $(e_1, e_2, \cdots, e_{n_1})$, $(f_1, f_2, \cdots, f_{n_2})$. Then this correspondence is an equivalence of the vector space $\mathfrak{L}(\mathfrak{R}_1, \mathfrak{R}_2)$ onto the vector space of $n_1 \times n_2$ matrices with elements in Φ; for we have seen that $A \rightarrow (\alpha)$ is a group isomorphism and we have just verified that $\gamma A \rightarrow \gamma(\alpha)$. Since the matrix space is n_1n_2-dimensional, it follows from this that $\mathfrak{L}(\mathfrak{R}_1, \mathfrak{R}_2)$ is n_1n_2-dimensional. This proves the following

Theorem 1. *Let $\mathfrak{R}_i$, $i = 1, 2$, be n_i-dimensional vector spaces over a field Φ and let $\mathfrak{L}(\mathfrak{R}_1, \mathfrak{R}_2)$ be the set of linear transformations of $\mathfrak{R}_1$ into $\mathfrak{R}_2$. Define $A + B$, and γA for γ in Φ as above. Then $\mathfrak{L}(\mathfrak{R}_1, \mathfrak{R}_2)$ is an n_1n_2-dimensional vector space relative to these compositions.*

We return now to the case of an arbitrary Δ, but we specialize by taking $\mathfrak{R}_1 = \mathfrak{R}_2 = \mathfrak{R}$. Also we take the $f_i = e_i$ so that (α) is now the matrix of A relative to the single basis $(e_1, e_2, \cdots, e_n)$. In this case we have a correspondence $A \rightarrow (\alpha)$ of the ring $\mathfrak{L} = \mathfrak{L}(\mathfrak{R}, \mathfrak{R})$ onto the set Δ_n of $n \times n$ matrices. The addition and multiplication compositions introduced in Δ_n turn this set into a ring. Also our results show that, if $A \rightarrow (\alpha)$, $B \rightarrow (\beta)$, then $A + B \rightarrow (\alpha) + (\beta)$ and $AB \rightarrow (\alpha)(\beta)$. Hence we have the important

Theorem 2. *Let $\mathfrak{R}$ be an n-dimensional vector space with the basis $(e_1, e_2, \cdots, e_n)$ over Δ. If A is a linear transformation in $\mathfrak{R}$, we associate with A its matrix (α) relative to the basis $(e_1, e_2, \cdots, e_n)$. Then $A \rightarrow (\alpha)$ is an isomorphism of the ring $\mathfrak{L}$ of linear transformations in $\mathfrak{R}$ onto the matrix ring Δ_n.*

EXERCISE

1. Prove associativity and distributivity of matrix multiplication by using the corresponding properties of multiplication of linear transformations.

5. Change of basis. Equivalence and similarity of matrices. Let (α) be the matrix of $A \in \mathfrak{L}(\mathfrak{R}_1, \mathfrak{R}_2)$ relative to the bases $(e_1, e_2, \cdots, e_{n_1})$, $(f_1, f_2, \cdots, f_{n_2})$. We now change the bases in $\mathfrak{R}_1$

and $\mathfrak{R}_2$, and we shall calculate the matrix of A relative to the new bases. Thus let $(u_1, u_2, \cdots, u_{n_1})$ be a second basis in $\mathfrak{R}_1$ where $u_i = \Sigma\mu_{ij}e_j$ and let $(v_1, v_2, \cdots, v_{n_2})$ be a second basis in $\mathfrak{R}_2$ where $v_p = \Sigma\nu_{pq}f_q$. The matrices (μ) and (ν) are non-singular, and we write their inverses as $(\mu)^{-1} = (\mu_{ij}^*)$, $(\nu)^{-1} = (\nu_{pq}^*)$. Now we have

$$u_iA = (\Sigma\mu_{ij}e_j)A = \Sigma\mu_{ij}(e_jA) = \Sigma\mu_{ij}\alpha_{jp}f_p$$

$$= \Sigma\mu_{ij}\alpha_{jp}\nu_{pq}{}^*v_q = \Sigma\tilde{\alpha}_{iq}v_q$$

where

$$\tilde{\alpha}_{iq} = \sum_{j,p} \mu_{ij}\alpha_{jp}\nu_{pq}{}^*.$$

Hence the new matrix of A is

$$(\tilde{\alpha}) = (\mu)(\alpha)(\nu)^{-1} \tag{15}$$

where (μ) gives the change of basis in $\mathfrak{R}_1$ and where (ν) gives the change of basis in $\mathfrak{R}_2$.

Now we shall call two $n_1 \times n_2$ matrices (α) and $(\tilde{\alpha})$ *equivalent (or associates)* if there exist matrices (μ) and (ν) in $L(\Delta, n_1)$ and $L(\Delta, n_2)$ respectively such that

$$(\tilde{\alpha}) = (\mu)(\alpha)(\nu). \tag{16}$$

Thus we see that any two matrices of the linear transformation A relative to different bases in the two spaces are equivalent. The converse is also clear. For let (α) and $(\tilde{\alpha})$ be related as in (16) and let A be the linear transformation whose matrix is (α) relative to $(e_1, e_2, \cdots, e_{n_1})$ and $(f_1, f_2, \cdots, f_{n_2})$. Then this linear transformation has the matrix $(\tilde{\alpha})$ relative to $(u_1, u_2, \cdots, u_{n_1})$, $(w_1, w_2, \cdots, w_{n_2})$ where

$$u_i = \Sigma\mu_{ij}e_j, \quad w_p = \Sigma\nu_{pq}{}^*f_q, \quad (\nu)^{-1} = (\nu_{pq}{}^*).$$

Assume next that $\mathfrak{R}_1 = \mathfrak{R}_2 = \mathfrak{R}$ and that $(e_1, e_2, \cdots, e_n)$ is a basis. Let (α) be the matrix of A relative to $(e_1, e_2, \cdots, e_n)$. Then $e_iA = \Sigma\alpha_{ij}e_j$, and our computation shows that the matrix of A relative to $(u_1, u_2, \cdots, u_n)$, $u_i = \Sigma\mu_{ij}e_j$ is

$$(\tilde{\alpha}) = (\mu)(\alpha)(\mu)^{-1}. \tag{17}$$

Two matrices in Δ_n that are related in this way are said to be *similar*. As in the case of equivalence it is clear that two matrices in Δ_n are similar if and only if both are matrices of the same linear transformation relative to (single) bases for $\mathfrak{R}$ over Δ.

As we shall see in the next section, it is easy to give necessary and sufficient conditions for equivalence of matrices. On the other hand, the problem of similarity requires a fairly elaborate analysis, which we shall undertake in the next chapter. At this point we illustrate the method that we shall use to handle this problem.

Example. We wish to prove that the matrices

$$(\alpha) = \begin{bmatrix} 0 & 1 & 0 & \cdots & 0 \\ 0 & 0 & 1 & \cdots & 0 \\ \cdot & \cdot & \cdot & \cdots & \cdot \\ 0 & 0 & 0 & \cdots & 1 \\ 1 & 0 & 0 & \cdots & 0 \end{bmatrix}, \quad (\beta) = \begin{bmatrix} \zeta_1 & & & & 0 \\ & \zeta_2 & & & \\ & & \cdot & & \\ & & & \cdot & \\ 0 & & & & \zeta_n \end{bmatrix}$$

where the ζ_i are the n distinct nth roots of unity, are similar in C_n, C the field of complex numbers. We use (α) to determine a linear transformation A in an n dimensional vector space $\mathfrak{R}$ over C. This is done by choosing a basis $(e_1, e_2, \cdots, e_n)$ in $\mathfrak{R}$ and defining $e_iA = \sum_j \alpha_{ij}e_j = e_{i+1}$ for $i < n$ and $= e_1$ for $i = n$. Then in order to prove our assertion, we must find a basis $(u_1, u_2, \cdots, u_n)$ such that $u_iA = u_i\zeta_i$ for the A that we have just defined. Without giving the details as to how one goes about finding such u_i, we shall show that the following u_i satisfy the requirements.

$$u_i = e_1 + \zeta_i^{-1}e_2 + \zeta_i^{-2}e_3 + \cdots + \zeta_i^{-(n-1)}e_n.$$

Thus

$$u_iA = e_2 + \zeta_i^{-1}e_3 + \cdots + \zeta_i^{-(n-1)}e_1 = \zeta_i u_i,$$

and the u_i form a basis since the matrix of $(u_1, u_2, \cdots, u_n)$ relative to $(e_1, e_2, \cdots, e_n)$ is the Vandermonde matrix

$$(\mu) = \begin{bmatrix} 1 & \zeta_1^{-1} & \zeta_1^{-2} & \cdots & \zeta_1^{-(n-1)} \\ 1 & \zeta_2^{-1} & \zeta_2^{-2} & \cdots & \zeta_2^{-(n-1)} \\ \cdot & \cdot & \cdot & \cdots & \cdot \\ 1 & \zeta_n^{-1} & \zeta_n^{-2} & \cdots & \zeta_n^{-(n-1)} \end{bmatrix}$$

in which the ζ_i^{-1} are distinct (cf. Exercise 1, p. 24). This proves the similarity and shows in fact that $(\beta) = (\mu)(\alpha)(\mu)^{-1}$ where (μ) is the above matrix.

EXERCISES

1. Prove that the relations of equivalence and of similarity are reflexive, symmetric and transitive.

2. Prove that, if Δ has characteristic 0, then the following two matrices are similar in Δ_n

$$(\alpha) = \begin{bmatrix} 1 & 1 & \cdots & 1 \\ 1 & 1 & \cdots & 1 \\ \cdot & \cdot & \cdots & \cdot \\ 1 & 1 & \cdots & 1 \end{bmatrix}, \quad (\beta) = \begin{bmatrix} n & 0 & \cdots & 0 \\ 0 & 0 & \cdots & 0 \\ \cdot & \cdot & \cdots & \cdot \\ 0 & 0 & \cdots & 0 \end{bmatrix}.$$

6. Rank space and null space of a linear transformation. If A is a linear transformation of $\mathfrak{R}_1$ into $\mathfrak{R}_2$, and $\mathfrak{S}_1$ is a subspace of $\mathfrak{R}_1$, then the image $\mathfrak{S}_1 A$ consisting of all vectors of the form $x_1 A$, x_1 in $\mathfrak{S}_1$, is a subspace of $\mathfrak{R}_2$. If $x_1, y_1 \varepsilon \mathfrak{S}_1$, then $x_1 + y_1 \varepsilon \mathfrak{S}_1$, and so $x_1 A + y_1 A = (x_1 + y_1) A \varepsilon \mathfrak{S}_1 A$. Also, if $x_1 \varepsilon \mathfrak{S}_1$, $\alpha x_1 \varepsilon \mathfrak{S}_1$; hence $\alpha(x_1 A) = (\alpha x_1) A \varepsilon \mathfrak{S}_1 A$. If the vectors $f_1, f_2, \cdots, f_m$ are generators for $\mathfrak{S}_1$, any $x_1 \varepsilon \mathfrak{S}_1$ has the form $\Sigma \xi_i f_i$. Hence any $x_1 A = \Sigma \xi_i (f_i A)$. Thus the image vectors $f_1 A, f_2 A, \cdots, f_m A$ are generators for $\mathfrak{S}_1 A$. If the f_i form a basis for $\mathfrak{S}_1$, $m = \dim \mathfrak{S}_1$. The images $f_i A$ need not form a basis for $\mathfrak{S}_1 A$, but since they are, in any case, generators, their number $m \geq \dim \mathfrak{S}_1 A$. Hence we see that the dimensionality of the image space never exceeds that of the original space.

We shall call the subspace $\mathfrak{R}_1 A$ of $\mathfrak{R}_2$ the *rank space* of A; its dimensionality the *rank* of A. If $(e_1, e_2, \cdots, e_{n_1})$ is a basis for $\mathfrak{R}_1$, $\mathfrak{R}_1 A = [e_1 A, e_2 A, \cdots, e_{n_1} A]$, the space spanned by the vectors $e_i A$. Hence the rank of A is the rank of the set $(e_1 A, e_2 A, \cdots, e_{n_1} A)$. If $(f_1, f_2, \cdots, f_{n_2})$ is a basis for $\mathfrak{R}_2$ and $e_i A = \sum_j \alpha_{ij} f_j$, $i = 1, 2, \cdots, n_1$, then the rank of the set $(e_1 A, e_2 A, \cdots, e_{n_1} A)$ is the same as the row rank of the matrix (α) of A determined by the bases $(e_1, e_2, \cdots, e_{n_1})$, $(f_1, f_2, \cdots, f_{n_2})$. This proves the following

Theorem 3. *The rank of a linear transformation A of a vector space equals the row rank of any matrix of A.*

We consider next the totality $\mathfrak{N}$ of vectors z in $\mathfrak{R}_1$ such that $zA = 0$. It is readily verified that $\mathfrak{N}$ is a subspace of $\mathfrak{R}_1$. We call it the *null space* of A and its dimensionality the *nullity* of A.

We prove now the main theorem relating the rank and the nullity of A.

Theorem 4. *Rank of* A + *nullity of* $A = n_1$, *the dimensionality of the space* $\mathfrak{R}_1$.

Proof. Let $(z_1, z_2, \cdots, z_\nu)$ be a basis for $\mathfrak{N}$. We can supplement this basis by $n_1 - \nu$ vectors x_i to obtain the basis $(x_1, \cdots, x_{n_1-\nu}; z_1, z_2, \cdots, z_\nu)$ for $\mathfrak{R}_1$. The vectors

$$x_1A, x_2A, \cdots, x_{n_1-\nu}A, z_1A, \cdots, z_\nu A$$

are generators for the rank space $\mathfrak{R}_1A \subseteq \mathfrak{R}_2$. Since the $z_iA = 0$, the vectors $x_1A, \cdots, x_{n_1-\nu}A$ are also generators for $\mathfrak{R}_1A$. But these vectors are linearly independent. Thus if

$$\beta_1(x_1A) + \beta_2(x_2A) + \cdots + \beta_{n_1-\nu}(x_{n_1-\nu}A) = 0,$$

then $(\Sigma\beta_ix_i)A = 0$ and $\Sigma\beta_ix_i \in \mathfrak{N} = [z_1, z_2, \cdots, z_\nu]$. Since the set $(x_1, \cdots, x_{n_1-\nu}, z_1, \cdots, z_\nu)$ is an independent set, this implies that the β_i are all 0. Hence if we set $y_i = x_iA$, $(y_1, y_2, \cdots, y_{n_1-\nu})$ is a basis for $\mathfrak{R}_1A$. Thus $\dim \mathfrak{R}_1A = n_1 - \nu = n_1 -$ nullity of A and this proves the theorem.

We now supplement the basis $(y_1, y_2, \cdots, y_{n_1-\nu})$ of $\mathfrak{R}_1A$ to a basis $(y_1, y_2, \cdots, y_{n_1-\nu}, w_1, w_2, \cdots, w_{n_2-n_1+\nu})$ for $\mathfrak{R}_2$. Then we have the relations

$$x_iA = y_i, \quad i = 1, 2, \cdots, \rho \equiv n_1 - \nu$$

$$z_jA = 0, \quad j = 1, \cdots, \nu.$$

These show that the matrix of A relative to the bases $(x_1, \cdots, x_\rho, z_1, \cdots, z_\nu)$, $(y_1, \cdots, y_\rho, w_1, \cdots, w_{n_2-\rho})$ is

$$\text{diag}\,\{\overbrace{1, \cdots, 1}^{\rho}, 0, \cdots, 0\}.* \tag{18}$$

If the matrices of these bases relative to the original bases $(e_1, e_2, \cdots, e_{n_1})$ and $(f_1, f_2, \cdots, f_{n_2})$ are respectively (μ) and (ν), then, as we have seen in the preceding section, $(\mu)(\alpha)(\nu)^{-1}$ is the matrix (18). The number ρ is the row rank of the matrix (α). This proves the following

* We use this notation, introduced in (14), for a matrix whose non-zero entries occur only in the (1,1), (2,2), etc., positions.

Theorem 5. *If* (α) *is an* $n_1 \times n_2$ *matrix with elements in a division ring* Δ *and if* (α) *has row rank* ρ, *then* (α) *is equivalent to the matrix given in equation* (18).

If (α) and $(\tilde{\alpha})$ are equivalent matrices, then we know that these can be taken to be matrices of the same linear transformation A. The row ranks of (α) and of $(\tilde{\alpha})$ coincide with the rank of A. Hence equivalent matrices have the same row rank. Conversely if (α) and $(\tilde{\alpha})$ have the same row rank ρ, then both of these matrices are equivalent to the same matrix (18). Hence they are equivalent.

Theorem 6. *Two* $n_1 \times n_2$ *matrices with elements in a division ring* Δ *are equivalent if and only if they have the same row rank.*

We consider now the conditions that a linear transformation A of $\mathfrak{R}_1$ into $\mathfrak{R}_2$ be an equivalence. Since A is a homomorphism, A is 1–1 if and only if its kernel $\mathfrak{N} = 0$. Clearly $\mathfrak{N}$ is the null-space of A. Hence A is a 1–1 transformation of $\mathfrak{R}_1$ onto $\mathfrak{R}_2$ if and only if: (1) $\mathfrak{N} = 0$; (2) $\mathfrak{R}_1 A = \mathfrak{R}_2$. In the special case in which $\mathfrak{R}_1 = \mathfrak{R}_2 = \mathfrak{R}$ either of these conditions is sufficient; for if $\mathfrak{N} = 0$, rank $A = \dim \mathfrak{R} = n$. Hence $\mathfrak{R}A = \mathfrak{R}$. On the other hand, if $\mathfrak{R}A = \mathfrak{R}$, rank $A = n$ and nullity $A = 0$. Hence $\mathfrak{N} = 0$.

It should be noted here, too, that, if A is an equivalence, then its inverse A^{-1} is also an equivalence. The verification is left to the reader. The equivalences of a vector space onto itself constitute a group relative to the resultant operation. If A is a linear transformation in $\mathfrak{R}$ and (α) is its matrix relative to the basis $(e_1, e_2, \cdots, e_n)$, then A is an equivalence in $\mathfrak{R}$ if and only if (α) is a unit. Hence we see that the group of equivalences in $\mathfrak{R}$ is isomorphic to the group $L(\Delta, n)$ of non-singular matrices in Δ_n. The former group is called the *full linear group* in the vector space $\mathfrak{R}$.

EXERCISES

1. Prove that, if $A \varepsilon \mathfrak{L}(\mathfrak{R}_1, \mathfrak{R}_2)$ and $\mathfrak{S}_1$ and $\mathfrak{U}_1$ are subspaces of $\mathfrak{R}_1$, then $(\mathfrak{S}_1 + \mathfrak{U}_1)A = \mathfrak{S}_1 A + \mathfrak{U}_1 A$ and $(\mathfrak{S}_1 \cap \mathfrak{U}_1)A \subseteq \mathfrak{S}_1 A \cap \mathfrak{U}_1 A$.

2. Prove that, if (α) and (β) are $m \times n$ matrices with elements in Δ, then (row) rank $[(\alpha) + (\beta)] \leq$ rank (α) + rank (β).

3. Prove that, if (α) is an $m \times n$ matrix and (β) is an $n \times p$ matrix with elements in Δ, then rank $(\alpha)(\beta) \leq \min$ (rank (α), rank (β)).

4. If

$$(\alpha) = \begin{bmatrix} 1 & -2 & 3 \\ 2 & 0 & 1 \end{bmatrix}$$

find units (μ) and (ν) such that $(\mu)(\alpha)(\nu)$ has the form (18).

5. (Fitting). If A is a linear transformation in $\mathfrak{R}$, let $\mathfrak{Z}$ be the set of vectors z such that $zA^m = 0$ for some m and let $\mathfrak{S}$ be the intersection of all the rank spaces $\mathfrak{R}A^k$, $k = 1, 2, \cdots$. Show that $\mathfrak{Z}$ and $\mathfrak{S}$ are subspaces and prove that $\mathfrak{R} = \mathfrak{Z} \oplus \mathfrak{S}$.

6. Show that A maps the two spaces $\mathfrak{Z}$ and $\mathfrak{S}$ of Ex. 5 into themselves; that A is nilpotent in $\mathfrak{Z}$ and an equivalence in $\mathfrak{S}$. Use this result to prove that any matrix in Δ_n is similar to a matrix of the form

$$\begin{bmatrix} (\beta) & 0 \\ 0 & (\gamma) \end{bmatrix}$$

where (β) is nilpotent and where (γ) is non-singular.

7. Systems of linear equations. We consider the left-handed system of linear equations

$$\begin{aligned} \xi_1\alpha_{11} + \xi_2\alpha_{21} + \cdots + \xi_n\alpha_{n_1} &= \delta_1 \\ \xi_1\alpha_{12} + \xi_2\alpha_{22} + \cdots + \xi_n\alpha_{n_2} &= \delta_2 \\ \cdots\cdots\cdots\cdots\cdots\cdots \\ \xi_1\alpha_{1m} + \xi_2\alpha_{2m} + \cdots + \xi_n\alpha_{nm} &= \delta_m. \end{aligned} \tag{19}$$

Here the α_{ij} and δ_i are given elements of the division ring Δ, and we seek solutions $\xi_i = \beta_i$ in Δ. If $\xi_i = \beta_i$ satisfies these equations, then we say that the n-tuple $(\beta_1, \beta_2, \cdots, \beta_n)$ is a solution. If $(\beta_1, \beta_2, \cdots, \beta_n)$ and $(\beta_1', \beta_2', \cdots, \beta_n')$ are two solutions of (19), then $(\gamma_1, \gamma_2, \cdots, \gamma_n)$ where $\gamma_i = \beta_i' - \beta_i$ is a solution of the system of *homogeneous* equations obtained by taking the $\delta_i = 0$ in (19). Conversely if $(\beta_1, \beta_2, \cdots, \beta_n)$ is a solution of (19) and $(\gamma_1, \gamma_2, \cdots, \gamma_n)$ is a solution of the homogeneous system, then $(\beta_1', \beta_2', \cdots, \beta_n')$ where $\beta_i' = \beta_i + \gamma_i$ is another solution of (19). This shows that in order to obtain the solutions, if any, of (19) we have to find a particular solution of this system and all solutions of the corresponding homogeneous system. We will then obtain all solutions of (19) by adding to the particular solution all solutions of the homogeneous system.

We therefore consider first the question of the existence of solutions for (19). We introduce the vectors $u_i = \Sigma\alpha_{ij}f_j$, $i = 1$,

$2, \cdots, n$, $v = \Sigma\delta_j f_j$ where $(f_1, f_2, \cdots, f_m)$ is a basis for the m dimensional space $\mathfrak{S}$. Then it is immediate that $(\beta_1, \beta_2, \cdots, \beta_n)$ is a solution of (19) if and only if the β_i satisfy

$$v = \beta_1 u_1 + \beta_2 u_2 + \cdots + \beta_n u_n. \tag{20}$$

Hence (20) is solvable if and only if v is linearly dependent on $u_1, u_2, \cdots, u_n$. On the other hand, v is linearly dependent on the u_i if and only if

$$\text{rank } (u_1, u_2, \cdots, u_n) = \text{rank } (u_1, u_2, \cdots, u_n, v),$$

and this in turn holds if and only if the row rank of the matrix (α) is the same as that of the augmented matrix

$$\begin{bmatrix} \alpha_{11} & \alpha_{12} & \cdots & \alpha_{1m} \\ \alpha_{21} & \alpha_{22} & \cdots & \alpha_{2m} \\ \cdot & \cdot & \cdots & \cdot \\ \alpha_{n1} & \alpha_{n2} & \cdots & \alpha_{nm} \\ \delta_1 & \delta_2 & \cdots & \delta_m \end{bmatrix}. \tag{21}$$

In particular, if $\Delta = \Phi$ is commutative, then we have the following

Theorem 7. *A system of linear equations* (19) *with* α_{ij} *and* δ_j *in a field* Φ *has a solution* $\xi_i = \beta_i$ *in* Φ *if and only if the matrix* (α) *of the coefficients and the augmented matrix* (21) *have the same determinantal ranks.*

We consider next the homogeneous system obtained by setting $\delta_j = 0$. To study this system we introduce also an n dimensional vector space $\mathfrak{R}$ with the basis $(e_1, e_2, \cdots, e_n)$, and we let A be the linear transformation whose matrix is (α) relative to the bases $(e_1, e_2, \cdots, e_n)$, $(f_1, f_2, \cdots, f_m)$. Then in the above notation the vectors $u_i = e_i A$ and $(\beta_1, \beta_2, \cdots, \beta_n)$ constitute a solution of the homogeneous system if and only if $\Sigma\beta_i u_i = 0$. Since $u_i = e_i A$, this amounts to the condition $(\Sigma\beta_i e_i)A = 0$. Thus $(\beta_1, \beta_2, \cdots, \beta_n)$ is a solution if and only if $\Sigma\beta_i e_i$ is in the null space $\mathfrak{N}$ of A. If ν is the nullity of A, we have a basis $(z_1, z_2, \cdots, z_\nu)$ for $\mathfrak{N}$ and, if $z_k = \Sigma\beta_i^{(k)} e_i$, $k = 1, 2, \cdots, \nu$, then

$$(\beta_1^{(1)}, \beta_2^{(1)}, \cdots, \beta_n^{(1)}), \cdots, (\beta_1^{(\nu)}, \beta_2^{(\nu)}, \cdots, \beta_n^{(\nu)})$$

is a set of (left) linearly independent solutions of the homogeneous system. Moreover, any solution $(\beta_1, \beta_2, \cdots, \beta_n)$ is a linear combination of these solutions. Because of the relation between rank and nullity, we know that $\nu = n - \rho$ where ρ is the row rank of the matrix (α). We therefore have the following result:

Theorem 8. *Let* $\sum_{i=1}^{n} \xi_i \alpha_{ij} = 0, j = 1, 2, \cdots, m,$ *be a left-handed system of homogeneous equations and let the row rank of* (α) *be* ρ. *Then there exist* $n - \rho$ *linearly independent solutions* $(\beta_1^{(k)}, \beta_2^{(k)}, \cdots, \beta_n^{(k)})$, $k = 1, 2, \cdots, n - \rho$, *such that any solution of the system is a left linear combination of these solutions.*

An immediate consequence of this theorem is the result, noted previously in an exercise (p. 13), that a system of m homogeneous equations in more than m unknowns has a non-trivial solution. We remark also that in the commutative case we can drop the modifier "left" in the above statement and replace rank by determinantal rank.

EXERCISE

Find all the solutions of the following system

$$\begin{aligned} 2\xi_1 - \xi_2 + \xi_3 - 3\xi_4 &= 0 \\ \xi_1 + \xi_2 - \xi_3 + 2\xi_4 &= 0 \\ 4\xi_1 + \xi_2 - \xi_3 + \xi_4 &= 0, \end{aligned}$$

Δ, the field of rational numbers.

8. Linear transformations in right vector spaces. If $\mathfrak{R}_1'$ and $\mathfrak{R}_2'$ are right vector spaces, a *linear transformation* of $\mathfrak{R}_1'$ into $\mathfrak{R}_2'$ is defined to be a mapping of $\mathfrak{R}_1'$ into $\mathfrak{R}_2'$ such that

$$(x' + y')A = x'A + y'A, \quad (x'\alpha)A = (x'A)\alpha \tag{22}$$

for all x', y' in $\mathfrak{R}_1'$ and all α in Δ. The discussion for left vector spaces can be carried over to the present situation with one or two notational changes. If $(e_1', e_2', \cdots, e_{n_1}')$ and $(f_1', f_2', \cdots, f_{n_2}')$ are bases in $\mathfrak{R}_1'$ and $\mathfrak{R}_2'$, respectively, we write

$$e_i'A = \sum_{j=1}^{n_2} f_j'\alpha_{ji}, \quad i = 1, 2, \cdots, n_1 \tag{23}$$

and we call

$$(24)\qquad \begin{bmatrix} \alpha_{11} & \alpha_{12} & \cdots & \alpha_{1n_2} \\ \alpha_{21} & \alpha_{22} & \cdots & \alpha_{2n_2} \\ \cdot & \cdot & \cdots & \cdot \\ \alpha_{n_11} & \alpha_{n_12} & \cdots & \alpha_{n_1n_2} \end{bmatrix}$$

the matrix of the linear transformation A relative to the given bases. It should be noted that (24) is the transposed of the matrix of the coefficients on the right-hand side of (23). As before, to the sum of linear transformations corresponds the sum of matrices. The situation for the product is, however, different from that of left vector spaces. Suppose that $(g_1', g_2', \cdots, g_{n_3}')$ is a basis for $\mathfrak{R}_3'$ and let B be a linear transformation of $\mathfrak{R}_2'$ into $\mathfrak{R}_3'$. Let (β) be the matrix of B relative to the bases $(f_1', f_2', \cdots, f_{n_2}')$, $(g_1', g_2', \cdots, g_{n_3}')$. Then

$$f_j'B = \Sigma g_k'\beta_{kj}, \quad j = 1, 2, \cdots, n_2.$$

Hence

$$\begin{aligned} e_i'AB &= \Big(\sum_j f_j'\alpha_{ji}\Big)B = \sum_j (f_j'B)\alpha_{ji} = \sum_{j,k} g_k'\beta_{kj}\alpha_{ji} \\ &= \Sigma g_k'\gamma_{ki} \end{aligned}$$

where $\gamma_{ki} = \sum_j \beta_{kj}\alpha_{ji}$. Thus the matrix of $C = AB$ is the product $(\beta)(\alpha)$ and not $(\alpha)(\beta)$ as before.

If $\mathfrak{L}' = \mathfrak{L}'(\mathfrak{R}', \mathfrak{R}')$ denotes the ring of linear transformations of $\mathfrak{R}'$ into itself, then the correspondence $A \to (\alpha)$ between the linear transformations and their matrices relative to a definite basis $(e_1', e_2', \cdots, e_n')$ is now an anti-automorphism. Thus it is 1–1, and to the sum of linear transformations corresponds the sum of the matrices, and to the product of linear transformations corresponds the product of the corresponding matrices taken in reverse order. The ring $\mathfrak{L}'$ is anti-isomorphic to the matrix ring Δ_n and hence also to the ring $\mathfrak{L}$ of linear transformations in an n-dimensional left vector space over Δ.

A change of basis from $(e_1', e_2', \cdots, e_{n_1}')$ to $(u_1', u_2', \cdots, u_{n_1}')$ where $u_i' = \Sigma e_j'\mu_{ji}$ and a change of basis from $(f_1', f_2', \cdots, f_{n_2}')$ to $(v_1', v_2', \cdots, v_{n_2}')$ where $v_p' = \Sigma f_q'\nu_{qp}$ is now reflected in a change of the matrix of A from (α) to

$$(\tilde{\alpha}) = (\nu)^{-1}(\alpha)(\mu).$$

If a single basis is used in $\mathfrak{R}_1' = \mathfrak{R}_2' = \mathfrak{R}'$, the matrix (α) of A is replaced on changing the basis by the similar one $(\mu)^{-1}(\alpha)(\mu)$ where (μ) gives the change of basis.

Our discussion of rank and nullity may be carried over without change. The rank of A now turns out to be the column rank of its matrices. This implies that equivalent matrices have the same column rank as well as the same row rank. Now we have seen that any matrix (α) is equivalent to one in the "normal" form (18). Also it is immediate from the definition that a matrix in normal form has the same row rank and column rank. It follows that the arbitrary matrix (α) has the same row rank and column rank. We state this result as

Theorem 9. *The row rank and the column rank of any matrix are the same.*

A second, and somewhat more geometric, proof of this result will be given in § 12. We note finally that the theory of right-handed systems of linear equations can be developed in a manner completely analogous to that of left-handed systems considered above. We have only to replace left vector spaces by right vector spaces.

EXERCISE

1. State and prove the analogue of Theorem 7 for right-handed systems.

9. Linear functions. We have defined a linear function on a vector space $\mathfrak{R}$ to be a mapping $x \to f(x)$ of $\mathfrak{R}$ into Δ such that

$$f(x + y) = f(x) + f(y), \quad f(\alpha x) = \alpha f(x).$$

If we define the sum of two such mappings in the usual way by

$$(f + g)(x) = f(x) + g(x)$$

and the product $f\mu$ for μ in Δ by

$$(f\mu)(x) = f(x)\mu,$$

then we obtain a right vector space $\mathfrak{R}^*$ over Δ. This can be verified directly. However, it may be more illuminating to integrate this result into the general theory of linear transformations developed in § 2.

For this purpose we recall that a linear function is just a linear transformation of $\mathfrak{R}$ into the one-dimensional vector space Δ. Now the right multiplication μ_r: $\xi \rightarrow \xi\mu$ is a linear transformation in the vector space Δ, since

$$(\xi + \eta)\mu_r = (\xi + \eta)\mu = \xi\mu + \eta\mu = \xi\mu_r + \eta\mu_r$$
$$(\alpha\xi)\mu_r = (\alpha\xi)\mu = \alpha(\xi\mu) = \alpha(\xi\mu_r).$$

Also we know that the set of linear transformations of $\mathfrak{R}$ into Δ is a commutative group under addition and that the resultant of a linear transformation of $\mathfrak{R}$ into Δ and a linear transformation of Δ into itself is a linear transformation of $\mathfrak{R}$ into Δ. Hence if f and $g \in \mathfrak{L}(\mathfrak{R}, \Delta)$, then $f + g$ and $f\mu_r \in \mathfrak{L}(\mathfrak{R}, \Delta)$. Since the definition of $f\mu$ given above is simply the resultant $f\mu_r$,

$$(f + g)\mu = (f + g)\mu_r = f\mu_r + g\mu_r = f\mu + g\mu$$
$$f(\mu + \nu) = f(\mu + \nu)_r = f(\mu_r + \nu_r) = f\mu_r + f\nu_r = f\mu + f\nu$$
$$f(\mu\nu) = f(\mu\nu)_r = f(\mu_r\nu_r) = (f\mu_r)\nu_r = (f\mu)\nu$$
$$f1 = f1_r = f.$$

This establishes the assertion made above that the set $\mathfrak{R}^*$ of linear functions is a right vector space over Δ. We call this space the *conjugate space* of the vector space $\mathfrak{R}$.

Now suppose $\mathfrak{R}$ is finite dimensional with the basis $(e_1, e_2, \cdots, e_n)$. A linear function f is completely determined by the values $f(e_i) = \alpha_i$. Moreover, for arbitrary $\alpha_i \in \Delta$ there exists a linear function such that $f(e_i) = \alpha_i$. If $x = \Sigma\xi_i e_i$, then $f(x) = \Sigma\xi_i\alpha_i$. We can also state these simple results in terms of matrices as follows. If f is a linear function and $f(e_i) = \alpha_i$, then we can use the basis 1 in Δ and write $f(e_i) = \alpha_i 1$. Then we see that the matrix of f relative to $(e_1, e_2, \cdots, e_n)$, (1) is

$$\begin{bmatrix} \alpha_1 \\ \alpha_2 \\ \cdot \\ \cdot \\ \cdot \\ \alpha_n \end{bmatrix} \tag{25}$$

We obtain in this way a 1–1 correspondence between $\mathfrak{R}^*$ and the set of $n \times 1$ matrices with elements in Δ. Sums correspond under this correspondence. Also it is immediate that, if f has the matrix (25), then $f\mu$ has the matrix

$$\begin{bmatrix} \alpha_1\mu \\ \alpha_2\mu \\ \cdot \\ \cdot \\ \cdot \\ \alpha_n\mu \end{bmatrix}.$$

From this it is clear that $\mathfrak{R}^*$ is equivalent to the right vector space of n-tuples (now written as columns). Evidently this implies that $\mathfrak{R}^*$ is n dimensional over Δ.

The last result can be made somewhat more transparent in the following way. If $(e_1, e_2, \cdots, e_n)$ is a basis for $\mathfrak{R}$ over Δ, we define a set $(e_1^*, e_2^*, \cdots, e_n^*)$ of linear functions by the equations

$$e_i^*(e_j) = \delta_{ij}, \quad i, j = 1, 2, \cdots, n, \tag{26}$$

where $\delta_{ij} = 0$ if $i \neq j$ and $= 1$ if $i = j$. Then the value of the linear function $\Sigma e_i^*\alpha_i$ for e_j is

$$\sum_i e_i^*\alpha_i(e_j) = \sum_i e_i^*(e_j)\alpha_i = \alpha_j, \quad j = 1, 2, \cdots, n$$

and this implies that the e_i^* form a basis for $\mathfrak{R}^*$ over Δ. Thus if f is any linear function and $f(e_j) = \alpha_j$, then f and $\Sigma e_i^*\alpha_i$ have the same values at the e_j. Hence $f = \Sigma e_i^*\alpha_i$. Also the e_i^* are linearly independent; for if $\Sigma e_i^*\alpha_i = 0$, then the displayed equations show that every $\alpha_j = 0$. We shall call the basis $(e_1^*, e_2^*, \cdots, e_n^*)$ the *complementary basis* to $(e_1, e_2, \cdots, e_n)$. Such bases will play an important role in the sequel.

EXERCISE

1. Prove that the right multiplications are the only linear transformations in the vector space Δ.

10. Duality between a finite dimensional space and its conjugate space. Let x be a fixed vector in the space $\mathfrak{R}$ and let f range over the conjugate space $\mathfrak{R}^*$ of $\mathfrak{R}$. Then the mapping

$f \to f(x)$ appears to be a mapping of $\mathfrak{R}^*$ into Δ. To emphasize that we are now dealing with a function of $f \varepsilon \mathfrak{R}^*$ we write $x(f)$ for $f(x)$. We assert that $x(f)$ is linear; for

$$x(f + g) = (f + g)(x) = f(x) + g(x) = x(f) + x(g)$$

$$x(f\alpha) = (f\alpha)(x) = f(x)\alpha = x(f)\alpha.$$

Thus we see that every vector $x \varepsilon \mathfrak{R}$ determines an element $x(f)$ of the (left) vector space $\mathfrak{R}^{**}$ of linear functions defined on $\mathfrak{R}^*$.

We consider next the properties of the mapping $x \to x(f)$ of $\mathfrak{R}$ into $\mathfrak{R}^{**}$. We note first that this mapping is a linear transformation of $\mathfrak{R}$ into $\mathfrak{R}^{**}$; for

$$(x + y)(f) = f(x + y) = f(x) + f(y) = x(f) + y(f)$$

and this means that the function associated with $x + y$ is the sum of the function $x(f)$ and $y(f)$. Also

$$(\alpha x)(f) = f(\alpha x) = \alpha f(x) = \alpha x(f),$$

so that the linear function associated with αx is α times the linear function associated with x. We prove next that $x \to x(f)$ is 1–1. For this it suffices to show that, if $x(f) = 0$ for all f, then $x = 0$. Now this is clear; for if $x \neq 0$, we can take x as the first vector e_1 of a basis $(e_1, e_2, \cdots, e_n)$. We have seen that there exists a linear function f such that $f(e_i) = \alpha_i$ for any given α_i. In particular, we can find an f such that $f(x) = f(e_1) \neq 0$.

We have now established that the mapping $x \to x(f)$ is a 1–1 linear transformation of $\mathfrak{R}$ into $\mathfrak{R}^{**}$. Hence if $\mathfrak{S}$ denotes the image space, $\dim \mathfrak{S} = \dim \mathfrak{R} = n$. On the other hand, we know that $\dim \mathfrak{R} = \dim \mathfrak{R}^* = \dim \mathfrak{R}^{**}$. Hence $\dim \mathfrak{S} = \dim \mathfrak{R}^{**}$. This, of course, implies that $\mathfrak{S} = \mathfrak{R}^{**}$. We are now led to the striking conclusion that every linear function on $\mathfrak{R}^*$ can be obtained as an $x(f)$ for some vector $x \varepsilon \mathfrak{R}$. This is the important *principle of duality* that $\mathfrak{R}$ can be identified with the conjugate space of (the space of linear functions on) $\mathfrak{R}^*$. As a first application of this result we prove the following

Theorem 10. *If $(e_1^*, e_2^*, \cdots, e_n^*)$ is a basis for $\mathfrak{R}^*$, then there exists a basis $(e_1, \cdots, e_n)$ for $\mathfrak{R}$ such that $e_i^*(e_j) = \delta_{ij}$.*

Proof. We know that we can find in $\mathfrak{R}^{**}$ a basis $(e_1^{**}, e_2^{**}, \cdots, e_n^{**})$ that is complementary to the given basis $(e_1^*, e_2^*, \cdots, e_n^*)$, that is, $e_i^{**}(e_j^*) = \delta_{ij}$. But there exist vectors $e_1, e_2, \cdots, e_n$ in $\mathfrak{R}$ such that $e_j^*(e_i) = e_i^{**}(e_j^*)$, $i, j = 1, 2, \cdots, n$. These vectors satisfy the conditions of the theorem.

We shall establish next a reciprocal relationship between the subspaces of $\mathfrak{R}$ and the subspaces of its conjugate space $\mathfrak{R}^*$. If $\mathfrak{S}$ is a subspace of $\mathfrak{R}$, we let $j(\mathfrak{S})$ be the totality of vectors $g \in \mathfrak{R}^*$ such that $g(y) = 0$ for all $y \in \mathfrak{S}$. The set $j(\mathfrak{S})$ is clearly a subspace of $\mathfrak{R}^*$, and we shall call it *the subspace of* $\mathfrak{R}^*$ *incident with* $\mathfrak{S}$. Similarly if $\mathfrak{S}^*$ is a subspace of $\mathfrak{R}^*$, we obtain the subspace $j(\mathfrak{S}^*)$ incident with $\mathfrak{S}^*$. This consists of the vectors y such that $g(y) = 0$ for all $g \in \mathfrak{S}^*$. Evidently if $\mathfrak{S}_1 \supseteq \mathfrak{S}_2$, then $j(\mathfrak{S}_1) \subseteq j(\mathfrak{S}_2)$, that is, the correspondence $\mathfrak{S} \rightarrow j(\mathfrak{S})$ is an order reversing correspondence of the lattice $L(\mathfrak{R})$ of subspaces of $\mathfrak{R}$ into the lattice $L(\mathfrak{R}^*)$ of subspaces of $\mathfrak{R}^*$. It is clear also that $j(j(\mathfrak{S})) \supseteq \mathfrak{S}$. We prove next the following

Lemma. *For any* $\mathfrak{S} \subseteq \mathfrak{R}$, *dim* $j(\mathfrak{S}) = n -$ *dim* $\mathfrak{S}$ *and for any* $\mathfrak{S}^* \subseteq \mathfrak{R}^*$, *dim* $j(\mathfrak{S}^*) = n -$ *dim* $\mathfrak{S}^*$.

Proof. Because of the duality between $\mathfrak{R}$ and $\mathfrak{R}^*$, it suffices to prove one of these statements. We choose the first. Let $\mathfrak{S}$ be a subspace of $\mathfrak{R}$ and let $(u_1, u_2, \cdots, u_n)$ be a basis for $\mathfrak{R}$ such that $(u_1, u_2, \cdots, u_r)$ is a basis for $\mathfrak{S}$. Let $(u_1^*, u_2^*, \cdots, u_n^*)$ be the complementary basis for $\mathfrak{R}^*$. Suppose now that $g = \Sigma u_i^*\beta_i \in j(\mathfrak{S})$. Then $(\Sigma u_i^*\beta_i)(u_j) = 0$ for $j = 1, 2, \cdots, r$. But $(\Sigma u_i^*\beta_i)(u_j) = \beta_j$. Hence $g = \sum_{r+1}^{n} u_i^*\beta_i$. Conversely, any linear form $g = \sum_{r+1}^{n} u_i^*\beta_i$ satisfies $g(u_j) = 0$ for $j = 1, 2, \cdots, r$. Hence also $g(y) = 0$ for every $y \in \mathfrak{S}$. Thus $j(\mathfrak{S}) = [u_{r+1}^*, u_{r+2}^*, \cdots, u_n^*]$. Hence dim $\mathfrak{S} = r$ while dim $j(\mathfrak{S}) = n - r$.

An immediate consequence of this lemma is that $j(j(\mathfrak{S})) = \mathfrak{S}$ and $j(j(\mathfrak{S}^*)) = \mathfrak{S}^*$ for any $\mathfrak{S}$ and $\mathfrak{S}^*$. Thus we have seen that $j(j(\mathfrak{S})) \supseteq \mathfrak{S}$. Moreover, dim $j(j(\mathfrak{S})) = n - (n - r)$ if dim $\mathfrak{S} = r$. Hence dim $j(j(\mathfrak{S}))$ = dim $\mathfrak{S}$, and this implies that $\mathfrak{S} = j(j(\mathfrak{S}))$.

We now see that the mapping $\mathfrak{S} \rightarrow j(\mathfrak{S})$ is a 1–1 mapping of $L(\mathfrak{R})$ onto $L(\mathfrak{R}^*)$. If $j(\mathfrak{S}_1) = j(\mathfrak{S}_2)$, then $\mathfrak{S}_1 = j(j(\mathfrak{S}_1)) =$

$j(j(\mathfrak{S}_2)) = \mathfrak{S}_2$. Moreover, if $\mathfrak{S}^*$ is any element of $L(\mathfrak{R}^*)$, then $\mathfrak{S}^* = j(\mathfrak{S})$ where $\mathfrak{S} = j(\mathfrak{S}^*)$.

EXERCISES

1. Prove that, if $\mathfrak{S} \to \bar{\mathfrak{S}}$ is a 1–1 order reversing correspondence of one lattice onto a second one, then $\overline{\mathfrak{S}_1 + \mathfrak{S}_2} = \bar{\mathfrak{S}}_1 \cap \bar{\mathfrak{S}}_2$ and $\overline{\mathfrak{S}_1 \cap \mathfrak{S}_2} = \bar{\mathfrak{S}}_1 + \bar{\mathfrak{S}}_2$.

2. Let f be a linear function on $\mathfrak{R}$ and let $\bar{f}$ denote its contraction on the subspace $\mathfrak{S}$ of $\mathfrak{R}$, that is, $\bar{f}$ is the mapping $y \to f(y)$ for y in $\mathfrak{S}$. Show that the mapping $f \to \bar{f}$ is a linear transformation of $\mathfrak{R}^*$ into the space $\mathfrak{S}^*$ of linear functions on $\mathfrak{S}$. Show that $j(\mathfrak{S})$ is the null-space of this transformation. Prove that $f \to \bar{f}$ is a mapping of $\mathfrak{R}^*$ onto $\mathfrak{S}^*$.

3. Prove that, if $e_1^*, e_2^*, \cdots, e_r^*$ are linearly independent linear functions then there exist vectors $e_1, e_2, \cdots, e_r$ such that $e_i^*(e_j) = \delta_{ij}$, $i, j = 1, 2, \cdots, r$.

11. Transpose of a linear transformation. Let A be a linear transformation of $\mathfrak{R}_1$ into $\mathfrak{R}_2$ and let f be a linear function defined on $\mathfrak{R}_2$. Then the mapping $x_1 \to f(x_1A)$ is a linear function defined on $\mathfrak{R}_1$; for this mapping is simply the resultant Af of A and f, that is,

$$(Af)(x_1) = f(x_1A). \tag{27}$$

We now let f vary over the conjugate space $\mathfrak{R}_2^*$ of $\mathfrak{R}_2$. Then we obtain a mapping $f \to Af$ of $\mathfrak{R}_2^*$ into $\mathfrak{R}_1^*$. We assert that it is linear. By the distributive law

$$A(f + g) = Af + Ag$$

and by the associative law

$$A(f\mu) = A(f\mu_r) = (Af)\mu_r = (Af)\mu$$

and these are the defining conditions. We shall call the linear transformation $f \to Af$ of $\mathfrak{R}_2^*$ into $\mathfrak{R}_1^*$ *the transpose* of A and we denote it as A^*. Hence $fA^* = Af$ where on the right Af is the resultant of A and f.

We consider next the properties of the mapping $A \to A^*$. First, if B is a second member of $\mathfrak{L}(\mathfrak{R}_1, \mathfrak{R}_2)$, then $A + B \in \mathfrak{L}(\mathfrak{R}_1, \mathfrak{R}_2)$,

$$f(A + B)^* = (A + B)f = Af + Bf = fA^* + fB^* = f(A^* + B^*).$$

Hence

$$(A + B)^* = A^* + B^*. \tag{28}$$

Now let $\mathfrak{R}_3$ be a third vector space and let C be a linear transformation of $\mathfrak{R}_2$ into $\mathfrak{R}_3$. Then $AC \in \mathfrak{L}(\mathfrak{R}_1, \mathfrak{R}_3)$. Hence if $h \in \mathfrak{R}_3^*$, $h(AC)^* = (AC)h \in \mathfrak{R}_1^*$. Moreover,

$$h(AC)^* = (AC)h = A(Ch) = A(hC^*) = (hC^*)A^* = h(C^*A^*)$$

and this proves the rule

$$(AC)^* = C^*A^*. \tag{29}$$

Our results also apply, of course, to right vector spaces. In particular, if A^* is a linear transformation of the right vector space of linear functions $\mathfrak{R}_2^*$ into the right vector space $\mathfrak{R}_1^*$, then we can associate with it a transpose transformation of $\mathfrak{R}_1^{**}$ into $\mathfrak{R}_2^{**}$, where $\mathfrak{R}_i^{**}$ is the left vector space of linear functions on $\mathfrak{R}_i^*$. On the other hand, we have the natural equivalences of $\mathfrak{R}_i^{**}$ onto $\mathfrak{R}_i$ and these enable us to define a linear transformation of $\mathfrak{R}_1$ into $\mathfrak{R}_2$ corresponding to the transpose of A^*. We shall call this transformation *the transpose* A^{**} of A^* *in* $\mathfrak{R}_1$ *to* $\mathfrak{R}_2$. We now determine the transformation A^{**} for the given linear transformation A^* of $\mathfrak{R}_2^*$ into $\mathfrak{R}_1^*$. Let f vary over $\mathfrak{R}_2^*$ and let x_1 be a definite vector in $\mathfrak{R}_1$. Then the mapping

$$f \rightarrow x_1(fA^*) \equiv (fA^*)(x_1) \in \Delta$$

is a linear function on $\mathfrak{R}_2^*$. Hence there is a uniquely determined vector $x_2 \in \mathfrak{R}_2$ such that

$$(fA^*)(x_1) = x_1(fA^*) = x_2(f) \equiv f(x_2) \tag{30}$$

holds for all $f \in \mathfrak{R}_2^*$. We now have a mapping A^{**} sending x_1 into x_2. The argument given above shows that this mapping is linear. This can also be verified directly.

We shall now show that the two correspondences $A \rightarrow A^*$ and $A^* \rightarrow A^{**}$ are inverses of each other, that is, if A^* is the transpose of $A \in \mathfrak{L}(\mathfrak{R}_1, \mathfrak{R}_2)$, then the transpose of A^* in $\mathfrak{R}_1$ to $\mathfrak{R}_2$ is A and, if A^{**} is the transpose in $\mathfrak{R}_1$ to $\mathfrak{R}_2$ of $A^* \in \mathfrak{L}(\mathfrak{R}_2^*, \mathfrak{R}_1^*)$, then the transpose of A^{**} is A^*. To see the first of these statements we note that $(fA^*)(x_1) = f(x_1A^{**})$ by the definition of A^{**}. Hence $f(x_1A) = f(x_1A^{**})$ and $f(x_1A - x_1A^{**}) = 0$ holds for all f. As we have seen (p. 54) this implies that $x_1A = x_1A^{**}$ for all x_1 so that $A = A^{**}$. The second statement is proved by using the relation

$$(fA^*)(x_1) = f(x_1A^{**}) = (fA^{***})(x_1).$$

Since this holds for all x_1, $fA^* = fA^{***}$ and $A^* = A^{***}$. It is now immediate that the mapping $A \to A^*$ is 1–1 of $\mathfrak{L}(\mathfrak{R}_1, \mathfrak{R}_2)$ onto $\mathfrak{L}(\mathfrak{R}_2^*, \mathfrak{R}_1^*)$.

In the particular case of $\mathfrak{L} = \mathfrak{L}(\mathfrak{R}, \mathfrak{R})$ and $\mathfrak{L}^* = \mathfrak{L}(\mathfrak{R}^*, \mathfrak{R}^*)$ the mapping $A \to A^*$ is an anti-isomorphism between these two rings. This is clear from the 1–1-ness and equations (28) and (29).

12. Matrices of the transpose. Again let A be a linear transformation of $\mathfrak{R}_1$ into $\mathfrak{R}_2$ and let $(e_1, e_2, \cdots, e_{n_1})$, $(f_1, f_2, \cdots, f_{n_2})$ be bases in these spaces. We choose complementary bases $(e_1^*, e_2^*, \cdots, e_{n_1}^*)$, $(f_1^*, f_2^*, \cdots, f_{n_2}^*)$ in the spaces $\mathfrak{R}_1^*$, $\mathfrak{R}_2^*$. Hence

$$e_i^*(e_j) = \delta_{ij}, \quad f_p^*(f_q) = \delta_{pq}.$$

Suppose now that (α) and (β), respectively, are the matrices of A and of its transpose relative to these bases. Then

$$e_iA = \sum_p \alpha_{ip}f_p, \quad f_q^*A^* = \Sigma e_j^*\beta_{jq}.$$

Since $f_q^*(e_iA) = f_q^*A^*(e_i)$,

$$f_q^*\left(\sum_p \alpha_{ip}f_p\right) = \sum_j (e_j^*\beta_{jq})(e_i).$$

Hence $\alpha_{iq} = \beta_{iq}$ and $(\alpha) = (\beta)$. In other words, *if complementary bases are used in $\mathfrak{R}_i$, and in $\mathfrak{R}_i^*$, $i = 1, 2$, then the matrices of A and of its transpose are equal.*

This result is perhaps a little unexpected. Its explanation lies in our definitions of matrices of linear transformations in left vector spaces and in right vector spaces. In the special case in which $\Delta = \Phi$ is a field the result usually appears in a somewhat different form. Here it is customary to consider all spaces as left vector spaces, as can be done by writing αx for $x\alpha$. Then the relations $f_q^*A^* = \sum_j e_j^*\alpha_{jq}$ that we have just derived may be written as

$$f_q^*A^* = \sum_j e_j^*\alpha_{jq} = \sum_j \alpha_{jq}e_j^* = \sum_j \alpha_{qj}'e_j^*$$

where $\alpha_{qj}' = \alpha_{jq}$. Thus the matrix of A^* considering the $\mathfrak{R}_i^*$ as left vector spaces is the transposed matrix $(\alpha)'$ of the matrix (α) of A. If $\mathfrak{R}_1 = \mathfrak{R}_2 = \mathfrak{R}$ we have the isomorphisms $A \to (\alpha)$ and $A^* \to (\alpha)'$ of the rings of linear transformations $\mathfrak{L} = \mathfrak{L}(\mathfrak{R}, \mathfrak{R})$ and $\mathfrak{L}(\mathfrak{R}^*, \mathfrak{R}^*)$ respectively onto the matrix ring Φ_n. Since $A \to A^*$ is an anti-isomorphism we see that the mapping $(\alpha) \to (\alpha)'$ is an anti-isomorphism in Φ_n, that is, $(\alpha) \to (\alpha)'$ is 1–1 and

$$[(\alpha) + (\beta)]' = (\alpha)' + (\beta)', \quad [(\alpha)(\beta)]' = (\beta)'(\alpha)'.$$

This, of course, can also be verified directly.

We return now to the general case of an arbitrary division ring, and we shall establish the following relation between rank spaces and null spaces of a linear transformation and its transpose.

Theorem 11. *The null space of A is the subspace incident with the rank space of the transpose A^* of A. The rank space of A is the subspace incident with the null space of A^*.*

Proof. It is sufficient to prove the first of these statements. Hence let z be a vector such that $zA = 0$. Then $f(zA) = 0$ and $(fA^*)(z) = 0$ for all f. Hence $z \, \varepsilon \, j(\mathfrak{R}_2^*A^*)$. The converse follows by retracing the steps.

We can now prove

Theorem 12. *Rank A = Rank A^*.*

Proof. Rank $A^* = \dim (\mathfrak{R}_2^*A^*) = \dim j(\mathfrak{N})$ where $\mathfrak{N}$ is the null space of A. Also $\dim j(\mathfrak{N}) = n_1 - \dim \mathfrak{N}$ where $n_1 = \dim \mathfrak{R}_1 = \dim \mathfrak{R}_1^*$. On the other hand, rank $A = n_1 - \dim \mathfrak{N}$. Hence rank A = rank A^*.

If (α) is any matrix, (α) can serve as the matrix of a linear transformation A and also of the transpose A^* of A. Then, as we have seen, rank A is the row rank of (α) while rank A^* is the column rank of (α). Theorem 12 therefore gives a geometric proof of the theorem that the two ranks of any matrix are equal.

13. Projections. We conclude this chapter by considering a type of linear transformation of a vector space that is intimately connected with direct decompositions of the space into subspaces. Suppose that $\mathfrak{R} = \mathfrak{R}_1 \oplus \mathfrak{R}_2 \oplus \cdots \oplus \mathfrak{R}_r$, that is, $\mathfrak{R}$ is a direct sum of the subspaces $\mathfrak{R}_i$ in the sense of § 11, Chapter 1. Then if x is any vector in $\mathfrak{R}$, x may be written as

$$x = x_1 + x_2 + \cdots + x_r \tag{31}$$

where $x_i \in \mathfrak{R}_i$. We know also that for a given i *the component* x_i *of* x in $\mathfrak{R}_i$ is unique. Hence the mapping E_i that sends x into x_i is single-valued. We wish to investigate the properties of these mappings. Let $y = y_1 + y_2 + \cdots + y_r$, $y_i \in \mathfrak{R}_i$; then

$$x + y = (x_1 + y_1) + (x_2 + y_2) + \cdots + (x_r + y_r)$$

where $x_i + y_i \in \mathfrak{R}_i$. Thus the component in $\mathfrak{R}_i$ of $x + y$ is $x_i + y_i$. Hence we have the following condition on E_i:

$$(x + y)E_i = xE_i + yE_i.$$

Also by (31)

$$\alpha x = \alpha x_1 + \alpha x_2 + \cdots + \alpha x_r.$$

Hence the component in $\mathfrak{R}_i$ of αx is αx_i. In other words, we have the relation

$$(\alpha x)E_i = \alpha(xE_i).$$

We therefore see that E_i is a linear transformation in $\mathfrak{R}$ over Δ.

We consider next the relations that hold among the E_i. We note first that (31) may be rewritten as

$$x = xE_1 + xE_2 + \cdots + xE_r. \tag{32}$$

Hence

$$E_1 + E_2 + \cdots + E_r = 1. \tag{33}$$

If $x_i \in \mathfrak{R}_i$, the representation (31) of $x = x_i$ reads $x_i = x_i$. Thus $x_iE_i = x_i$, $x_iE_j = 0$ if $j \neq i$. Now for any x, $xE_i = x_i \in \mathfrak{R}_i$. Hence we see that $xE_i^2 = x_iE_i = x_i = xE_i$ and $xE_iE_j = x_iE_j = 0$. We therefore have the following equations.

$$E_i^2 = E_i, \quad E_iE_j = 0 \quad \text{if} \quad i \neq j. \tag{34}$$

Now, in general, we shall call a linear transformation E that is idempotent in the sense that it is equal to its square ($E^2 = E$) a *projection*. A set of projections will be called *orthogonal* if the product of any two distinct ones in the set is 0. Finally we shall say that the set of orthogonal projections $E_1, E_2, \cdots, E_r$ is *supplementary* if their sum is the identity mapping. Then the results that we have just obtained concerning the mappings E_i determined by the decomposition $\mathfrak{R} = \mathfrak{R}_1 \oplus \mathfrak{R}_2 \oplus \cdots \oplus \mathfrak{R}_r$ are that these mappings form a supplementary set of orthogonal projections. We shall now show that every finite supplementary set of orthogonal projections is obtained in this way. Thus let E_1, E_2, $\cdots$, E_r be such a set. Let $\mathfrak{R}_i = \mathfrak{R}E_i$, the rank space of E_i. Then if x is any vector in $\mathfrak{R}$

$$x = x1 = x(\Sigma E_i) = xE_1 + xE_2 + \cdots + xE_r$$

$$\varepsilon\, \mathfrak{R}E_1 + \mathfrak{R}E_2 + \cdots + \mathfrak{R}E_r.$$

Hence $\mathfrak{R} = \mathfrak{R}_1 + \mathfrak{R}_2 + \cdots + \mathfrak{R}_r$. Next let

$$z_i = z_1 + \cdots + z_{i-1} + z_{i+1} + \cdots + z_r \tag{35}$$

where $z_j \,\varepsilon\, \mathfrak{R}_j$. Since z_i has the form y_iE_i, $z_iE_i = y_iE_i^2 = y_iE_i = z_i$. Similarly if $j \neq i$, $z_jE_i = y_jE_jE_i = 0$. Hence if we operate on the left-hand side of (35) with E_i, we obtain z_i and, if we operate on the right-hand side, we obtain 0. Hence $z_i = 0$. This proves that $\mathfrak{R}_i \cap (\mathfrak{R}_1 + \cdots + \mathfrak{R}_{i-1} + \mathfrak{R}_{i+1} + \cdots + \mathfrak{R}_r) = 0$ for $i = 1, 2, \cdots, r$. Thus $\mathfrak{R} = \mathfrak{R}_1 \oplus \mathfrak{R}_2 \oplus \cdots \oplus \mathfrak{R}_r$. Since $x = \Sigma xE_i$ and $xE_i = x_i \,\varepsilon\, \mathfrak{R}_i$, the projections determined by this decomposition are the mappings $x \to xE_i$, that is, the given linear transformations E_i. We have therefore established a 1–1 correspondence between direct decomposition of the vector space and finite sets of supplementary orthogonal projections.

We obtain next canonical bases for the set of supplementary orthogonal projections $E_1, E_2, \cdots, E_r$. Let $\mathfrak{R} = \mathfrak{R}_1 \oplus \cdots \oplus \mathfrak{R}_r$ as above. Then we can obtain a basis $(f_1, f_2, \cdots, f_n)$ for $\mathfrak{R}$ such that $(f_1, \cdots, f_{\rho_1})$ is a basis for $\mathfrak{R}_1$, $(f_{\rho_1+1}, \cdots, f_{\rho_1+\rho_2})$ is a basis for $\mathfrak{R}_2$, etc. Thus $\rho_i = \text{rank } E_i$. Since $E_jE_i = 0$ for $j \neq i$, the linear transformation E_i annihilates all the f's outside the i-th subset $(f_{\rho_1+\cdots+\rho_{i-1}+1}, \cdots, f_{\rho_1+\cdots+\rho_i})$. Since $E_i^2 = E_i$, any vector in $\mathfrak{R}_i = \mathfrak{R}E_i$ is sent into itself by E_i. It follows that the matrix of E_i relative to the basis $(f_1, f_2, \cdots, f_n)$ is the diagonal matrix

$$(\delta_i) = \text{diag}\,\{0, \cdots, 0, \overbrace{1, \cdots, 1}^{\rho_i}, 0, \cdots, 0\} \tag{36}$$

in which all elements are 0 except those in $\rho_1 + \cdots + \rho_{i-1} + 1$st to $\rho_1 + \cdots + \rho_i$th rows.

We now drop the assumption that $\Sigma E_i = 1$. Thus we suppose that $E_1, E_2, \cdots, E_r$ is any finite set of orthogonal projections. Set $E = \Sigma E_i$ and $E_{r+1} = 1 - E$. Since $E_iE_j = 0$ for $i \neq j, \leq r$, $E^2 = \Sigma E_i^2 = \Sigma E_i = E$. Hence also $E_{r+1}^2 = (1 - E)^2 = 1 - 2E + E^2 = 1 - 2E + E = 1 - E = E_{r+1}$. Moreover, if $j \leq r$, $E_jE = E_j\Sigma E_i = E_j^2 = E_j$ and similarly $EE_j = E_j$. This implies $E_jE_{r+1} = E_j(1 - E) = E_j - E_jE = E_j - E_j = 0$ and $E_{r+1}E_j = 0$. We have therefore shown that the set $E_1, E_2, \cdots, E_{r+1}$ is a supplementary orthogonal set of projections. Our discussion shows that there exists a basis $(f_1, f_2, \cdots, f_n)$ of $\mathfrak{R}$ such that the matrix of E_i relative to this basis has the form (36), $i = 1, 2, \cdots, r$.

These remarks apply in particular to a single idempotent linear transformation. If E_1 is of this type, then E_1 and $E_2 = 1 - E_1$ are orthogonal and supplementary. Relative to a suitable basis the matrix of E_1 has the form diag $\{1, 1, \cdots, 1, 0, \cdots, 0\}$. The number of 1's is the rank of E_1.

EXERCISES

1. Prove the following theorem on matrices: If $(\epsilon_1), (\epsilon_2), \cdots, (\epsilon_r)$ is a set of matrices such that

$$(\epsilon_i)^2 = (\epsilon_i), \quad (\epsilon_i)(\epsilon_j) = 0 \quad \text{if} \quad i \neq j, \tag{37}$$

then there exists a matrix (μ) in $L(\Delta, n)$ such that $(\mu)(\epsilon_i)(\mu)^{-1}$ has the form (36).

2. Prove that two projections E and F have the same rank spaces if and only if $EF = E$ and $FE = F$. Prove that they have the same null spaces if and only if $EF = F$, $FE = E$.

3. Show that, if $E_1, E_2, \cdots, E_k$ are projections with the same rank spaces and $\alpha_1, \alpha_2, \cdots, \alpha_k$ are elements of Δ such that $\Sigma\alpha_i = 1$, then $\Sigma\alpha_iE_i$ is a projection with the same rank space as the E_i.

4. Assume that the characteristic of Δ is not 2. (This means that, if $\alpha \neq 0$, then $2\alpha = \alpha + \alpha \neq 0$ in Δ.) Prove that, if E_1 and E_2 are projections whose sum is a projection, then E_1 and E_2 are orthogonal.

5. Show that, if E_1 and E_2 are commutative projections, then E_1E_2 and $E_1 + E_2 - E_1E_2$ are projections.

Chapter III

THE THEORY OF A SINGLE LINEAR TRANSFORMATION

This chapter is devoted to the study of a single linear transformation in a vector space over a field. We shall obtain a decomposition of the vector spaces into so-called cyclic subspaces relative to a given linear transformation. By choosing appropriate bases in these spaces we obtain certain canonical matrices for the transformation. These results yield necessary and sufficient conditions for similarity of matrices. Following Krull we shall derive the fundamental decomposition theorems by making use of the theory of finitely generated $\mathfrak{o}$-modules, $\mathfrak{o}$ a principal ideal domain. We shall also prove in this chapter the Hamilton-Cayley-Frobenius theorems on the characteristic and minimum polynomials of a matrix. Finally we study the algebra of linear transformations that commute with a given transformation.

1. The minimum polynomial of a linear transformation. Throughout this chapter we assume that the underlying division ring is a field. We use the notation Φ for this field. Let $\mathfrak{R}$ be a vector space over Φ and let $\mathfrak{L} \equiv \mathfrak{L}(\mathfrak{R}, \mathfrak{R})$ be the set of linear transformations of $\mathfrak{R}$ into itself. We recall that $\mathfrak{L}$ can be regarded as an algebra over Φ: that is, 1) $\mathfrak{L}$ is a ring; 2) $\mathfrak{L}$ is a vector space over Φ with addition the same as the ring addition; and 3)

$$\gamma(AB) = (\gamma A)B = A(\gamma B)$$

holds for A, B in $\mathfrak{L}$ and γ in Φ. The product γA is by definition the resultant $\gamma_l A = A\gamma_l$. We recall also that $\mathfrak{L}$ is n^2 dimensional over Φ.

Now let A be an arbitrary linear transformation in $\mathfrak{R}$ and let $[1, A, A^2, \cdots]$ be the subspace of $\mathfrak{L}$ spanned by the powers of A. If A is not a multiple of 1, the pair $(1, A)$ is a linearly independent set. Again if A^2 is not linearly dependent on $(1, A)$, then, since $(1, A)$ is an independent set, so also is $(1, A, A^2)$. Continuing in this way we obtain the linearly independent sets (1), $(1, A)$, $(1, A, A^2)$, $(1, A, A^2, A^3)$, $\cdots$.

Since $\mathfrak{L}$ is finite dimensional, this process must break off after a finite number of steps. This, of course, occurs when we reach the first power A^m of A which is linearly dependent on preceding powers. Then $(1, A, A^2, \cdots, A^{m-1})$ is linearly independent, but $A^m \,\varepsilon\, [1, A, A^2, \cdots, A^{m-1}]$. Hence we have a relation of the form

$$A^m = \mu_0 1 + \mu_1 A + \cdots + \mu_{m-1} A^{m-1} \tag{1}$$

where the $\mu_i \,\varepsilon\, \Phi$. Multiplication of (1) by A yields, in virtue of the distributive law and the algebra condition 3) above, the equation

$$A^{m+1} = \mu_0 A + \mu_1 A^2 + \cdots + \mu_{m-1} A^m.$$

Since $A^m \,\varepsilon\, [1, A, A^2, \cdots, A^{m-1}]$, this shows that $A^{m+1} \,\varepsilon\, [1, A, A^2, \cdots, A^{m-1}]$. Similarly we can prove that $A^{m+2}, A^{m+3}, \cdots$ are in $[1, A, A^2, \cdots, A^{m-1}]$. Hence

$$[1, A, A^2, \cdots] = [1, A, A^2, \cdots, A^{m-1}],$$

and since the displayed set of A^k is a linearly independent one, we see that $[1, A, A^2, \cdots]$ has the basis $(1, A, A^2, \cdots, A^{m-1})$.

The relation (1) has an important interpretation which we shall now examine. We note first that the mapping

$$\alpha \to \alpha 1 = \alpha_l \tag{2}$$

is an isomorphism of Φ into $\mathfrak{L}$. Thus

$$\begin{aligned} (\alpha + \beta)1 &= \alpha 1 + \beta 1 \\ (\alpha\beta)1 = (\alpha\beta)(1^2) &= \alpha(\beta(1^2)) = \alpha(1(\beta 1)) = (\alpha 1)(\beta 1) \end{aligned} \tag{3}$$

and $\alpha 1 \neq 0$ if $\alpha \neq 0$ so that the mapping is 1–1. It follows that the image set $\Phi 1$ is a subfield of $\mathfrak{L}$ isomorphic to Φ.

Now let $\Phi[\lambda]$ denote the integral domain of polynomials in the transcendental element (indeterminate) λ with coefficients in Φ. The linear transformation A commutes with every element $\alpha 1 (= \alpha_l)$. It follows from this that the mapping

$$(4)\quad f(\lambda) = \alpha_0 + \alpha_1 \lambda + \alpha_2 \lambda^2 + \cdots \to f(A) \equiv \alpha_0 1 + \alpha_1 A + \alpha_2 A^2 + \cdots$$

is a homomorphism of $\Phi[\lambda]$ into $\mathfrak{L}$. This homomorphism is an extension of the isomorphism (2), and it is characterized by the property that it maps λ into A. Let $\mathfrak{R}$ be the kernel of our homomorphism, that is, $\mathfrak{R}$ is the ideal of polynomials $\nu(\lambda)$ such that $\nu(A) = 0$. Now it is known that $\Phi[\lambda]$ is a principal ideal domain.* By this we mean that every ideal in $\Phi[\lambda]$ consists of the multiples of a particular polynomial. In particular $\mathfrak{R} = (\mu(\lambda))$ is the set of multiples of $\mu(\lambda)$. The generator $\mu(\lambda)$ is uniquely determined if we specify that its leading coefficient is 1. By the relation (1)

$$A^m - \mu_{m-1} A^{m-1} - \mu_{m-2} A^{m-2} - \cdots - \mu_0 1 = 0.$$

Hence we see that the polynomial

$$(5)\qquad \lambda^m - \mu_{m-1} \lambda^{m-1} - \mu_{m-2} \lambda^{m-2} - \cdots - \mu_0$$

which is $\neq 0$ is in $\mathfrak{R}$. The result that we established before therefore shows that $\mathfrak{R} \neq 0$. We now note that the polynomial (5) as determined by (1) is the generator $\mu(\lambda)$ of $\mathfrak{R}$; for otherwise there exists a polynomial of degree lower than m such that $A^r + \nu_{r-1} A^{r-1} + \cdots + \nu_0 = 0$, and this contradicts the linear independence of $1, A, \cdots, A^r$.

We shall call the polynomial $\mu(\lambda)$ the *minimum polynomial* of the linear transformation A. It is characterized by the following properties: 1) leading coefficient $= 1$; 2) $\mu(A) = 0$; 3) if $\nu(\lambda)$ is any polynomial such that $\nu(A) = 0$, then $\nu(\lambda)$ is a multiple of $\mu(\lambda)$.

The isomorphism between $\mathfrak{L}$ and the matrix ring Φ_n enables us to carry over to matrices the results that we have just derived. Let (α) be the matrix of A relative to $(e_1, e_2, \cdots, e_n)$. Then we know that the matrix of $\beta_0 1 + \beta_1 A + \cdots + \beta_k A^k$ rela-

* See, for instance, Volume I, p. 100, of these Lectures.

tive to this basis is $\beta_0 1 + \beta_1(\alpha) + \cdots + \beta_k(\alpha)^k$. It follows that, if the polynomial $\mu(\lambda)$ given by (5) is the minimum polynomial of A, then

$$\mu((\alpha)) \equiv (\alpha)^m - \mu_{m-1}(\alpha)^{m-1} - \cdots - \mu_0 1 = 0.$$

Conversely, a relation in (α) implies a corresponding relation for A. This implies that $\mu(\lambda)$ is the polynomial of least degree that has (α) as a root. For this reason we shall also call $\mu(\lambda)$ the *minimum polynomial of the matrix* (α).

EXERCISES

1. Prove the rule $(\alpha\beta)(xy) = (\alpha x)(\beta y)$ for any algebra.
2. What is the minimum polynomial of an idempotent linear transformation $\neq 0$?
3. Let $\mathfrak{R}$ be the vector space of polynomials of degree $\leq n - 1$ with real coefficients. Find the minimum polynomials of the differentiation operator D and of the linear transformation A such that $f(\lambda)A = f(\lambda + 1)$.

2. Cyclic subspaces. Of special interest in the study of a linear transformation A is the discovery of subspaces $\mathfrak{S}$ that are *invariant* relative to A in the sense that they are mapped into themselves by A. If $\mathfrak{S}$ is such a subspace, then A induces a mapping in $\mathfrak{S}$ and it is obvious that this mapping is a linear transformation.

The simplest type of invariant subspace is an invariant subspace that is generated by a single vector. Let u be a particular vector in $\mathfrak{R}$ and let $\mathfrak{S}$ denote the smallest invariant subspace containing u. Since $u \in \mathfrak{S}$, so is uA; hence also $uA^2 = (uA)A$, $uA^3 = (uA^2)A, \cdots$. Thus

$$[u, uA, uA^2, \cdots] \subseteq \mathfrak{S}.$$

On the other hand, if v is any vector in the space $[u, uA, \cdots]$, then

$$v = \alpha_0 u + \alpha_1 uA + \cdots + \alpha_r uA^r$$
$$= uf(A)$$

where $f(\lambda) = \alpha_0 + \alpha_1\lambda + \cdots + \alpha_r\lambda^r$. Hence $vA = uf(A)A \in [u, uA, \cdots]$. Thus this space is invariant, and $[u, uA, uA^2, \cdots]$ is therefore the smallest invariant subspace of $\mathfrak{R}$ containing the vector u. We now denote $[u, uA, \cdots]$ by $\{u\}$, and we call this space *the cyclic space* (relative to A) generated by the vector u.

Suppose now that $u \neq 0$, and let uA^r, $r \geq 1$, be the first vector in the sequence $u, uA, uA^2, \cdots$ that is linearly dependent on the preceding. Then, as in the above discussion of the powers of A, we can conclude that

$$\{u\} = [u, uA, \cdots, uA^{r-1}] \tag{6}$$

and that $(u, uA, \cdots, uA^{r-1})$ is a basis for this space. Also we have a relation of the form

$$uA^r = \nu_0 u + \nu_1 uA + \cdots + \nu_{r-1} uA^{r-1}$$

or, what is the same thing, $u\mu_u(A) = 0$ where

$$\mu_u(\lambda) = \lambda^r - \nu_{r-1}\lambda^{r-1} - \cdots - \nu_0. \tag{7}$$

From the definition of r it is clear that $\mu_u(\lambda)$ is a polynomial $\neq 0$ of least degree having the property $u\mu_u(A) = 0$. We shall call $\mu_u(\lambda)$ *the order* of the vector u.

Consider now the totality $\mathfrak{J}_u$ of polynomials $\nu(\lambda)$ such that $u\nu(A) = 0$. It is readily seen that $\mathfrak{J}_u$ is an ideal in $\Phi[\lambda]$. The polynomial $\mu_u(\lambda)$ of positive degree is in $\mathfrak{J}_u$ and is a non-zero polynomial of least degree in this ideal. It follows that $\mathfrak{J}_u = (\mu_u(\lambda))$. We observe finally that $\mu_u(\lambda)$ is the minimum polynomial $\bar{\mu}(\lambda)$ of the linear transformation induced by A in $\{u\}$. For if v is any vector in $\{u\}$, $v = uf(A)$ so that

$$v\mu_u(A) = uf(A)\mu_u(A) = u\mu_u(A)f(A) = 0f(A) = 0.$$

Hence $\mu_u(A) = 0$ in $\{u\}$ and therefore $\bar{\mu}(\lambda) \mid \mu_u(\lambda)$. On the other hand, $u\bar{\mu}(A) = 0$. Hence $\bar{\mu}(\lambda) \in \mathfrak{J}_u$ and so $\mu_u(\lambda) \mid \bar{\mu}(\lambda)$. These two relations imply that $\bar{\mu}(\lambda) = \mu_u(\lambda)$.

If $u = 0$, $\{u\} = 0$. In this case $\mathfrak{J}_u = (1)$, and we shall say that the order is 1.

If $\mu(\lambda)$ is the minimum polynomial of A, then $\mu(A) = 0$ and $u\mu(A) = 0$. It follows that the minimum polynomial of A is a multiple of every order $\mu_u(\lambda)$.

3. Existence of a vector whose order is the minimum polynomial. Let $e_1, e_2, \cdots, e_r$ be a set of vectors that *generate* $\mathfrak{R}$

relative to A in the sense that every vector x can be written in the form $x = \sum_1^r e_i\phi_i(A)$, for suitable $\phi_i(\lambda)$ in $\Phi[\lambda]$. Such finite sets exist; for any ordinary basis $(e_1, e_2, \cdots, e_n)$ has this property. Let $\bar{\mu}(\lambda)$ be the least common multiple $[\mu_{e_1}(\lambda), \mu_{e_2}(\lambda), \cdots, \mu_{e_r}(\lambda)]$ of the orders $\mu_{e_i}(\lambda)$. Since each $\mu_{e_i}(\lambda)$ is a factor of the minimum polynomial $\mu(\lambda)$ of A, $\bar{\mu}(\lambda) \mid \mu(\lambda)$. On the other hand, if $x = \Sigma e_i\phi_i(A)$, then

$$x\bar{\mu}(A) = \Sigma e_i\phi_i(A)\bar{\mu}(A) = \Sigma e_i\bar{\mu}(A)\phi_i(A) = 0.$$

Hence $\bar{\mu}(A) = 0$ and this implies that $\mu(\lambda) \mid \bar{\mu}(\lambda)$. Hence $\mu(\lambda) = \bar{\mu}(\lambda)$. Thus the minimum polynomial of A is the least common multiple of the orders of any set of generators. This generalizes the result established above in the cyclic case.

We now write the $\mu_{e_i}(\lambda)$ in terms of the same primes, say

$$\mu_{e_i}(\lambda) = \pi_1(\lambda)^{k_{1i}}\pi_2(\lambda)^{k_{2i}} \cdots \pi_s(\lambda)^{k_{si}}$$

where the π's are distinct and $k_{ji} \geq 0$. We may also suppose that the leading coefficients of the π's are all 1. Then if $k_j = \max(k_{j1}, k_{j2}, \cdots, k_{jr})$, the polynomial

$$\pi_1(\lambda)^{k_1}\pi_2(\lambda)^{k_2} \cdots \pi_s(\lambda)^{k_s} = \bar{\mu}(\lambda) = \mu(\lambda).$$

Now, in general, if the order $\mu_x(\lambda) = \mu_1(\lambda)\mu_2(\lambda)$, then the order of $y = x\mu_1(A)$ is $\mu_2(\lambda)$. The proof is immediate. Applying this fact we see that, if $k_1 = k_{1i_1}$, then the order of

$$f_1 = e_{i_1}\pi_2(A)^{k_{2i_1}}\pi_3(A)^{k_{3i_1}} \cdots \pi_s(A)^{k_{si_1}}$$

is $\pi_1(\lambda)^{k_1}$. Similarly, we can find a vector f_j, $j = 2, \cdots, s$, of order $\pi_j(\lambda)^{k_j}$. We shall now show that the vector

$$f = f_1 + f_2 + \cdots + f_s$$

has order $\mu_f(\lambda) = \mu(\lambda)$. This will follow from the following more general result:

Lemma. *If the orders $\mu_i(\lambda)$ of f_i, $i = 1, 2, \cdots, s$, are relatively prime in pairs, then the order of $f = f_1 + f_2 + \cdots + f_s$ is the product $\nu(\lambda) = \Pi\mu_i(\lambda)$.*

Proof. Since $f_i\mu_i(A) = 0$ for each i, $f\nu(A) = \Sigma f_i\nu(A) = 0$. Hence $\mu_f(\lambda) \mid \nu(\lambda)$. We note next that if $\rho_1(\lambda) = \mu_f(\lambda)\mu_2(\lambda)\mu_3(\lambda) \cdots \mu_s(\lambda)$, then $f\rho_1(A) = 0$ and $f_j\rho_1(A) = 0$ for $j = 2, 3, \cdots, s$. Hence, also $f_1\rho_1(A) = 0$. This implies that $\mu_1(\lambda) \mid \rho_1(\lambda)$. Since $\mu_1(\lambda)$ is prime to $\mu_2(\lambda), \cdots, \mu_s(\lambda)$, it follows that $\mu_1(\lambda) \mid \mu_f(\lambda)$. Similarly $\mu_2(\lambda) \mid \mu_f(\lambda)$, $\mu_3(\lambda) \mid \mu_f(\lambda), \cdots$. Hence $\nu(\lambda) = \Pi\mu_i(\lambda)$ is a factor of $\mu_f(\lambda)$. Thus $\mu_f(\lambda) = \nu(\lambda)$. This completes the proof.

The above discussion now gives the following

Theorem 1. *There exists a vector in $\mathfrak{R}$ whose order is the minimum polynomial of A.*

We know that the degree of the order of any vector u is the dimensionality of the space $\{u\}$. Hence $\deg \mu_u(\lambda) \leq n$. Thus we have the

Corollary. *The degree of the minimum polynomial of A is $\leq n$.*

EXERCISE

1. Use the method of this section to prove that, if m is the maximum order of the elements of a finite commutative group G, then $x^m = 1$ for all x in G.

4. Cyclic linear transformations. The space $\mathfrak{R}$ is called *cyclic* (relative to A) and A is called *cyclic* (sometimes *non-derogatory*) if there exists a single vector e that generates $\mathfrak{R}$, that is, $\mathfrak{R} = \{e\}$. We know that in this case $\mu(\lambda) = \mu_e(\lambda)$, and this shows that the orders of any two generators of a cyclic space are the same. It shows also that the minimum polynomial of a cyclic linear transformation has degree n; for if $\mathfrak{R} = \{e\}$, $\deg \mu_e(\lambda) = \dim \mathfrak{R} = n$. Hence $\deg \mu(\lambda) = n$.

Conversely, suppose A is any linear transformation whose minimum polynomial has degree n. Then by Theorem 1 there exists a vector e such that $\mu_e(\lambda) = \mu(\lambda)$. Then $\deg \mu_e(\lambda) = n$ and so $\dim \{e\} = n$. This, of course, means that $\{e\} = \mathfrak{R}$. We therefore have the following criterion.

Theorem 2. *A linear transformation is cyclic if and only if its minimum polynomial has degree n.*

Suppose now that $\mathfrak{R} = \{e\}$ and let

$$\mu(\lambda) = \mu_e(\lambda) = \lambda^n - \mu_{n-1}\lambda^{n-1} - \cdots - \mu_0.$$

We know that the vectors $(e, eA, \cdots, eA^{n-1})$ constitute a basis. We wish to determine the matrix of A relative to this basis. Since

$$\begin{aligned} eA &= eA \\ (eA)A &= eA^2 \\ (eA^{n-2})A &= eA^{n-1} \\ (eA^{n-1})A &= \mu_0 e + \mu_1(eA) + \cdots + \mu_{n-1}(eA^{n-1}), \end{aligned}$$

the matrix we seek is

$$\begin{bmatrix} 0 & 1 & 0 & \cdot & \cdot & \cdot & 0 \\ 0 & 0 & 1 & 0 & \cdot & \cdot & 0 \\ \cdot & \cdot & \cdot & \cdot & \cdot & \cdot & \cdot \\ \cdot & \cdot & \cdot & \cdot & \cdot & \cdot & \cdot \\ 0 & \cdot & \cdot & \cdot & \cdot & 0 & 1 \\ \mu_0 & \mu_1 & \mu_2 & \cdot & \cdot & \cdot & \mu_{n-1} \end{bmatrix}. \tag{8}$$

If $\mu(\lambda) = \lambda^n - \mu_{n-1}\lambda^{n-1} - \cdots - \mu_0$ is any polynomial, then the $n \times n$ matrix given in (8) is called *the companion matrix* of $\mu(\lambda)$. If, as in the present case, A is cyclic with minimum polynomial $\mu(\lambda)$, then (8) is called the *Jordan canonical matrix* of A. Clearly the minimum polynomial may be read off from this matrix.

We proceed now to derive a canonical matrix that displays the prime powers of $\mu(\lambda)$. We prove first

Theorem 3. *Let $\mathfrak{R} = \{e\}$ and $\mu(\lambda) = \mu_1(\lambda) \cdots \mu_s(\lambda)$ where the $\mu_i(\lambda)$ are relatively prime in pairs. Then $\mathfrak{R} = \{e_1\} \oplus \{e_2\} \oplus \cdots \oplus \{e_s\}$ where $\mu_{e_i}(\lambda) = \mu_i(\lambda)$.*

Let $\nu_i(\lambda) = \mu(\lambda)(\mu_i(\lambda))^{-1}$ and let $e_i = e\nu_i(A)$. Then $\mu_{e_i}(\lambda) = \mu_i(\lambda)$. We know also that the order of $e' = e_1 + e_2 + \cdots + e_s$ is $\mu(\lambda)$. Hence $\{e'\} = \mathfrak{R}$. On the other hand, $\{e'\} \subseteq \{e_1\} + \{e_2\} + \cdots + \{e_s\}$ so that $\mathfrak{R} = \{e_1\} + \{e_2\} + \cdots + \{e_s\}$. Since $\dim \mathfrak{R} = \deg \mu(\lambda) = \Sigma \deg \mu_i(\lambda) = \Sigma \dim \{e_i\}$, the subspaces $\{e_i\}$ are independent. Hence $\mathfrak{R} = \{e_1\} \oplus \{e_2\} \oplus \cdots \oplus \{e_s\}$.

Now let $\mu(\lambda) = \pi_1(\lambda)^{k_1}\pi_2(\lambda)^{k_2} \cdots \pi_s(\lambda)^{k_s}$ where the π_i are distinct primes. Then by Theorem 3, $\mathfrak{R} = \{e_1\} \oplus \{e_2\} \oplus \cdots \oplus \{e_s\}$ where $\mu_{e_i}(\lambda) = \pi_i(\lambda)^{k_i}$.

For the sake of simplicity of notation we assume for the moment that $s = 1$. Then $\mu(\lambda) = \pi(\lambda)^k$ where $\pi(\lambda)$ is prime, and if

$$\pi(\lambda) = \lambda^q - \rho_{q-1}\lambda^{q-1} - \cdots - \rho_0,$$

$n = \deg \mu(\lambda) = kq$. We introduce the vectors

$$f_1 = e\pi(A)^{k-1}, \quad f_2 = e\pi(A)^{k-1}A, \quad \cdots, \quad f_q = e\pi(A)^{k-1}A^{q-1};$$

$$f_{q+1} = e\pi(A)^{k-2}, \quad f_{q+2} = e\pi(A)^{k-2}A, \quad \cdots, \quad f_{2q} = e\pi(A)^{k-2}A^{q-1};$$

. . . .

$$f_{(k-1)q+1} = e, \quad f_{(k-1)q+2} = eA, \quad \cdots, \quad f_{kq} = eA^{q-1}.$$

Thus each f is of the form $e\phi(A)$ where $\deg \phi(\lambda) < kq$. Moreover, the degree of the $\phi(\lambda)$'s associated with distinct f's are different. It follows that the f's are linearly independent, since a non-trivial relation of the form $\Sigma\delta_i f_i = 0$ entails the existence of a polynomial $\nu(\lambda) \neq 0$ of degree $< kq$ such that $e\nu(A) = 0$. This contradicts the fact that $\deg \mu(\lambda) = kq$ and establishes the linear independence of the f's. Since there are altogether kq f's, $(f_1, f_2, \cdots, f_{kq})$ is a basis for $\mathfrak{R}$ over Φ.

What is the matrix of A relative to $(f_1, f_2, \cdots, f_{kq})$? We have the following relations:

$$f_1 A = f_2$$

$$f_2 A = f_3$$

. . . .

$$f_{q-1}A = f_q$$

$$\begin{aligned} f_q A &= e\pi(A)^{k-1}A^q = e\pi(A)^{k-1}[A^q - \pi(A)] \\ &= e\pi(A)^{k-1}[\rho_0 1 + \rho_1 A + \cdots + \rho_{q-1}A^{q-1}] \\ &= \rho_0 f_1 + \rho_1 f_2 + \cdots + \rho_{q-1} f_q \end{aligned}$$

$$f_{q+1}A = f_{q+2}$$

$$f_{q+2}A = f_{q+3}$$

. . . .

$$
\begin{aligned}
f_{2q-1}A &= f_{2q} \\
f_{2q}A &= e\pi(A)^{k-2}A^q = e\pi(A)^{k-2}[A^q - \pi(A)] \\
&\quad + e\pi(A)^{k-1} \\
&= \rho_0 f_{q+1} + \rho_1 f_{q+2} + \cdots + \rho_{q-1} f_{2q} + f_1 \\
&\cdots\cdots\cdots \\
f_{(k-1)q+1}A &= f_{(k-1)q+2} \\
&\cdots\cdots\cdots \\
f_{(k-1)q+q-1}A &= f_{kq} \\
f_{kq}A &= \rho_0 f_{(k-1)q+1} + \rho_1 f_{(k-1)q+2} + \cdots + \rho_{q-1} f_{kq} \\
&\quad + f_{(k-2)q+1}.
\end{aligned}
$$

Hence the matrix of A relative to the basis $(f_1, f_2, \cdots, f_{kq})$ has the form

$$
B = \begin{bmatrix} P & & & & \\ N & P & & & \\ & N & & & \\ & & \ddots & & \\ & & & & \\ & & & N & P \end{bmatrix} \tag{9}
$$

where P is the companion matrix of $\pi(\lambda)$ and N is the $q \times q$ matrix

$$
N = \begin{bmatrix} 0 & 0 & \cdots & 0 \\ 0 & 0 & \cdots & 0 \\ \cdot & \cdot & \cdots & \cdot \\ 1 & 0 & \cdots & 0 \end{bmatrix}. \tag{10}
$$

We return now to the general case in which $\mu(\lambda) = \pi_1(\lambda)^{k_1}\pi_2(\lambda)^{k_2} \cdots \pi_s(\lambda)^{k_s}$ and $\mathfrak{R} = \{e_1\} \oplus \{e_2\} \oplus \cdots \oplus \{e_s\}$. We choose in each $\{e_i\}$ a basis of the type just indicated. Together these bases give a basis for $\mathfrak{R}$. The matrix of A relative to this basis is

$$\begin{bmatrix} B_1 & & & \\ & B_2 & & \\ & & \ddots & \\ & & & B_s \end{bmatrix} \tag{11}$$

where each B_i is determined by $\pi_i(\lambda)^{k_i}$ in the same way that B is determined by $\pi(\lambda)^k$.

We shall call (11) the *classical canonical matrix* of the cyclic linear transformation A. The prime powers $\pi_i(\lambda)^{k_i}$ that occur in the factorization of the minimum polynomial will be called the *elementary divisors* of the cyclic linear transformation. To the elementary divisor $\pi_i(\lambda)^{k_i}$ is associated the block B_i in the classical canonical form. As an example, let Φ be the field of rational numbers, and let $\mu(\lambda) = (\lambda - 1)^3(\lambda^2 - 2)^2$. The classical canonical matrix of the cyclic linear transformation with minimum polynomial $\mu(\lambda)$ is

$$\left[\begin{array}{ccc|cc|cc} 1 & 0 & 0 & & & & \\ 1 & 1 & 0 & & & & \\ 0 & 1 & 1 & & & & \\ \hline & & & 0 & 1 & 0 & 0 \\ & & & 2 & 0 & 0 & 0 \\ \hline & & & 0 & 0 & 0 & 1 \\ & & & 1 & 0 & 2 & 0 \end{array}\right].$$

EXERCISES

1. Show that, if $\mu(\lambda)$ is any polynomial with leading coefficient 1, then there exists a matrix (α) that has $\mu(\lambda)$ as its minimum polynomial.
2. Show that, if A is a linear transformation that has the diagonal matrix

$$\operatorname{diag}\,\{\alpha_1, \alpha_2, \cdots, \alpha_n\}$$

where the α_i are distinct, then the minimum polynomial of A is $\Pi(\lambda - \alpha_i)$. Hence prove A cyclic.

3. Use 2 to prove that, if C is the field of complex numbers and $\zeta_1, \zeta_2, \cdots, \zeta_n$ are the n distinct roots of unity, then

$$\begin{bmatrix} \zeta_1 & & & \\ & \zeta_2 & & \\ & & \ddots & \\ & & & \zeta_n \end{bmatrix} \quad \text{and} \quad \begin{bmatrix} 0 & 1 & 0 & \cdots & 0 \\ 0 & 0 & 1 & \cdots & 0 \\ \cdot & \cdot & \cdot & \cdots & \cdot \\ 0 & 0 & 0 & \cdot\ 0 & 1 \\ 1 & 0 & 0 & \cdots & 0 \end{bmatrix}$$

are similar in C_n.

4. Show that, if A is a linear transformation with matrix

$$\begin{bmatrix} 0 & 1 & \alpha_{13} & \cdot & \cdot & \cdot & \alpha_{1n} \\ 0 & 0 & 1 & \alpha_{24} & \cdot & \cdot & \alpha_{2n} \\ \cdot & \cdot & \cdot & \cdot & \cdot & \cdot & \cdot \\ \cdot & \cdot & \cdot & \cdot & \cdot & \cdot & \cdot \\ 0 & 0 & 0 & \cdot & \cdot & 0 & 1 \\ 0 & 0 & 0 & \cdot & \cdot & 0 & 0 \end{bmatrix},$$

then A is cyclic. Use this to prove that the given matrix is similar to

$$\begin{bmatrix} 0 & 1 & 0 & \cdots & 0 \\ 0 & 0 & 1 & \cdots & 0 \\ \cdot & \cdot & \cdot & \cdots & \cdot \\ 0 & 0 & 0 & \cdot\ 0 & 1 \\ 0 & 0 & 0 & \cdot\ 0 & 0 \end{bmatrix}.$$

5. What is the classical canonical matrix of the cyclic linear transformation whose minimum polynomial is $(\lambda - 1)^3(\lambda^2 + 1)^2(\lambda + 1)^2$? Assume Φ to be the field of real numbers.

6. Prove that, if $\mathfrak{R} = \{e_1\} \oplus \{e_2\} \oplus \cdots \oplus \{e_s\}$ and the orders $\mu_{e_i}(\lambda)$ are relatively prime in pairs, then $\mathfrak{R}$ is cyclic.

5. The $\Phi[\lambda]$-module determined by a linear transformation. For the further study of a linear transformation we shall make use of another idea, namely, that the homomorphism between $\Phi[\lambda]$ and the ring $\mathfrak{A} = [1, A, A^2, \cdots]$ can be used to turn the vector space $\mathfrak{R}$ into a $\Phi[\lambda]$-module. For this purpose we define the product $\phi(\lambda)x$, x in $\mathfrak{R}$, to be the vector $x\phi(A)$. Then

$$\phi(\lambda)(x + y) = (x + y)\phi(A) = x\phi(A) + y\phi(A) = \phi(\lambda)x + \phi(\lambda)y,$$

and since the linear transformations corresponding to $\phi(\lambda) + \psi(\lambda)$ and $\phi(\lambda)\psi(\lambda)$ are respectively $\phi(A) + \psi(A)$ and $\phi(A)\psi(A) = \psi(A)\phi(A)$,

$$[\phi(\lambda) + \psi(\lambda)]x = x[\phi(A) + \psi(A)] = x\phi(A) + x\psi(A)$$
$$= \phi(\lambda)x + \psi(\lambda)x$$
$$[\phi(\lambda)\psi(\lambda)]x = x[\psi(A)\phi(A)] = (x\psi(A))\phi(A) = \phi(\lambda)(\psi(\lambda)x)$$
$$1x = x1 = x.$$

Hence the basic module conditions hold.* There are several advantages to be gained in adopting the module point of view. In the first place we replace the ring $\mathfrak{A}$ by the ring $\Phi[\lambda]$. The former may have quite a complicated algebraic structure; it may contain nilpotent elements and it may have zero divisors. On the other hand, we know that the polynomial ring $\Phi[\lambda]$ is an integral domain, and we have available the arithmetic theory of this ring. The basic arithmetic fact concerning $\Phi[\lambda]$ is that every ideal in this ring is a principal ideal.

A second advantage of the module point of view is that it leads us to the consideration of other $\Phi[\lambda]$-modules that are of a simpler structure than $\mathfrak{R}$ and that can be used to study $\mathfrak{R}$. In fact we shall reduce our study of $\mathfrak{R}$ to that of free modules and of submodules of free modules. This is accomplished in the following way: Let $(e_1, e_2, \cdots, e_n)$ be a basis for $\mathfrak{R}$ over Φ. Then, of course, the e_i form a set of generators for $\mathfrak{R}$ relative to $\Phi[\lambda]$. The $\Phi[\lambda]$-module $\mathfrak{R}$ is not a free module since for any $e \in \mathfrak{R}$ there exists a non-zero polynomial $\mu_e(\lambda)$ such that $\mu_e(\lambda)e = e\mu_e(A) = 0$. Hence there exist relations of the form $\sum_1^n \gamma_i(\lambda)e_i = 0$ in which the $\gamma_i(\lambda)$ are not all 0. In order to study these relations it is natural to introduce the free $\Phi[\lambda]$-module $\mathfrak{F}$ with basis $(t_1, t_2, \cdots, t_n)$. If $v = \Sigma\phi_i(\lambda)t_i$ is an element of $\mathfrak{F}$, then we associate with v the element $vT = \Sigma\phi_i(\lambda)e_i$ in $\mathfrak{R}$. This correspondence is a $\Phi[\lambda]$-homomorphism of $\mathfrak{F}$ on $\mathfrak{R}$. Let $\mathfrak{N}$ denote the kernel of this homomorphism. It is immediate that $\mathfrak{N}$ is a submodule of $\mathfrak{F}$. By definition $\Sigma\gamma_i(\lambda)t_i \in \mathfrak{N}$ if and only if the relation $\Sigma\gamma_i(\lambda)e_i = 0$ holds

* It should be noted that the commutativity of multiplication in $\mathfrak{A}$ was used in the verification of the third axiom. We can also regard $\mathfrak{R}$ as a right $\Phi[\lambda]$-module. Here $x\phi(\lambda) = x\phi(A)$. This is probably the more natural point of view. However, we have adopted the above method in order to be consistent with our previous emphasis on left modules.

among the generators e_i. Thus in a sense $\mathfrak{N}$ corresponds to the set of relations that obtain among the e_i.

We can easily determine the submodule $\mathfrak{N}$. Let $e_iA = \Sigma\alpha_{ij}e_j$ so that (α) is the matrix of A relative to the given basis. Then it is clear that the elements

$$
(12)\qquad
\begin{aligned}
v_1 &= (\lambda - \alpha_{11})t_1 - \alpha_{12}t_2 - \cdots - \alpha_{1n}t_n\\
v_2 &= -\alpha_{21}t_1 + (\lambda - \alpha_{22})t_2 - \cdots - \alpha_{2n}t_n\\
&\cdots\cdots\cdots\cdots\cdots\cdots\\
v_n &= -\alpha_{n1}t_1 - \alpha_{n2}t_2 - \cdots + (\lambda - \alpha_{nn})t_n
\end{aligned}
$$

are in $\mathfrak{N}$. We shall now show that $\mathfrak{N}$ is free and that the v's form a basis for this module. The definition $v_i = \lambda t_i - \Sigma\alpha_{ij}t_j$ gives $\lambda t_i = v_i + \Sigma\alpha_{ij}t_j$. These relations can be used to express any element $v = \Sigma\phi_i(\lambda)t_i$ in the form $\Sigma\psi_i(\lambda)v_i + \Sigma\rho_i t_i$ where the $\rho_i \varepsilon \Phi$. Clearly if $v \varepsilon \mathfrak{N}$, then $\Sigma\rho_i t_i \varepsilon \mathfrak{N}$. Hence $\Sigma\rho_i e_i = (\Sigma\rho_i t_i)T = 0$. But the e_i are Φ-independent, and this implies that every $\rho_i = 0$. Hence $v = \Sigma\psi_i(\lambda)v_i$, which shows that the v_i are generators for $\mathfrak{N}$. Suppose next that $\Sigma\psi_i(\lambda)v_i = 0$. Then

$$\Sigma\psi_i(\lambda)\lambda t_i = \Sigma\psi_i(\lambda)\alpha_{ij}t_j$$

and

$$\psi_i(\lambda)\lambda = \sum_j \psi_j(\lambda)\alpha_{ji}.$$

If any $\psi_i \neq 0$, let ψ_r be one of maximum degree. Then clearly the relation $\psi_r(\lambda)\lambda = \sum_j \psi_j(\lambda)\alpha_{jr}$ is impossible. This proves that the v_i form a basis for $\mathfrak{N}$.

6. Finitely generated $\mathfrak{o}$-modules, $\mathfrak{o}$, a principal ideal domain. There is still another advantage of the module theoretic method: It is easily generalized. In place of $\Phi[\lambda]$ we can consider any principal ideal domain $\mathfrak{o}$ and any finitely generated $\mathfrak{o}$-module $\mathfrak{R}$.* With a slight increase in difficulty we shall obtain in this way other important applications. The most noteworthy of these is the theory of ordinary finitely generated commutative groups.

* A good deal of what we shall do can also be done for $\mathfrak{o}$-modules for which $\mathfrak{o}$ is a non-commutative principal ideal domain. This theory can be applied to the theory of a single linear transformation in a vector space over a division ring. See the author's *Theory of Rings*, New York, 1943, Chapter 3.

As we have noted in Chapter 1, this theory can be obtained by regarding commutative groups as $\mathfrak{o}$-modules for $\mathfrak{o}$, the ring of rational integers.

Hence let $\mathfrak{R}$ be any finitely generated $\mathfrak{o}$-module and let $(e_1, e_2, \cdots, e_n)$ be a set of generators. Following the procedure that we used above we introduce the free module $\mathfrak{F}$ with basis $(t_1, t_2, \cdots, t_n)$ and the mapping T: $\Sigma\phi_i t_i \to \Sigma\phi_i e_i$. The mapping T is an $\mathfrak{o}$-homomorphism. Hence the kernel $\mathfrak{N}$ is a submodule of $\mathfrak{F}$. We shall now show that any submodule of a free $\mathfrak{o}$-module, $\mathfrak{o}$ principal, is free. It is convenient to use the convention that the module 0 is a free module with a vacuous basis.

Thus let $\mathfrak{N}$ be any submodule of the free module $\mathfrak{F}$ with the basis $(t_1, t_2, \cdots, t_n)$. We shall show by induction on n that $\mathfrak{N}$ is free and has a basis of $m \leq n$ elements. This is clear if $n = 0$. Hence we assume it true for modules $\mathfrak{F}'$ with bases of $n - 1$ elements. We consider now the totality $\mathfrak{J}$ of elements β_1 of $\mathfrak{o}$ such that $\beta_1 t_1 + \beta_2 t_2 + \cdots + \beta_n t_n \,\varepsilon\, \mathfrak{N}$. It is immediate that $\mathfrak{J}$ is an ideal; hence $\mathfrak{J} = (\delta_1)$. If $\delta_1 = 0$ every element of $\mathfrak{N}$ is of the form $\beta_2 t_2 + \cdots + \beta_n t_n$, and therefore it belongs to the free module $\mathfrak{F}'$ with the basis $(t_2, t_3, \cdots, t_n)$. In this case the result is clear. Now assume that $\delta_1 \neq 0$ and let $v_1 = \delta_1 t_1 + \sum_2^n \beta_j' t_j$ be an element of $\mathfrak{N}$ that has the "leading coefficient" δ_1. Then if v is any element of $\mathfrak{N}$, $v = \Sigma\beta_i t_i$ and $\beta_1 = \mu_1\delta_1$. Hence $v' = v - \mu_1 v_1 \,\varepsilon\, \mathfrak{F}'$. Of course this element also belongs to $\mathfrak{N}$. Now the intersection $\mathfrak{N}' = \mathfrak{N} \cap \mathfrak{F}'$ is a submodule of $\mathfrak{F}'$, and so we can assume that it is a free module with the basis $v_2, v_3, \cdots, v_m$, $m \leq n$. Then $v' = \sum_{k=2}^{m} \mu_k v_k$ and $v = \mu_1 v_1 + \Sigma\mu_k v_k$. Hence $(v_1, v_2, \cdots, v_m)$ is a set of generators for $\mathfrak{N}$. Suppose next that $\sum_1^m \rho_i v_i = 0$. Then if we replace v_1 by $\delta_1 t_1 + \Sigma\beta_j' t_j$ and the other v's by their expressions in terms of $t_2, t_3, \cdots, t_n$ we obtain

$$\rho_1\delta_1 t_1 + \rho_2' t_2 + \cdots + \rho_m' t_m = 0.$$

Since $\delta_1 \neq 0$, this implies that $\rho_1 = 0$. Hence we have $\sum_2^n \rho_k v_k = 0$ from which we conclude that all the ρ's are 0 since the elements $(v_2, v_3, \cdots, v_m)$ form a basis for $\mathfrak{N}'$. This concludes the proof of the following

Theorem 4. *If $\mathfrak{F}$ is a free $\mathfrak{o}$-module with a basis of n elements and $\mathfrak{o}$ is a principal ideal domain, then any submodule of $\mathfrak{F}$ is free and has a basis of $m \leq n$ elements.*

7. Normalization of the generators of $\mathfrak{F}$ and of $\mathfrak{N}$. We shall suppose now that the elements $(v_1, v_2, \cdots, v_m)$ are merely generators for the submodule $\mathfrak{N}$ of the free $\mathfrak{o}$-module $\mathfrak{F}$ with basis $(t_1, t_2, \cdots, t_n)$. We write

$$
\begin{aligned}
v_1 &= \sigma_{11}t_1 + \sigma_{12}t_2 + \cdots + \sigma_{1n}t_n \\
v_2 &= \sigma_{21}t_1 + \sigma_{22}t_2 + \cdots + \sigma_{2n}t_n \\
&\cdots\cdots\cdots\cdots\cdots \\
v_m &= \sigma_{m1}t_1 + \sigma_{m2}t_2 + \cdots + \sigma_{mn}t_n.
\end{aligned}
\tag{13}
$$

The matrix (σ) with elements σ_{ij} in $\mathfrak{o}$ is uniquely determined by the ordered sets $(v_1, v_2, \cdots, v_m)$ and $(t_1, t_2, \cdots, t_n)$.

Let (μ_{ij}) be a unit in the matrix ring and let the inverse matrix be (μ_{ij}^*). Define $t_i' = \Sigma\mu_{ij}t_j$. Then we assert that the t_i' constitute another basis for $\mathfrak{F}$. In the first place we have the relations

$$\Sigma\mu_{ki}^*t_i' = \sum_{i,j} \mu_{ki}^*\mu_{ij}t_j = \Sigma\delta_{kj}t_j = t_k.$$

Hence any element $\Sigma\phi_it_i$ may be written in the form $\Sigma\psi_it_i'$ and the t_i' are generators. Next suppose that $\Sigma\gamma_it_i' = 0$. Then $\sum_{i,j} \gamma_i\mu_{ij}t_j = 0$. Hence $\sum_i \gamma_i\mu_{ij} = 0$. But this implies that

$$\gamma_k = \sum_{i,j} \gamma_i\mu_{ij}\mu_{jk}^* = 0,$$

and this proves that the t_i' form a basis.

In a similar fashion we see that if (ν_{pq}) is a unit in $\mathfrak{o}_m$, then the elements $v_p' = \Sigma\nu_{pq}v_q$ constitute a second set of generators of $\mathfrak{N}$.

Suppose now that the basis $(t_1, t_2, \cdots, t_n)$ is replaced by $(t_1', t_2', \cdots, t_n')$ where $t_i' = \Sigma\mu_{ij}t_j$, (μ) a unit, and that the set of generators $(v_1, v_2, \cdots, v_m)$ of $\mathfrak{N}$ is replaced by $(v_1', v_2', \cdots, v_m')$ where $v_p' = \Sigma\nu_{pq}v_q$. What is the matrix of $(v_1', v_2', \cdots, v_m')$

relative to $(t_1', t_2', \cdots, t_n')$? We have the relations

$$v_p' = \sum_q \nu_{pq} v_q = \sum_{q,i} \nu_{pq}\sigma_{qi} t_i = \sum_{q,i,j} \nu_{pq}\sigma_{qi}\mu_{ij}{}^* t_j'$$
$$= \Sigma \tau_{pj} t_j',$$

where $\tau_{pj} = \sum_{q,i} \nu_{pq}\sigma_{qi}\mu_{ij}{}^*$. Thus the new matrix is related to (σ) by

$$(\tau) = (\nu)(\sigma)(\mu)^{-1}. \tag{14}$$

We are now led to the problem of making the "right" choices for the units (ν) and $(\mu)^{-1}$ so that the matrix (τ) has a simple "normal" form.

8. Equivalence of matrices with elements in a principal ideal domain. Two $m \times n$ matrices with elements in $\mathfrak{o}$ are said to be *equivalent* if there exist a unit (μ) in $\mathfrak{o}_n$ and a unit (ν) in $\mathfrak{o}_m$ such that

$$(\tau) = (\nu)(\sigma)(\mu).$$

This relation is an equivalence. The special case in which $\mathfrak{o}$ is a field has been considered previously. Here we learned that a necessary and sufficient condition that (σ) and (τ) be equivalent is that they have the same determinantal rank. A consequence of this result for the general case of an arbitrary principal ideal domain is that a necessary condition for equivalence is equality of ranks. For let $(\tau) = (\nu)(\sigma)(\mu)$, where (μ) and (ν) are units. Then if P is the quotient field of $\mathfrak{o}$, then the given equation may be regarded as a relation between matrices with elements in P. Clearly $(\mu) \in L(\text{P}, n)$ and $(\nu) \in L(\text{P}, m)$. Hence by the result in the field case (τ) and (σ) have the same determinantal ranks.

We consider now the problem of selecting among the matrices equivalent to a given $m \times n$ matrix (σ) one that has a particularly simple form. The result that we wish to obtain is the following

Theorem 5. *If (σ) is an $m \times n$ matrix with elements in a principal ideal domain $\mathfrak{o}$, there exists a matrix equivalent to (σ) which has the "diagonal" form (cf. (18), p. 45)*

$$\text{diag}\,\{\delta_1, \delta_2, \cdots, \delta_r, 0, \cdots, 0\}$$

where the δ_i are $\neq 0$ and $\delta_i \mid \delta_j$ (δ_i is a factor of δ_j) if $i < j$.

(a) We give first a constructive proof of this theorem which is valid for the case where $\mathfrak{o}$ is Euclidean. We recall that a commutative integral domain is called Euclidean if there is defined a *degree* $\delta(a)$ for every a in the domain, such that $\delta(a)$ is a non-negative integer and such that

1. $$\delta(a) = 0 \quad \text{if and only if} \quad a = 0.$$

2. $$\delta(ab) = \delta(a)\delta(b).$$

3. If a is arbitrary and $b \neq 0$, then there exist elements q and r with $\delta(r) < \delta(b)$ such that $a = bq + r$.

We refer the reader to Volume I, p. 123, for the proof of the fact that any ideal in a *Euclidean* domain is principal. We proceed now with the proof of the theorem.

We note first that the square matrices

$$T_{ij}(\beta) = \begin{bmatrix} 1 & & & & & & \\ & \cdot & & & \cdot & & \\ & & 1 & \cdots & \beta & \cdots & \\ & & & \cdot & \cdot & & \\ & & & & 1 & & \\ & & & & & \cdot & \\ & & & & & & 1 \end{bmatrix} \begin{matrix} \\ \\ i \\ \\ \\ \\ \\ \end{matrix} \quad (\beta \text{ in column } j)$$

$$D_i(\gamma) = \begin{bmatrix} 1 & & & & & \\ & \cdot & & \cdot & & \\ & & 1 & & & \\ & & & \gamma & \cdots & \\ & & & & 1 & \\ & & & & & \cdot \\ & & & & & & 1 \end{bmatrix} \begin{matrix} \\ \\ \\ i \\ \\ \\ \\ \end{matrix} \quad (\gamma \text{ in column } i)$$

where $i \neq j$ and γ is a unit in $\mathfrak{o}$ and

$$P_{ij} = \begin{array}{c} \\ \left[\begin{array}{ccccccccccc}
1 & & & \cdot & & & & \cdot & & & \\
 & \cdot & & \cdot & & & & \cdot & & & \\
 & & 1 & & & & & & & & \\
 & & & 0 & \cdot & \cdot & \cdot & 1 & \cdots & & \\
 & & & \cdot & 1 & & & \cdot & & & \\
 & & & \cdot & & \cdot & & \cdot & & & \\
 & & & & & & 1 & & & & \\
 & & & 1 & \cdot & \cdot & \cdot & 0 & \cdots & & \\
 & & & & & & & & 1 & & \\
 & & & & & & & & & \cdot & \\
 & & & & & & & & & & 1
\end{array}\right] \end{array}
\begin{array}{c} \\ \\ \\ \\ i \\ \\ \\ \\ j \\ \\ \\ \\ \end{array}$$

(the columns through the entries $0, \cdot, \cdot, 1$ and $1, \cdot, \cdot, 0$ being the ith and jth columns)

are units; for $T_{ij}(\beta)^{-1} = T_{ij}(-\beta)$, $D_i(\gamma)^{-1} = D_i(\gamma^{-1})$ and $P_{ij}^{-1} = P_{ij}$. It is easy to verify that

I. Left multiplication of (σ) by an $m \times m$ matrix $T_{ij}(\beta)$ gives a matrix whose ith row is obtained by multiplying the jth row of (σ) by β and adding it to the ith row and whose other rows are the same as in (σ).

Right multiplication of (σ) by an $n \times n$ matrix $T_{ij}(\beta)$ gives a matrix whose jth column is β times the ith column of (σ) plus the jth column of (σ) and whose other columns are as in (σ).

II. Left multiplication of (σ) by an $m \times m$ matrix $D_i(\gamma)$ amounts to the operation of multiplying the ith row of (σ) by γ and leaving the other rows unchanged.

Right multiplication of (σ) by an $n \times n$ matrix $D_i(\gamma)$ amounts to the operation of multiplying the ith column of (σ) by γ and leaving the other columns unchanged.

III. Left multiplication of (σ) by an $m \times m$ matrix P_{ij} amounts to interchanging the ith and jth rows of (σ) and leaving the remaining rows unchanged.

Multiplication of (σ) on the right by an $n \times n$ matrix P_{ij} amounts to the interchange of the ith and jth columns of (σ).

We refer to the matrices T_{ij}, D_i, P_{ij} as *elementary matrices*, and multiplication by these matrices as *elementary transformations* of the respective types I, II and III. Any matrix that is obtained from (σ) by a finite sequence of elementary operations is equivalent to (σ). We now proceed to the main part of the proof.

If $(\sigma) = 0$, we have nothing to prove. Otherwise let σ_{ij} be a non-zero element of least degree in (σ). Elementary transformations of type III will move this element to the (1,1) position. If $\sigma_{1k} = \sigma_{11}\beta_k + \rho_{1k}$ where the degree $\delta(\rho_{1k}) < \delta(\sigma_{11})$, then we may use an elementary transformation of type I to replace σ_{1k} by ρ_{1k}. If $\rho_{1k} \neq 0$, we have a new matrix equivalent to (σ) in which there are non-zero elements of degree less than the minimum non-zero degree in (σ). We repeat the original process with this new matrix. Similarly if $\sigma_{k1} = \sigma_{11}\beta_k' + \rho_{k1}$ where $\delta(\rho_{k1}) < \delta(\sigma_{11})$, then either $\rho_{k1} = 0$ or we obtain a matrix equivalent to (σ) that has a non-zero element of degree less than the minimum $\delta(\sigma_{ij}) \neq 0$. Since the minimum degree $\neq 0$ is constantly decreasing, we eventually reach a matrix in which $\sigma_{1k} = \sigma_{11}\beta_k$ and $\sigma_{k1} = \sigma_{11}\beta_k'$ for all $k = 1, 2, \cdots, n$. Then elementary transformations of type I yield an equivalent matrix of the form

$$\begin{bmatrix} \rho_{11} & 0 & \cdots & 0 \\ 0 & \rho_{22} & \cdots & \rho_{2n} \\ \cdot & \cdot & \cdots & \cdot \\ 0 & \rho_{m2} & \cdots & \rho_{mn} \end{bmatrix}, \quad \rho_{11} = \sigma_{11}. \tag{15}$$

This process can now be applied to the submatrix (ρ_{ij}) $i = 2, \cdots, m$; $j = 2, \cdots, n$. The necessary transformations do not affect the first row or column, and yield an equivalent matrix of the form

$$\begin{bmatrix} \tau_{11} & 0 & \cdot & \cdots & 0 \\ 0 & \tau_{22} & 0 & \cdots & 0 \\ 0 & 0 & \tau_{33} & \cdots & \tau_{3n} \\ \cdot & \cdot & \cdot & \cdots & \cdot \\ 0 & 0 & \tau_{m3} & \cdots & \tau_{mn} \end{bmatrix}.$$

Continuing in this way we obtain finally an equivalent matrix of diagonal form, say diag $\{\epsilon_1, \epsilon_2, \cdots, \epsilon_r, 0, \cdots, 0\}$ where the $\epsilon_i \neq 0$.

If $\epsilon_i = \epsilon_1\beta + \rho_i$ with $\delta(\rho_i) < \delta(\epsilon_1)$ and $\rho_i \neq 0$, then we may replace our diagonal matrix by the equivalent matrix

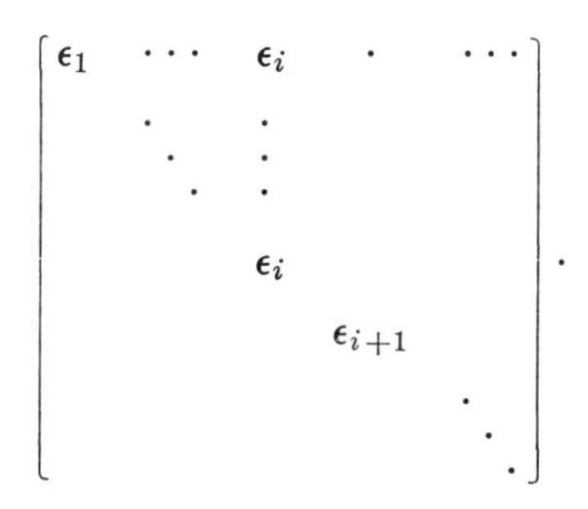

This is equivalent to a matrix in which the element in the $(1, i)$ position is ρ_i. We then repeat our earlier process to "diagonalize" this matrix. We obtain a new matrix diag $\{\eta_1, \eta_2, \cdots, \eta_r, 0, \cdots, 0\}$ in which $\delta(\eta_1) < \delta(\epsilon_1)$. A number of repetitions of this process leads eventually to an equivalent diagonal matrix, diag $\{\delta_1, \delta_2, \cdots, \delta_r, 0, \cdots, 0\}$ in which the non-zero elements δ_i satisfy the required divisibility conditions.

(b) The argument in the general case is similar to this. Here we use the *length* of an element, that is, the number r of primes π_i that occur on a factorization of the element as $\pi_1\pi_2 \cdots \pi_r$ as a substitute for the degree used above. This number is an invariant because of the unique factorization theorem in $\mathfrak{o}$ (cf. Volume I, p. 122). Suppose that σ_{11} has smallest length for the non-zero elements and suppose that $\sigma_{11} \nmid \sigma_{1k}$.* Write $\alpha = \sigma_{11}$, $\beta = \sigma_{1k}$ and let δ be a highest common factor of α and β. Then there exist elements ξ, η such that $\alpha\xi + \beta\eta = \delta$. Set $\zeta = \beta\delta^{-1}$, $\theta = -\alpha\delta^{-1}$. Then we have the matrix relation

$$\begin{bmatrix} -\theta & \zeta \\ \eta & -\xi \end{bmatrix} \begin{bmatrix} \xi & \zeta \\ \eta & \theta \end{bmatrix} = \begin{bmatrix} 1 & 0 \\ 0 & 1 \end{bmatrix}.$$

Thus these matrices are units; consequently

* $a \nmid b$ means "a is not a factor of b."

$$
U = \begin{bmatrix}
\xi & 0 & \cdots & 0 & \zeta & 0 & \cdots & \\
0 & 1 & & & & & & \\
\vdots & & \ddots & \vdots & & & & \\
& & & 1 & & & & \\
\eta & 0 & \cdots & 0 & \theta & \cdot & \cdots & \\
& & & & & 1 & & \\
& & & & & & \ddots & \\
& & & & & & & 1
\end{bmatrix}
$$

(column k marked above ζ; row k marked at the right of the η row)

is a unit. The matrix $(\sigma)U$ has as its first row the vector $(\delta, \sigma_{12}, \sigma_{13}, \cdots, \sigma_{1(k-1)}, 0, \sigma_{1(k+1)}, \cdots, \sigma_{1n})$. Since $\alpha \not| \beta$, the length of δ is less than that of $\alpha = \sigma_{11}$. Eventually this process and a similar one applied to the first column yield a matrix equivalent to (σ) in which σ_{11} is a factor of each σ_{1k} and of each σ_{k1}. As before, this leads to a matrix of the form (15). The remainder of the argument is a repetition of the one given above.

A matrix equivalent to (σ) that has the form given in Theorem 5 is called a *normal form* for the matrix (σ). The diagonal elements of a normal form are called *invariant factors* of (σ). We shall show in § 10 that they are, in fact, invariants of the given matrix.

EXERCISES

1. Obtain a normal form for the matrix

$$
\begin{bmatrix}
6 & 2 & 3 & 0 \\
2 & 3 & -4 & 1 \\
-3 & 3 & 1 & 2 \\
-1 & 2 & -3 & 5
\end{bmatrix},
$$

$\mathfrak{o} = I$ the ring of integers.

2. Obtain a normal form for the matrix

$$
(\sigma) = \begin{bmatrix}
\lambda - 17 & 8 & 12 & -14 \\
-46 & \lambda + 22 & 35 & -41 \\
2 & -1 & \lambda - 4 & 4 \\
-4 & 2 & 2 & \lambda - 3
\end{bmatrix},
$$

$\mathfrak{o} = R_0[\lambda]$, R_0 the field of rational numbers. Also find units (μ) and (ν) such that $(\nu)(\sigma)(\mu) =$ normal form.

3. Prove that a diagonal square matrix, diag $\{\delta_1, \delta_2, \cdots, \delta_n\}$ is a unit if and only if every δ_i is a unit.

4. Use 3. to prove that, if $\mathfrak{o}$ is Euclidean, then any unit in $\mathfrak{o}_n$ is a product of elementary matrices.

5. Show that, if d is a greatest common divisor of the elements $a_1, a_2, \cdots, a_n$, then there exists a unit (μ) in $\mathfrak{o}_n$ such that $(a_1, \cdots, a_n)(\mu) = (d, 0, \cdots, 0)$.

6. Show that, if the elements $a_{11}, a_{12}, \cdots, a_{1n}$ are relatively prime (have greatest common divisor $= 1$), then there exist elements a_{ij}, $i = 2, \cdots, n, j = 1, 2, \cdots, n$, such that (a) is a unit.

9. Structure of finitely generated $\mathfrak{o}$-modules. In order to state the structure theorems it will be well to introduce some general concepts concerning $\mathfrak{o}$-modules. Some of these have already been encountered in the special case of modules determined by a linear transformation.

Let x be any element of the $\mathfrak{o}$-module $\mathfrak{R}$. Then the set $\mathfrak{J}_x$ of elements $\beta \in \mathfrak{o}$ such that $\beta x = 0$ is an ideal. We call this ideal the *order ideal* of the element x. If $\mathfrak{o}$ is a principal ideal domain, $\mathfrak{J}_x = (\mu_x)$. Moreover, if $\mathfrak{o} = \Phi[\lambda]$ and $\mathfrak{J}_x \neq (0)$, then we may normalize μ_x to have leading coefficient 1. The μ_x thus obtained is the order that we have considered before.

The totality $\{x\}$ of multiples αx of the fixed element x in $\mathfrak{R}$ is a submodule. We call $\{x\}$ the *cyclic submodule* generated by x and we call $\mathfrak{R}$ *cyclic* if there exists an e such that $\mathfrak{R} = \{e\}$. If $\mathfrak{J}_x = (0)$, the submodule $\{x\}$ is a free module. In the contrary case $\mathfrak{J}_x \neq (0)$ we say that x is of *finite order*.

If $y = \alpha x$ and $\beta \in \mathfrak{J}_x$, then $\beta y = \beta\alpha x = \alpha\beta x = 0$. Hence $\beta \in \mathfrak{J}_y$ and so $\mathfrak{J}_x \subseteq \mathfrak{J}_y$. If y is a second generator of $\{x\}$, then also $\mathfrak{J}_y \subseteq \mathfrak{J}_x$. Hence $\mathfrak{J}_x = \mathfrak{J}_y$. Thus the order ideal depends only on the cyclic submodule and not on the particular generator of this submodule.

A module $\mathfrak{R}$ is said to be a *direct sum* of the submodules $\mathfrak{R}_1, \mathfrak{R}_2, \cdots, \mathfrak{R}_s$ if

$$\mathfrak{R} = \mathfrak{R}_1 + \mathfrak{R}_2 + \cdots + \mathfrak{R}_s \tag{16}$$

in the sense that $\mathfrak{R}$ is the smallest submodule containing all of the $\mathfrak{R}_i$, and

$$\mathfrak{R}_i \cap (\mathfrak{R}_1 + \cdots + \mathfrak{R}_{i-1} + \mathfrak{R}_{i+1} + \cdots + \mathfrak{R}_s) = 0 \tag{17}$$

for $i = 1, 2, \cdots, s$. If these conditions hold, we write $\mathfrak{R} = \mathfrak{R}_1 \oplus \mathfrak{R}_2 \oplus \cdots \oplus \mathfrak{R}_s$. As in the case of a vector space $\mathfrak{R} = \mathfrak{R}_1 \oplus \mathfrak{R}_2 \oplus \cdots \oplus \mathfrak{R}_s$ if and only if every x in $\mathfrak{R}$ may be represented in one and only one way in the form $x_1 + x_2 + \cdots + x_s$ where $x_i \in \mathfrak{R}_i$.

We may now state the following fundamental structure theorem for $\mathfrak{o}$-modules, $\mathfrak{o}$ a principal ideal domain.

Theorem 6. *If $\mathfrak{o}$ is a principal ideal domain, any finitely generated $\mathfrak{o}$-module is a direct sum of cyclic submodules.*

Proof. Let $(e_1, e_2, \cdots, e_n)$ be a set of generators for $\mathfrak{R}$. As above, we introduce the free module $\mathfrak{F}$ with basis $(t_1, t_2, \cdots, t_n)$ and the $\mathfrak{o}$-homomorphism T: $\Sigma\phi_i t_i \to \Sigma\phi_i e_i$ of $\mathfrak{F}$ onto $\mathfrak{R}$. Let $\mathfrak{N}$ be the kernel. Then we have seen that $\mathfrak{N}$ has a basis of $m \leq n$ elements. For our purpose here we shall require only the weaker result that $\mathfrak{N}$ is finitely generated. Let $(v_1, v_2, \cdots, v_m)$ be a set of generators and let $v_i = \sum_{j=1}^{n} \sigma_{ij} t_j$, $i = 1, 2, \cdots, m$, be the relations connecting the v's and the t's as before. We now replace the t's by $(t_1', t_2', \cdots, t_n')$ and the v's by $(v_1', v_2', \cdots, v_m')$ where $t_i' = \Sigma\mu_{ij} t_j$, $v_p' = \Sigma\nu_{pq} v_q$ and (μ) and (ν) are units. Then the matrix of the new set of generators $(v_1', v_2', \cdots, v_m')$ of $\mathfrak{N}$ relative to the new basis $(t_1', t_2', \cdots, t_n')$ of $\mathfrak{F}$ is $(\nu)(\sigma)(\mu)^{-1}$. We choose the units (μ) and (ν) so that $(\nu)(\sigma)(\mu)^{-1}$ has the normal form diag $\{\delta_1, \delta_2, \cdots, \delta_r, 0, 0, \cdots, 0\}$. The relations connecting the new generators now read

$$(18) \quad v_1' = \delta_1 t_1', \quad v_2' = \delta_2 t_2', \quad \cdots, \quad v_r' = \delta_r t_r', \qquad v_{r+1}' = 0, \quad \cdots, \quad v_m' = 0.$$

Now define $e_i' = \Sigma\mu_{ij} e_j$. Then since (μ) is a unit these elements are also generators of $\mathfrak{R}$. We assert that

$$(19) \qquad \mathfrak{R} = \{e_1'\} \oplus \{e_2'\} \oplus \cdots \oplus \{e_n'\}$$

and that the order ideal of $\{e_i'\}$ for $i \leq r$ is (δ_i), while for $i > r$ it is (0). Since the e_i' are generators, (16) holds. To prove (17) we must show that, if $\sum_1^n \beta_i e_i' = 0$, then each $\beta_i e_i' = 0$. Since $t_i' = \Sigma\mu_{ij} t_j$, $t_i'T = e_i'$. Hence $\Sigma\beta_i e_i' = 0$ implies that $\Sigma\beta_i t_i' \in \mathfrak{N}$.

Since the v_i' are generators of $\mathfrak{N}$, this means that $\sum_1^n \beta_i t_i' = \sum_1^r \gamma_j v_j'$. Hence

$$\beta_1 t_1' + \beta_2 t_2' + \cdots + \beta_n t_n' = \gamma_1 \delta_1 t_1' + \gamma_2 \delta_2 t_2' + \cdots + \gamma_r \delta_r t_r'.$$

Since the t_i' form a basis for $\mathfrak{F}$, this implies that

$$\beta_i = \gamma_i \delta_i \quad \text{for } i \leq r \quad \text{and} \quad \beta_i = 0 \quad \text{for } i > r.$$

Thus certainly $\beta_i e_i' = 0$ for $i > r$. Moreover, since $v_i' \varepsilon \mathfrak{N}$ and $v_i' = \delta_i t_i'$ for $i \leq r$, $v_i' T = \delta_i e_i' = 0$. Hence also $\beta_i e_i' = \gamma_i \delta_i e_i' = 0$ for $i \leq r$. This proves our first assertion and completes the proof of Theorem 6.

We have also shown that $\beta_i e_i' = 0$ for $i > r$ implies that $\beta_i = 0$. Hence the order ideal $\mathfrak{J}_{e_i'} = (0)$. Moreover, we saw that, if $i \leq r$, then $\beta_i e_i' = 0$ if and only if β_i is a multiple of δ_i. Hence for these i's, $\mathfrak{J}_{e_i'} = (\delta_i)$. This proves the second assertion made above.

There is another interpretation of our results that we now give. We observe first that a direct sum of cyclic submodules such as $\{e_{r+1}'\}$, $\{e_{r+2}'\}$, $\cdots$, $\{e_n'\}$ in which the generators e_j' have order ideal $= (0)$ is a free module. For if $\beta_{r+1} e_{r+1}' + \beta_{r+2} e_{r+2}' + \cdots + \beta_n e_n' = 0$, then each $\beta_j e_j' = 0$ and since $\mathfrak{J}_{e_j'} = (0)$, $\beta_j = 0$. We note next that the subset $\mathfrak{S}$ of $\mathfrak{R}$ of elements of finite order is a submodule. For if $y_1, y_2 \varepsilon \mathfrak{S}$ and $\beta_i y_i = 0$ for $\beta_i \neq 0$, then $\beta_1 \beta_2 (y_1 - y_2) = 0$ and $\beta_1 \beta_2 \neq 0$. Also if $y \varepsilon \mathfrak{S}$, then $\alpha y \varepsilon \mathfrak{S}$ for any α. We shall now prove that the submodule $\bar{\mathfrak{S}} = \{e_1'\} \oplus \{e_2'\} \oplus \cdots \oplus \{e_r'\}$ that consists of all linear combinations $\alpha_1 e_1' + \alpha_2 e_2' + \cdots + \alpha_r e_r'$ is precisely the submodule $\mathfrak{S}$ of elements of finite order in $\mathfrak{R}$. Since each e_i', $i = 1, 2, \cdots, r$, is in $\mathfrak{S}$, $\bar{\mathfrak{S}} \subseteq \mathfrak{S}$. On the other hand, let $y = \alpha_1 e_1' + \alpha_2 e_2' + \cdots + \alpha_n e_n' \varepsilon \mathfrak{S}$. Then there exists a $\beta \neq 0$ such that

$$\beta \alpha_1 e_1' + \beta \alpha_2 e_2' + \cdots + \beta \alpha_n e_n' = 0.$$

We have seen that this implies that $\beta \alpha_j = 0$ for $j > r$. Hence $\alpha_j = 0$ for $j > r$ and $y \varepsilon \bar{\mathfrak{S}}$. Thus $\mathfrak{S} = \bar{\mathfrak{S}}$. Now it is clear that $\mathfrak{R} = \mathfrak{S} \oplus \{e_{r+1}'\} \oplus \cdots \oplus \{e_n'\} = \mathfrak{S} \oplus \mathfrak{E}$ where $\mathfrak{E} = \{e_{r+1}'\} \oplus \cdots \oplus \{e_n'\}$ is free. This proves the following

Theorem 7. *If $\mathfrak{o}$ is a principal ideal domain, any finitely generated $\mathfrak{o}$-module is a direct sum of the submodule of elements of finite order and a free module.*

We note finally that in the decomposition $\mathfrak{R} = \{e_1'\} \oplus \{e_2'\} \oplus \cdots \oplus \{e_n'\}$ we may drop the terms $\{e_i'\}$ whose order ideals are $(\delta_j) = (1)$. If there are h such terms, we write $e_{h+j}' = f_j$, $j = 1, 2, \cdots, t$, where $t = n - h$. We shall also denote the number $r - h$ of f's of finite order by u and the number $n - r$ of the remaining f's by v. Finally we change the notation δ_{h+j} to δ_j. Then

$$\mathfrak{R} = \{f_1\} \oplus \{f_2\} \oplus \cdots \oplus \{f_t\} \tag{20}$$

where

$$\mathfrak{I}_{f_j} = (\delta_j), \quad j = 1, \cdots, u; \quad \mathfrak{I}_{f_k} = (0), \tag{21}$$
$$k = u + 1, \cdots, t \quad \text{and} \quad \delta_j \mid \delta_{j'} \quad \text{if} \quad j \leq j'.$$

For the case of an ordinary commutative group our results specialize to the following theorems.

Theorem 8. *Any finitely generated commutative group is a direct sum of cyclic groups.*

Theorem 9. *Any finitely generated commutative group is a direct sum of a finite group and a free group.*

The finite group is $\mathfrak{S} = \{f_1\} \oplus \{f_2\} \oplus \cdots \oplus \{f_u\}$. Its order is $\delta_1\delta_2 \cdots \delta_u$ if δ_j is normalized to be positive.

10. Invariance theorems. In this section we prove that the order ideals $(\delta_1), (\delta_2), \cdots, (\delta_u)$ and the number $v = t - u$ of f's in the free part $\{f_{u+1}\} \oplus \{f_{u+2}\} \oplus \cdots \oplus \{f_t\}$ are invariants. We discuss first some more ideas concerning $\mathfrak{o}$-modules which will be needed for the proof.

Let $\mathfrak{R}$ be an $\mathfrak{o}$-module and let $\mathfrak{S}$ be a submodule and $\mathfrak{R}/\mathfrak{S}$ the factor group of cosets $x + \mathfrak{S}$. If $\alpha \,\varepsilon\, \mathfrak{o}$ and $x + \mathfrak{S} \,\varepsilon\, \mathfrak{R}/\mathfrak{S}$, we define $\alpha(x + \mathfrak{S}) = \alpha x + \mathfrak{S}$. Then since $\mathfrak{S}$ is closed under scalar multiplication, the result $\alpha x + \mathfrak{S}$ given here does not depend on the choice of the representative x in the coset $x + \mathfrak{S}$. It is readily seen that $\mathfrak{R}/\mathfrak{S}$ becomes an $\mathfrak{o}$-module relative to this definition of scalar multiplication. The proof is the same as for vector spaces (cf. § 9, Chapter I). We note also that, if $\mathfrak{R}$ is finitely generated with generators $e_1, e_2, \cdots, e_n$, then $\mathfrak{R}/\mathfrak{S}$ is finitely generated with generators $e_1 + \mathfrak{S}, e_2 + \mathfrak{S}, \cdots, e_n + \mathfrak{S}$.

We suppose next that $\mathfrak{B}$ is an ideal in $\mathfrak{o}$ having the property that $\beta x = 0$ for all $\beta \in \mathfrak{B}$ and all $x \in \mathfrak{R}$. Let $\bar{\alpha} = \alpha + \mathfrak{B}$ be a coset in the difference ring $\bar{\mathfrak{o}} = \mathfrak{o}/\mathfrak{B}$. Then we define $\bar{\alpha}x = \alpha x$. If $\bar{\alpha} = \bar{\alpha}_1$, then $\alpha - \alpha_1 = \beta \in \mathfrak{B}$. Hence $\alpha x = (\alpha_1 + \beta)x = \alpha_1 x$. Thus $\bar{\alpha}x$ is a single-valued function of $\bar{\alpha}$ in $\bar{\mathfrak{o}}$ and x in $\mathfrak{R}$. It is easy to verify that $\mathfrak{R}$ is an $\bar{\mathfrak{o}}$-module relative to this function.

We give now the invariance proof which we shall break up into several stages. First let $\mathfrak{R} = \mathfrak{F}$ be a free module with basis $(t_1, t_2, \cdots, t_n)$. Let $\mathfrak{N}$ be a submodule and suppose now that $(v_1, v_2, \cdots, v_m)$ represents a basis for $\mathfrak{N}$. Then the new elements t_i', v_p' used in the proof of Theorem 6 form bases for $\mathfrak{F}$ and $\mathfrak{N}$ respectively. We have the relations (18) connecting these elements. Now since no element of a basis can be 0, clearly $r = m$ in (18). Hence the number of t's is not less than the number of v's. Consequently $m \leq n$. If $\mathfrak{N} = \mathfrak{F}$, we may reverse the roles of the t' and the v' and prove that $m = n$. This proves that the number of elements in any basis of a free $\mathfrak{o}$-module, $\mathfrak{o}$ a principal ideal domain is an invariant. As for vector spaces we call this number the *dimensionality* of the free module.

Assume next that every element of $\mathfrak{R} = \mathfrak{S}$ has finite order. Let

$$\mathfrak{S} = \{f_1'\} \oplus \{f_2'\} \oplus \cdots \oplus \{f_{t'}'\}$$

be a second decomposition into cyclic submodules such that $\mathfrak{Z}_{f_j'} = (\delta_j') \neq (0)$ and $\delta_j' \mid \delta_k'$ for $j \leq k$. We wish to prove that $t = t'$ and that $(\delta_j) = (\delta_j')$ for $j = 1, 2, \cdots, t$. We define the *length* of a decomposition $\mathfrak{S} = \{f_1\} \oplus \{f_2\} \oplus \cdots \oplus \{f_t\}$ to be Σs_i, where s_i is the length of δ_i. We suppose that the first decomposition has minimum length, and we shall prove the theorem by induction on this minimum length.

Let π be a fixed prime and let $\mathfrak{S}'$ denote the subset of elements y in $\mathfrak{S}$ such that $\pi y = 0$. $\mathfrak{S}'$ is a submodule of $\mathfrak{S}$. Suppose that $y = \Sigma \gamma_i f_i \in \mathfrak{S}'$. Then $\Sigma \pi \gamma_i f_i = 0$. Hence $\pi \gamma_i$ is divisible by δ_i. This implies that either $\delta_i \mid \gamma_i$ or $\pi \mid \delta_i$. In the former case $\gamma_i f_i$

$= 0$, and in the latter γ_i is a multiple of $\epsilon_i = \delta_i \pi^{-1}$. Thus y has the form $\sum_h^t \rho_j \epsilon_j f_j$, where h is the smallest integer such that δ_h is divisible by π. This shows that

$$\mathfrak{S}' = \{\epsilon_h f_h\} \oplus \{\epsilon_{h+1} f_{h+1}\} \oplus \cdots \oplus \{\epsilon_t f_t\}.$$

Similarly,

$$\mathfrak{S}' = \{\epsilon_{h'}' f_{h'}'\} \oplus \cdots \oplus \{\epsilon_{t'}' f_{t'}'\}.$$

Since every element of $\mathfrak{S}'$ satisfies the equation $\pi y = 0$, $\mathfrak{S}'$ may be regarded as an $\bar{\mathfrak{o}}$-module, $\bar{\mathfrak{o}} = \mathfrak{o}/(\pi)$. Since π is prime, we know that $\bar{\mathfrak{o}}$ is a field. (Vol. I, Ex. 2, p. 122.) Hence $\mathfrak{S}'$ is a vector space over $\bar{\mathfrak{o}}$. Hence by the theorem on invariance of dimensionality (or by the above result for free modules) the number of base elements in the two decompositions is the same. Thus $t - h = t' - h'$.

Now choose π to be a divisor of δ_1. Then $h = 1$ and the equation $t - h = t' - h'$ shows that $t' \geq t$. Similarly if π is chosen to be a divisor of δ_1', then we obtain $t \geq t'$. Hence $t = t'$. We now see that, if π is chosen to be a divisor of δ_1, then $h = 1$ implies that $h' = 1$. Thus π is also a divisor of δ_1'.

We consider next the submodule $\pi\mathfrak{S}$. The argument just used shows that

$$\begin{aligned}(22) \qquad \pi\mathfrak{S} &= \{\pi f_k\} \oplus \{\pi f_{k+1}\} \oplus \cdots \oplus \{\pi f_t\} \\ &= \{\pi f_{k'}'\} \oplus \{\pi f_{k'+1}'\} \oplus \cdots \oplus \{\pi f_t'\}\end{aligned}$$

where k and k' are respectively the smallest integers such that δ_k and $\delta_{k'}$ are not associates of π. Then the order ideals of πf_l are (ϵ_l) and those of $\pi f_{l'}'$ are $(\epsilon_{l'}')$, and the ϵ_l and the $\epsilon_{l'}'$ satisfy the divisibility conditions. We can now use the induction hypothesis on the length of the module to conclude that $t - k = t - k'$ and that $(\epsilon_k) = (\epsilon_{k'}')$, $(\epsilon_{k+1}) = (\epsilon_{k'+1}')$, $\cdots$. These relations imply that also $(\delta_1) = (\delta_1')$, $(\delta_2) = (\delta_2')$, $\cdots$. This proves the invariance of the (δ_i) for the module $\mathfrak{S}$.

We consider finally the general case. Then if $\mathfrak{S}$ is the submodule of elements of finite order, the result just proved shows that the number u of f_i of finite order is the same as the number u' of f_i' of finite order. Moreover, we have $(\delta_i) = (\delta_i')$ for the corresponding order ideals. Now consider the module $\mathfrak{R}/\mathfrak{S}$. It is easy to see that this module is free with basis $f_{u+1} + \mathfrak{S}, \cdots, f_t + \mathfrak{S}$ and also with basis $f_{u+1}' + \mathfrak{S}, \cdots, f_{t'}' + \mathfrak{S}$. Hence by the result proved for free modules, $t - u = t' - u$. This completes the proof of the following

Theorem 10. *Let* $\mathfrak{R} = \{f_1\} \oplus \cdots \oplus \{f_t\} = \{f_1'\} \oplus \cdots \oplus \{f_{t'}'\}$ *be two decompositions of an* $\mathfrak{o}$*-module,* $\mathfrak{o}$ *a principal ideal domain, into cyclic modules* $\neq 0$. *Assume that the order ideals satisfy conditions* (21) *and analogous ones for the* f_i'. *Then* $t = t'$ *and* $\mathfrak{J}_{f_i} = \mathfrak{J}_{f_i'}$, $i = 1, 2, \cdots, t$.

The module theorem just proved can be used to prove the uniqueness up to unit factors of the invariant factors of a matrix. Thus we have the following

Theorem 11. *If diag* $\{\delta_1, \delta_2, \cdots, \delta_r, 0, \cdots, 0\}$ *and diag* $\{\delta_1', \delta_2', \cdots, \delta_{r'}', 0, \cdots, 0\}$ *are equivalent* $m \times n$ *matrices with elements in a principal ideal domain* $\mathfrak{o}$ *and* $\delta_i \mid \delta_j$ *for* $i \leq j$, $\delta_k' \mid \delta_l'$ *for* $k \leq l$, *then* $r = r'$ *and* δ_i *is an associate of* δ_i' *for* $i = 1, 2, \cdots, r$.

Proof. Let $\mathfrak{F}$ be a free module with basis $(t_1, t_2, \cdots, t_n)$, and let $\mathfrak{N}$ be the submodule generated by $v_1 = \delta_1 t_1, \cdots, v_r = \delta_r t_r, v_{r+1} = 0, \cdots, v_m = 0$. Then $\mathfrak{F}/\mathfrak{N}$ is a direct sum of cyclic modules whose order ideals are $(\delta_1), (\delta_2), \cdots, (\delta_r)$ and a free module of $n - r$ dimensions. On the other hand, the assumption of equivalence implies that we can find a new basis $(t_1', t_2', \cdots, t_n')$ for $\mathfrak{N}$ and new generators $v_1' = \delta_1' t_1', v_2' = \delta_2' t_2', \cdots, v_{r'}' = \delta_{r'}' t_{r'}', v_{r'+1}' = 0, \cdots, v_m' = 0$ for $\mathfrak{N}$. This gives a decomposition of $\mathfrak{F}/\mathfrak{N}$ as a direct sum of cyclic modules with order ideals $(\delta_1'), (\delta_2'), \cdots, (\delta_{r'}')$ and a free module of $n - r'$ dimensions. Since the divisibility conditions hold, we conclude from Theorem 10 that $r = r'$ and $(\delta_i) = (\delta_i')$ for the $(\delta_i) \neq (1)$. This implies $(\delta_i) = (\delta_i')$ for all i.

We shall give next a second and purely matrix proof of this result. At the same time we shall obtain some useful formulas for calculating the invariant factors.

We observe first that, if (ν) is any $m \times n$ matrix, then the rows of $(\nu)(\sigma)$ are linear combinations of the rows of (σ). Hence for any j, the j-rowed minors of $(\nu)(\sigma)$ are linear combinations of the

j-rowed minors of (σ). Similarly the j-rowed minors of $(\sigma)(\mu)$ are linear combinations of those of (σ). Combining these two results we see that the j-rowed minors of $(\nu)(\sigma)(\mu)$ are linear combinations of those of (σ). Now let $\Delta_j(\sigma)$ be a highest common factor of the j-rowed minors of (σ). Then our result shows that, if $(\tau) = (\nu)(\sigma)(\mu)$, then $\Delta_j(\sigma) \mid \Delta_j(\tau)$. If (μ) and (ν) are units, then also $\Delta_j(\tau) \mid \Delta_j(\sigma)$. Hence $\Delta_j(\tau)$ and $\Delta_j(\sigma)$ are associates. We apply this now to the equivalent normal forms

$$\operatorname{diag}\{\delta_1, \delta_2, \cdots, \delta_r, 0, \cdots, 0\}$$

$$\operatorname{diag}\{\delta_1', \delta_2', \cdots, \delta_{r'}', 0, \cdots, 0\}.$$

Denote the Δ_j for these matrices by Δ_j and Δ_j' respectively. Then because of the divisibility conditions on the δ's and the δ''s, we may take

$$\Delta_j = \delta_1\delta_2 \cdots \delta_j \quad \text{and} \quad \Delta_j' = \delta_1'\delta_2' \cdots \delta_j'.$$

Since Δ_j is an associate of Δ_j' for each j, it follows that $r = r'$. Also $\Delta_1 = \delta_1$ is an associate of $\Delta_1' = \delta_1'$. Since $\delta_1\delta_2$ is an associate of $\delta_1'\delta_2'$ we also have that δ_2 and δ_2' are associates. Continuing in this way we see that δ_i and δ_i' are associates for each i. We have also proved the following

Theorem 12. *Let $\Delta_j(\sigma)$ be a highest common factor of the j-rowed minors of (σ) and suppose that $\Delta_j(\sigma) \neq 0$ for $j \leqq r$. Then the elements $\delta_1 = \Delta_1(\sigma)$, $\delta_2 = \Delta_2(\sigma)\Delta_1(\sigma)^{-1}, \cdots, \delta_r = \Delta_r(\sigma)\Delta_{r-1}(\sigma)^{-1}$ constitute a set of invariant factors for (σ).*

11. Decomposition of a vector space relative to a linear transformation. We return now to the consideration of a linear transformation A in a vector space $\mathfrak{R}$ over Φ. We apply the above results to the $\Phi[\lambda]$-module $\mathfrak{R}$ determined by A. Since every vector x has an order $\mu_x(\lambda) \neq 0$, $\mathfrak{R} = \{f_1\} \oplus \{f_2\} \oplus \cdots \oplus \{f_t\}$ where $\mathfrak{J}_{f_j} = (\delta_j) \neq (0), \neq (1)$ and $\delta_j \mid \delta_h$ if $j \leq h$. The invariant factor ideals (δ_j) are uniquely determined.

If $(e_1, e_2, \cdots, e_n)$ is a basis for $\mathfrak{R}$ over Φ and $e_iA = \Sigma\alpha_{ij}e_j$, then (α) is the matrix of A relative to this basis, and the elements $v_i = \Sigma\alpha_{ij}t_j - \lambda t_i$ form a basis for the kernel $\mathfrak{N}$ of the homomorphism T between the free module $\mathfrak{F}$ and $\mathfrak{R}$. Hence the matrix that must be normalized to obtain the invariant factor ideals is

$$(23) \quad (\sigma) = \lambda 1 - (\alpha) = \begin{bmatrix} \lambda - \alpha_{11} & -\alpha_{12} & \cdots & -\alpha_{1n} \\ -\alpha_{21} & \lambda - \alpha_{22} & \cdots & -\alpha_{2n} \\ \cdot & \cdot & \cdots & \cdot \\ -\alpha_{n1} & -\alpha_{n2} & \cdots & \lambda - \alpha_{nn} \end{bmatrix}.$$

The normal form of this matrix is

$$(24) \qquad \operatorname{diag}\{1, 1, \cdots, 1, \delta_1, \delta_2, \cdots, \delta_t\}$$

where it is assumed that the leading coefficient of $\delta_i = \delta_i(\lambda)$ is 1. Our results show us also how to obtain a set of f's such that $\mathfrak{R} = \{f_1\} \oplus \{f_2\} \oplus \cdots \oplus \{f_t\}$. If (μ) and (ν) are units such that $(\nu)(\sigma)(\mu)$ is the normal form, and $e_i' = \Sigma \mu_{ij}{}^* e_j$, $(\mu^*) = (\mu)^{-1}$, then we may take $f_i = e_{n-t+i}'$.

We obtain a Φ-basis for $\mathfrak{R}$ by stringing together Φ-bases in the cyclic subspaces $\{f_i\}$. If the degree of δ_i is n_i, then $(f_i, f_iA, \cdots, f_iA^{n_i-1})$ is a basis for $\{f_i\}$. Hence

$$(f_1, f_1A, \cdots, f_1A^{n_1-1};\ f_2, f_2A, \cdots, f_2A^{n_2-1};\ \cdots;\ \cdots,\ f_tA^{n_t-1})$$

is a basis for $\mathfrak{R}$ over Φ. The matrix of A relative to this basis has the form

$$(25) \qquad \begin{bmatrix} B_1 & & & \\ & B_2 & & \\ & & \ddots & \\ & & & B_t \end{bmatrix}$$

where the diagonal block B_i is the companion matrix of $\delta_i(\lambda)$. The matrix (25), which is completely determined by the invariant factor ideals, is called the *Jordan canonical matrix* of A.

A more refined canonical form can be obtained by applying the following considerations. Let

$$\delta_i(\lambda) = \pi_{i1}(\lambda)^{k_{i1}} \pi_{i2}(\lambda)^{k_{i2}} \cdots \pi_{is_i}(\lambda)^{k_{is_i}}$$

be the factorization of δ_i into powers of distinct prime factors. Then by Theorem 3, p. 70, $\{f_i\} = \{f_{i1}\} \oplus \{f_{i2}\} \oplus \cdots \oplus \{f_{is_i}\}$

where the order of f_{ij} is $\pi_{ij}(\lambda)^{k_{ij}}$. We choose now a basis in each $\{f_{ij}\}$ of the type given on p. 71. These bases together give a basis for $\mathfrak{R}$ and the matrix of A relative to this basis is

$$\begin{bmatrix} C_1 & & & \\ & C_2 & & \\ & & \ddots & \\ & & & C_t \end{bmatrix} \tag{26}$$

where C_i has the same dimensions as B_i in (25) and

$$C_i = \begin{bmatrix} C_{i1} & & & \\ & C_{i2} & & \\ & & \ddots & \\ & & & C_{is_i} \end{bmatrix} \tag{27}$$

where

$$C_{ij} = \begin{bmatrix} P_{ij} & & & & \\ N_{ij} & P_{ij} & & & \\ & N_{ij} & \ddots & & \\ & & \ddots & \ddots & \\ & & & N_{ij} & P_{ij} \end{bmatrix} \quad k_{ij} \text{ blocks} \tag{28}$$

and P_{ij} is the companion matrix of $\pi_{ij}(\lambda)$ and N_{ij} has the form (10). The matrix thus obtained will be called a *classical canonical matrix* of A. It displays the prime factors $\pi_{ij}(\lambda)$ and the exponents of these primes in the factorization of the $\delta_i(\lambda)$. The ideals $(\pi_{ij}(\lambda)^{k_{ij}})$ will be called the *elementary divisor ideals* of A. Also the polynomials $\pi_{ij}(\lambda)^{k_{ij}}$ will be called the *elementary divisors* of $(\sigma) = \lambda 1 - (\alpha)$.

For example, suppose that the invariant factors $\neq 1$ of $\lambda 1 - (\alpha)$ are $\lambda^3 - \lambda^2 - \lambda + 1 = (\lambda - 1)^2(\lambda + 1)$ and $\lambda^6 - 3\lambda^4 + 3\lambda^2 - 1 = (\lambda - 1)^3(\lambda + 1)^3$. Then the Jordan canonical matrix is

$$\begin{bmatrix} \begin{array}{|ccc|} \hline 0 & 1 & 0 \\ 0 & 0 & 1 \\ -1 & 1 & 1 \\ \hline \end{array} & \\ & \begin{array}{|cccccc|} \hline 0 & 1 & 0 & 0 & 0 & 0 \\ 0 & 0 & 1 & 0 & 0 & 0 \\ 0 & 0 & 0 & 1 & 0 & 0 \\ 0 & 0 & 0 & 0 & 1 & 0 \\ 0 & 0 & 0 & 0 & 0 & 1 \\ 1 & 0 & -3 & 0 & 3 & 0 \\ \hline \end{array} \end{bmatrix}.$$

The elementary divisors here are $(\lambda - 1)^2$, $(\lambda + 1)$, $(\lambda - 1)^3$, $(\lambda + 1)^3$. Hence a classical canonical matrix is

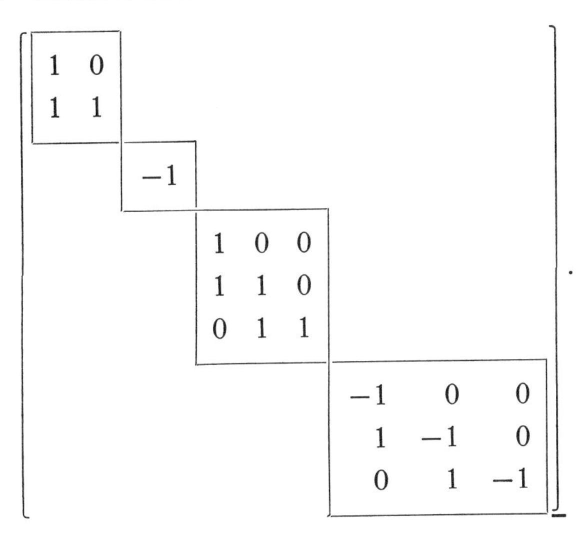

To illustrate the method of obtaining the canonical matrices, let A be the linear transformation such that

$$\begin{aligned} e_1A &= -e_1 - 2e_2 + 6e_3 \\ e_2A &= -e_1 \qquad\;\; + 3e_3 \\ e_3A &= -e_1 - \;\, e_2 + 4e_3. \end{aligned}$$

The matrix (α) here is

$$\begin{bmatrix} -1 & -2 & 6 \\ -1 & 0 & 3 \\ -1 & -1 & 4 \end{bmatrix} \quad \text{and} \quad \lambda 1 - (\alpha) = \begin{bmatrix} \lambda + 1 & 2 & -6 \\ 1 & \lambda & -3 \\ 1 & 1 & \lambda - 4 \end{bmatrix}.$$

We have

$$\begin{bmatrix} 0 & 1 & 0 \\ 0 & -1 & 1 \\ 1 & -\lambda + 2 & -3 \end{bmatrix} [\lambda 1 - (\alpha)] \begin{bmatrix} 1 & 3 & -3 + \lambda \\ 0 & 0 & -1 \\ 0 & 1 & -1 \end{bmatrix} = \begin{bmatrix} 1 & 0 & 0 \\ 0 & \lambda - 1 & 0 \\ 0 & 0 & (\lambda - 1)^2 \end{bmatrix}.$$

Thus

$$(\mu) = \begin{bmatrix} 1 & 3 & -3 + \lambda \\ 0 & 0 & -1 \\ 0 & 1 & -1 \end{bmatrix}, \quad (\mu)^{-1} = \begin{bmatrix} 1 & \lambda & -3 \\ 0 & -1 & 1 \\ 0 & -1 & 0 \end{bmatrix}$$

and

$$\begin{aligned} e_1' &= e_1 + \lambda e_2 - 3e_3 = e_1 + e_2 A - 3e_3 = 0 \\ e_2' &= -e_2 + e_3 \\ e_3' &= -e_2. \end{aligned}$$

To obtain the Jordan matrix we use the basis $f_1 = e_2', f_2 = e_3'$, $f_3 = e_3'A = e_1 - 3e_3$. Hence the Jordan matrix is

$$\begin{bmatrix} \boxed{1} & & \\ & \boxed{\begin{matrix} 0 & 1 \\ -1 & 2 \end{matrix}} \end{bmatrix}.$$

The matrix that transforms (α) to this matrix is the matrix of (f_1, f_2, f_3) relative to (e_1, e_2, e_3) and this matrix is

$$\begin{bmatrix} 0 & -1 & 1 \\ 0 & -1 & 0 \\ 1 & 0 & -3 \end{bmatrix}.$$

We can check that

$$\begin{bmatrix} 0 & -1 & 1 \\ 0 & -1 & 0 \\ 1 & 0 & -3 \end{bmatrix} \begin{bmatrix} -1 & -2 & 6 \\ -1 & 0 & 3 \\ -1 & -1 & 4 \end{bmatrix} \begin{bmatrix} 0 & -1 & 1 \\ 0 & -1 & 0 \\ 1 & 0 & -3 \end{bmatrix}^{-1} = \begin{bmatrix} 1 & 0 & 0 \\ 0 & 0 & 1 \\ 0 & -1 & 2 \end{bmatrix}.$$

We consider now the classical canonical matrices for the case where Φ is the field of complex numbers, or, more generally, any algebraically closed field.* Here the only irreducible polynomials of positive degree are the linear ones. Hence the elementary divisors have the form $(\lambda - \rho)^k$. The block (C_{ij} in (28)) corresponding to this elementary divisor is

$$\begin{bmatrix} \rho & & & & \\ 1 & \rho & & & \\ & \cdot & \cdot & & \\ & & \cdot & \cdot & \\ & & & \cdot & \cdot \\ & & & 1 & \rho \end{bmatrix} \Bigg\} k. \tag{29}$$

The classical canonical form has blocks of this type strung down the main diagonal.

Suppose next that Φ is the field of real numbers. In this case the irreducible polynomials of positive degree are the linear ones and the quadratic ones $\lambda^2 - \beta\lambda - \gamma$ where $\beta^2 + 4\gamma < 0$. The elementary divisors are of the forms $(\lambda - \rho)^k$, $(\lambda^2 - \beta\lambda - \gamma)^k$. The block corresponding to $(\lambda - \rho)^k$ is (29) and that corresponding to $(\lambda^2 - \beta\lambda - \gamma)^k$ is

(30)

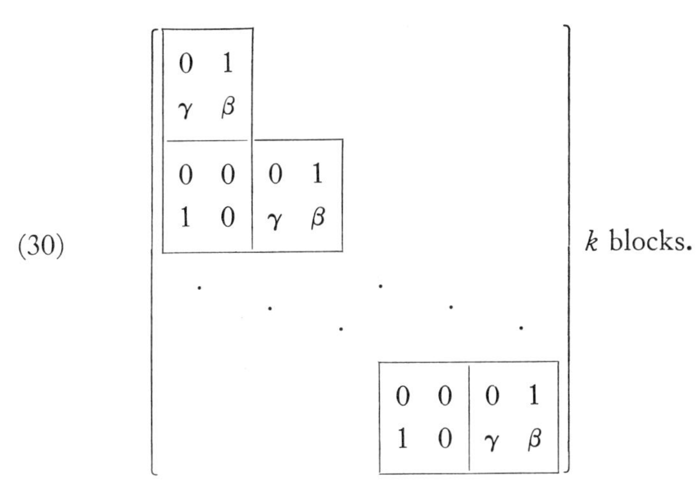

k blocks.

* A field is said to be *algebraically closed* if every polynomial with coefficients in the field has a root in the field. An equivalent definition is that every polynomial with coefficients in the field factors into linear factors.

EXERCISES

1. Find a classical canonical matrix similar to

$$\begin{bmatrix} 17 & -8 & -12 & 14 \\ 46 & -22 & -35 & 41 \\ -2 & 1 & 4 & -4 \\ 4 & -2 & -2 & 3 \end{bmatrix}$$

and find a matrix that transforms this matrix into the canonical matrix.

2. Prove the following theorem:

A necessary and sufficient condition that two matrices (α) *and* (β) *in* Φ_n *be similar is that* $\lambda 1 - (\alpha)$ *and* $\lambda 1 - (\beta)$ *have the same invariant factors in* $\Phi[\lambda]_n$.

3. Prove that any matrix is similar to its transposed.

4. Given the following elementary divisors: $(\lambda - 1)^3$, $(\lambda - 1)$, $(\lambda^2 + 1)^4$, $(\lambda^2 + 1)^2$, $(\lambda^2 + 1)$, $(\lambda + 2)$, what are the invariant factors?

5. Prove that if f is any vector whose order is the minimum polynomial of A, then there exists an invariant subspace $\mathfrak{S}$ ($\mathfrak{S}A \subseteq \mathfrak{S}$) such that $\mathfrak{R} = \{f\} \oplus \mathfrak{S}$.

6. Let $\mathfrak{S}$ be an invariant subspace such that the invariant factors of the transformations induced by A in $\mathfrak{S}$ and in $\mathfrak{R}/\mathfrak{S}$ together give all the invariant factors of A. Prove that there exists a second invariant subspace $\mathfrak{U}$ such that $\mathfrak{R} = \mathfrak{S} \oplus \mathfrak{U}$.

12. The characteristic and minimum polynomials. Again let $\mathfrak{R} = \{f_1\} \oplus \{f_2\} \oplus \cdots \oplus \{f_t\}$ where the order ideals of the f_i are the invariant factor ideals (δ_i). Thus the order of f_i is δ_i and $\delta_i \mid \delta_j$ if $i \leq j$. We know that the minimum polynomial $\mu(\lambda)$ of A is the least common multiple of the orders of the generators f_i. By the divisibility conditions this l.c.m. is $\delta_t(\lambda)$. Hence $\mu(\lambda) = \delta_t(\lambda)$.

Now let (α) be any matrix of A. We know that, if $\Delta_n(\lambda) = \det(\lambda 1 - (\alpha))$ and $\Delta_{n-1}(\lambda)$ is the highest common factor of the $(n - 1)$-rowed minors of $\lambda 1 - (\alpha)$, then the following relations hold

$$\Delta_n(\lambda) = \delta_1(\lambda)\delta_2(\lambda) \cdots \delta_t(\lambda) \tag{31}$$

$$\mu(\lambda) = \delta_t(\lambda) = \Delta_n(\lambda)[\Delta_{n-1}(\lambda)]^{-1}. \tag{32}$$

The polynomial $\Delta_n(\lambda)$ is called the *characteristic polynomial* of (α) (or of A). If we refer to (23) we can see that

$$\Delta_n(\lambda) = \lambda^n - \alpha_1\lambda^{n-1} + \alpha_2\lambda^{n-2} - \cdots + (-1)^n\alpha_n \tag{33}$$

where α_i is the sum of the diagonal minors of order i in (α).

Of particular importance are the first and last α's. These are

$$(34) \qquad \alpha_1 = \alpha_{11} + \alpha_{22} + \cdots + \alpha_{nn}, \quad \alpha_n = \det(\alpha).$$

The former is called *the trace* of (α). The main properties of this function of (α) will be considered in § 14.

Now we know that, if $\nu(\lambda)$ is any polynomial such that $\nu(A) = 0$, then $\nu((\alpha)) = 0$. Also if $\mu(\lambda)$ is the minimum polynomial of A, then $\mu(\lambda)$ is the minimum polynomial of (α). Hence our results may be stated as the following theorem on matrices.

Theorem 13. *Let (α) be a matrix in Φ_n and let $\Delta_n(\lambda) = \det(\lambda 1 - (\alpha))$ and $\mu(\lambda) = \Delta_n(\lambda)[\Delta_{n-1}(\lambda)]^{-1}$ where $\Delta_{n-1}(\lambda)$ is the highest common factor of the $(n-1)$-rowed minors of $\lambda 1 - (\alpha)$. Then* 1) *$\Delta_n((\alpha)) = \mu((\alpha)) = 0$;* 2) *if $\nu(\lambda)$ is any polynomial such that $\nu((\alpha)) = 0$, then $\mu(\lambda) \mid \nu(\lambda)$;* 3) *$\mu(\lambda)$ and $\Delta_n(\lambda)$ have the same prime factors.*

The first two statements are clear from what we have proved about A. The last statement follows from (31) and the fact that all the $\delta_i(\lambda)$ are factors of $\delta_t(\lambda) = \mu(\lambda)$.

Theorem 13 is a composite of the theorem of Hamilton-Cayley on the characteristic polynomial and Frobenius' theorem on the minimum polynomial. Direct matrix proofs of these results will be given in the next section.

EXERCISES

1. Prove that A is cyclic if and only if $\lambda 1 - (\alpha)$ has only one invariant factor $\neq 1$.

2. Prove that, if (α) is nilpotent, then the invariant factors of $\lambda 1 - (\alpha)$ all have the form λ^m. Hence prove that any nilpotent matrix is similar to a matrix of the form

$$\begin{bmatrix} N_1 & & & \\ & N_2 & & \\ & & \ddots & \\ & & & N_q \end{bmatrix}$$

where N_i has the form

$$N_i = \begin{bmatrix} 0 & & & \\ 1 & \ddots & & \\ & \ddots & \ddots & \\ & & 1 & 0 \end{bmatrix}.$$

3. Prove that a matrix with elements in the field of complex numbers is similar to a diagonal matrix if and only if its minimum polynomial has no multiple roots.

4. Show that, if (α) is idempotent, then the elementary divisors of $\lambda 1 - (\alpha)$ are either λ or $\lambda - 1$. Use this to prove the result on p. 62 for the case Δ a field.

5. Show that, if u is a vector $\neq 0$ such that $uA = \rho u$, then ρ is a root of the characteristic polynomial. Conversely, show that, if ρ is a root belonging to Φ of the characteristic polynomial, then there exists a vector $u \neq 0$ such that $uA = \rho u$.

6. A vector $u \neq 0$ such that $uA = \rho u$ is called a *characteristic* vector of A. Prove that such vectors always exist if Φ is an algebraically closed field. Prove also that, if Φ is the field of real numbers, then any linear transformation in an odd dimensional space $\mathfrak{R}$ over Φ possesses characteristic vectors.

13. Direct proof of Theorem 13. From the point of view of matrix theory the proof that we have given of Theorem 13 is somewhat roundabout. In this section we shall generalize the results contained in this theorem, and we shall give direct proofs of these results. We suppose first that $\mathfrak{o}$ is any commutative ring with an identity. Let $\mathfrak{o}[\lambda]$ be the polynomial ring in the indeterminate λ and consider the matrix ring $\mathfrak{o}[\lambda]_n$. This ring contains the subring $\mathfrak{o}_n$ of matrices with elements in $\mathfrak{o}$. Also it contains the matrix

$$\lambda 1 = \text{diag}\, \{\lambda, \lambda, \cdots \lambda\}$$

which evidently belongs to the center. Now the essential observation for our purposes is that $\mathfrak{o}[\lambda]_n = \mathfrak{o}_n[\lambda 1]$ and $\lambda 1$ is transcendental relative to $\mathfrak{o}_n$. To see this, let $(\alpha(\lambda))$ be an arbitrary matrix in $\mathfrak{o}[\lambda]_n$ and write

$$\alpha_{ij}(\lambda) = \alpha_{ij0} + \alpha_{ij1}\lambda + \alpha_{ij2}\lambda^2 + \cdots. \tag{35}$$

Then if we recall that the product $(\alpha)\lambda 1$ is obtained by multiplying all of the elements of (α) by λ, we see that

$$(\alpha(\lambda)) = (\alpha)_0 + (\alpha)_1\lambda 1 + (\alpha)_2(\lambda 1)^2 + \cdots \tag{36}$$

where $(\alpha)_k$ is the matrix that has α_{ijk} in its (i, j) position. Next suppose that $(\alpha)_0 + (\alpha)_1\lambda 1 + (\alpha)_2(\lambda 1)^2 + \cdots = 0$. Then the (i, j) element of the left-hand side is given by (35). Since this element is 0, $\alpha_{ijk} = 0$. Hence every $(\alpha)_k = 0$, and this proves the transcendency of $\lambda 1$ relative to $\mathfrak{o}_n$.

If

$$\phi(\lambda) = \beta_0 + \beta_1\lambda + \beta_2\lambda^2 + \cdots$$

is a polynomial in $\mathfrak{o}[\lambda]$, then by $\phi((\alpha))$ we shall understand as usual the matrix

$$\beta_0 1 + \beta_1(\alpha) + \beta_2(\alpha)^2 + \cdots$$

where in general $\beta_i(\gamma)$ is obtained by multiplying all the elements of (γ) by β_i. Thus $\phi((\alpha))$ is obtained by replacing $\lambda 1$ by (α) in $\phi(\lambda)1 = \beta_0 1 + \beta_1(\lambda 1) + \beta_2(\lambda 1)^2 + \cdots$.

If $(\alpha) \in \mathfrak{o}_n$ we define the *characteristic polynomial* $\Delta_n(\lambda)$ to be the polynomial $\det(\lambda 1 - (\alpha))$ belonging to $\mathfrak{o}[\lambda]$. We shall now prove the following

Theorem 14. (Hamilton-Cayley) *If* $(\alpha) \in \mathfrak{o}_n$, $\mathfrak{o}$ *a commutative ring with an identity, and* $\Delta_n(\lambda)$ *is the characteristic polynomial, then* $\Delta_n((\alpha)) = 0$.

Proof. We recall the identity (Vol. I, p. 59)

$$[\lambda 1 - (\alpha)] \operatorname{adj} [\lambda 1 - (\alpha)] = \det(\lambda 1 - (\alpha))1 = \Delta_n(\lambda)1. \tag{37}$$

The matrix $\operatorname{adj}[\lambda 1 - (\alpha)] \in \mathfrak{o}[\lambda]_n = \mathfrak{o}_n[\lambda 1]$, and the degrees of its elements are $\leq n - 1$. Hence

$$\operatorname{adj}[\lambda 1 - (\alpha)] = (\beta)_0(\lambda 1)^{n-1} + (\beta)_1(\lambda 1)^{n-2} + \cdots + (\beta)_{n-1}.$$

Also if $\Delta_n(\lambda) = \lambda^n - \alpha_1\lambda^{n-1} + \cdots + (-1)^n\alpha_n$, then

$$\Delta_n(\lambda)1 = (\lambda 1)^n - \alpha_1(\lambda 1)^{n-1} + \cdots + (-1)^n\alpha_n 1.$$

Hence the identity (37) is equivalent to

$$\begin{aligned}[\lambda 1 - (\alpha)][(\beta)_0(\lambda 1)^{n-1} + (\beta)_1(\lambda 1)^{n-2} + \cdots + (\beta)_{n-1}]\\ = (\lambda 1)^n - \alpha_1(\lambda 1)^{n-1} + \cdots + (-1)^n\alpha_n 1.\end{aligned} \tag{38}$$

Thus we see that $\lambda 1 - (\alpha)$ is a factor of $\Delta_n(\lambda)1$ in $\mathfrak{o}_n[\lambda 1]$. By the factor theorem (Volume I, p. 99) this implies that

$$(\alpha)^n - \alpha_1(\alpha)^{n-1} + \cdots + (-1)^n\alpha_n 1 = 0$$

as required.

We suppose next that $\mathfrak{o}$ is a Gaussian integral domain, that is, a commutative integral domain with an identity in which the unique factorization theorem holds. For such domains we have the following

Theorem 15. (Frobenius) *Let* $(\alpha) \varepsilon \mathfrak{o}_n$, $\mathfrak{o}$ *a Gaussian domain, and let* $\Delta_n(\lambda) = det\,(\lambda 1 - (\alpha))$ *and* $\mu(\lambda) = \Delta_n(\lambda)[\theta(\lambda)]^{-1}$ *where* $\theta(\lambda)$ *is the highest common factor of the* $(n - 1)$*-rowed minors of* $\lambda 1 - (\alpha)$. *Then* 1) $\mu((\alpha)) = 0$, 2) *if* $\nu(\lambda) \varepsilon \mathfrak{o}[\lambda]$ *and* $\nu((\alpha)) = 0$, *then* $\mu(\lambda) \mid \nu(\lambda)$ *in* $\mathfrak{o}[\lambda]$, 3) *any irreducible factor of* $\Delta_n(\lambda)$ *is a factor of* $\mu(\lambda)$.

Proof. The existence of $\theta(\lambda)$ is assured by the fact that $\mathfrak{o}[\lambda]$ is a Gaussian domain. Since some of the $(n - 1)$-rowed minors (e.g., the diagonal minors) have leading coefficient 1, we may suppose that $\theta(\lambda)$ has leading coefficient 1. Clearly $\theta(\lambda)$ is a factor of $\Delta_n(\lambda)$ since it is a factor of all the $(n - 1)$-rowed minors. The quotient $\mu(\lambda)$ has leading coefficient 1. Now let $(\gamma(\lambda))$ denote the matrix in $\mathfrak{o}[\lambda]_n$ that is obtained by dividing out the elements of adj $[\lambda 1 - (\alpha)]$ by $\theta(\lambda)$. Then by (37) we have

$$[\lambda 1 - (\alpha)](\gamma(\lambda)) = \mu(\lambda) 1. \tag{39}$$

By the argument used in the preceding proof, this relation implies that $\mu((\alpha)) = 0$. This proves 1).

Now let $\mu^*(\lambda)$ be a polynomial $\neq 0$ of least degree such that $\mu^*((\alpha)) = 0$. We may assume that $\mu^*(\lambda)$ is primitive. Suppose that $\nu(\lambda)$ is any polynomial such that $\nu((\alpha)) = 0$. If P is the quotient field of $\mathfrak{o}$, we may write

$$\nu(\lambda) = q(\lambda)\mu^*(\lambda) + r(\lambda)$$

in P$[\lambda]$ where deg $r(\lambda) <$ deg $\mu^*(\lambda)$. Multiplication by a suitable element $\eta \neq 0$ in $\mathfrak{o}$ gives a relation

$$\eta\nu(\lambda) = q_1(\lambda)\mu^*(\lambda) + r_1(\lambda)$$

where $q_1(\lambda) = \eta q(\lambda)$ and $r_1(\lambda) = \eta r(\lambda) \varepsilon \mathfrak{o}[\lambda]$. Substitution of (α) in this relation gives $r_1((\alpha)) = 0$. Hence by assumption of minimality for the degree of $\mu^*(\lambda)$, $r_1(\lambda) = 0$. Thus $\mu^*(\lambda) \mid \eta\nu(\lambda)$. Since $\mu^*(\lambda)$ is primitive, $\mu^*(\lambda) \mid \nu(\lambda)$. In particular, $\mu^*(\lambda) \mid \mu(\lambda)$, and so, since the leading coefficient of $\mu(\lambda)$ is 1, we may suppose that $\mu^*(\lambda)$ has this property too.

We now write $\mu(\lambda) = \mu^*(\lambda)\rho(\lambda)$. Since $\mu^*((\alpha)) = 0$, by reversing the argument used to prove 1), we see that

$$\mu^*(\lambda)1 = [\lambda 1 - (\alpha)](\delta(\lambda))$$

where $(\delta(\lambda)) \varepsilon\, \mathfrak{o}[\lambda]_n$. Hence

$$\Delta_n(\lambda)1 = \mu^*(\lambda)\rho(\lambda)\theta(\lambda)1 = [\lambda 1 - (\alpha)](\delta(\lambda))\rho(\lambda)\theta(\lambda)1.$$

If we compare this with (37) and use the fact that $\lambda 1 - (\alpha)$ is not a zero-divisor * in $\mathfrak{o}[\lambda]_n$, we obtain

$$\text{adj}\,[\lambda 1 - (\alpha)] = (\delta(\lambda))\rho(\lambda)\theta(\lambda)1.$$

Thus all the $(n - 1)$-rowed minors of $\lambda 1 - (\alpha)$ are divisible by $\rho(\lambda)\theta(\lambda)$. Since $\theta(\lambda)$ was assumed to be the highest common factor of the $(n - 1)$-rowed minors, it follows that $\rho(\lambda) = 1$. Thus $\mu(\lambda) = \mu^*(\lambda)$. Statement 2) now follows from what we proved above.

To prove 3) we take the determinants of both sides of (39). This gives

$$\Delta_n(\lambda) \det (\gamma(\lambda)) = [\mu(\lambda)]^n.$$

Statement 3) is an immediate consequence.

EXERCISE

1. Show that the characteristic and minimum polynomials of

$$\begin{bmatrix} \alpha_0 & \alpha_1 & \alpha_2 & \alpha_3 \\ -\alpha_1 & \alpha_0 & -\alpha_3 & \alpha_2 \\ -\alpha_2 & \alpha_3 & \alpha_0 & -\alpha_1 \\ -\alpha_3 & -\alpha_2 & \alpha_1 & \alpha_0 \end{bmatrix}$$

are, respectively,

$$(\lambda^2 - 2\alpha_0\lambda + (\alpha_0{}^2 + \alpha_1{}^2 + \alpha_2{}^2 + \alpha_3{}^2))^2$$

and

$$\lambda^2 - 2\alpha_0\lambda + (\alpha_0{}^2 + \alpha_1{}^2 + \alpha_2{}^2 + \alpha_3{}^2).$$

14. Formal properties of the trace and the characteristic polynomial. We have defined the trace of (α), tr (α), to be the nega-

* Its determinant is $\Delta_n[\lambda] \neq 0$.

tive of the coefficient of λ^{n-1} in the characteristic polynomial of (α). From this it follows that

$$\text{tr}\,(\alpha) = \alpha_{11} + \alpha_{22} + \cdots + \alpha_{nn}. \tag{40}$$

If the characteristic polynomial $\Delta_n(\lambda) = \Pi(\lambda - \rho_i)$ in some extension field, then the ρ_i are called the *characteristic roots* of (α) in this field. The coefficients of $\Delta_n(\lambda)$ are the so-called elementary symmetric functions of the ρ_i. In particular

$$\text{tr}\,(\alpha) = \rho_1 + \rho_2 + \rho_3 + \cdots + \rho_n. \tag{41}$$

From (40) it is clear that the trace function is linear:

$$\text{tr}\,[(\alpha) + (\beta)] = \text{tr}\,(\alpha) + \text{tr}\,(\beta). \tag{42}$$

$$\text{tr}\,(\rho(\alpha)) = \rho\,\text{tr}\,(\alpha).$$

We can also verify the following property

$$\text{tr}\,(\alpha)(\beta) = \text{tr}\,(\beta)(\alpha). \tag{43}$$

For the (i, i) element of $(\alpha)(\beta)$ is $\sum_j \alpha_{ij}\beta_{ji}$. Hence

$$\text{tr}\,(\alpha)(\beta) = \sum_{i,j} \alpha_{ij}\beta_{ji}$$

and this is symmetric in (α) and (β).

The result we have just noted holds also for the other coefficients of the characteristic polynomial. Thus $(\alpha)(\beta)$ and $(\beta)(\alpha)$ have the same characteristic polynomials. This, too, can be verified directly. However, the following indirect proof has some elements of interest of its own.

We shall suppose more generally that (α) is an $m \times n$ matrix and (β) an $n \times m$ matrix with elements in a field. Then multiplication in either order is possible, and the resulting matrices $(\alpha)(\beta)$ and $(\beta)(\alpha)$ are respectively $m \times m$ and $n \times n$ matrices. Assuming that $n \geq m$, we shall show that the characteristic polynomial of $(\beta)(\alpha) = \lambda^{n-m}$ times the characteristic polynomial of $(\alpha)(\beta)$. We assume first that (α) has the following form

$$(\alpha) = \text{diag}\,\{\overbrace{1, \cdots, 1}^{r}, 0, \cdots, 0\} \tag{44}$$

Then if

$$(\beta) = \begin{bmatrix} \beta_{11} & \beta_{12} & \cdots & \beta_{1m} \\ \beta_{21} & \beta_{22} & \cdots & \beta_{2m} \\ \cdot & \cdot & \cdots & \cdot \\ \beta_{n1} & \beta_{n2} & \cdots & \beta_{nm} \end{bmatrix}$$

$$(\alpha)(\beta) = \begin{bmatrix} \beta_{11} & \beta_{12} & \cdots & \beta_{1m} \\ \cdot & \cdot & \cdots & \cdot \\ \beta_{r1} & \beta_{r2} & \cdots & \beta_{rm} \\ 0 & 0 & \cdots & 0 \\ \cdot & \cdot & \cdots & \cdot \\ 0 & 0 & \cdots & 0 \end{bmatrix}$$

$$(\beta)(\alpha) = \begin{bmatrix} \beta_{11} & \cdots & \beta_{1r} & 0 & \cdots & 0 \\ \beta_{21} & \cdots & \beta_{2r} & 0 & \cdots & 0 \\ \cdot & \cdots & \cdot & \cdot & \cdots & \cdot \\ \cdot & \cdots & \cdot & \cdot & \cdots & \cdot \\ \beta_{n1} & \cdots & \beta_{nr} & 0 & \cdots & 0 \end{bmatrix}.$$

Hence if $g(\lambda)$ denotes the characteristic polynomial of

$$\begin{bmatrix} \beta_{11} & \beta_{12} & \cdots & \beta_{1r} \\ \cdot & \cdot & \cdots & \cdot \\ \cdot & \cdot & \cdots & \cdot \\ \beta_{r1} & \beta_{r2} & \cdots & \beta_{rr} \end{bmatrix},$$

then the characteristic polynomials of $(\alpha)(\beta)$ and of $(\beta)(\alpha)$ are respectively $\lambda^{m-r}g(\lambda)$ and $\lambda^{n-r}g(\lambda)$. This proves our assertion.

We suppose now that (α) is arbitrary. There exists a matrix (μ) in $L(\Phi, m)$ and a matrix (ν) in $L(\Phi, n)$ such that $(\mu)(\alpha)(\nu) = (\alpha)_1$ has the form (44). Set $(\nu)^{-1}(\beta)(\mu)^{-1} = (\beta)_1$. Then by what we have just proved, the characteristic polynomial of $(\beta)_1(\alpha)_1$ is λ^{n-m} times that of $(\alpha)_1(\beta)_1$. On the other hand,

$$(\alpha)_1(\beta)_1 = (\mu)(\alpha)(\nu)(\nu)^{-1}(\beta)(\mu)^{-1} = (\mu)(\alpha)(\beta)(\mu)^{-1}$$

is similar to $(\alpha)(\beta)$ and

$$(\beta)_1(\alpha)_1 = (\nu)^{-1}(\beta)(\mu)^{-1}(\mu)(\alpha)(\nu) = (\nu)^{-1}(\beta)(\alpha)(\nu)$$

is similar to $(\beta)(\alpha)$. Hence $(\alpha)(\beta)$ and $(\alpha)_1(\beta)_1$ have the same characteristic polynomials, and $(\beta)(\alpha)$ and $(\beta)_1(\alpha)_1$ have the same characteristic polynomials. The asserted result therefore holds for $(\alpha)(\beta)$ and $(\beta)(\alpha)$. This proves the following

Theorem 16. *Let (α) be an $m \times n$ matrix and (β) an $n \times m$ matrix with elements in a field. Then if $n \geq m$, the characteristic polynomial of $(\beta)(\alpha)$ is λ^{n-m} times the characteristic polynomial of $(\alpha)(\beta)$.*

EXERCISE

1. (Flanders) Show that the elementary divisors not divisible by λ of $(\alpha)(\beta)$ and $(\beta)(\alpha)$ are the same. Also obtain a relation for the elementary divisors which are powers of λ.

15. The ring of $\mathfrak{o}$-endomorphisms of a cyclic $\mathfrak{o}$-module. In the remainder of this chapter we shall consider the problem of determining the linear transformations that commute with a given linear transformation and the generalization of this problem to modules.

If A is a linear transformation in $\mathfrak{R}$ over Φ, the totality $\mathfrak{B}$ of linear transformations that commute with A is a subalgebra of $\mathfrak{L}$. If $B \varepsilon \mathfrak{B}$, then $\alpha_l B = B\alpha_l$ and $AB = BA$. Hence B commutes with every linear transformation $\beta_{0l} + \beta_{1l}A + \beta_{2l}A^2 + \cdots$ belonging to the subalgebra $\mathfrak{A} = \Phi_l[A]$ generated by 1 and A. Conversely, if B is any endomorphism in the group $\mathfrak{R}$ that commutes with every element in $\mathfrak{A}$, then $B \varepsilon \mathfrak{B}$; for $\alpha_l B = B\alpha_l$ for all α so that B is a linear transformation and $BA = AB$.

If $\mathfrak{R}$ is regarded as a $\Phi[\lambda]$-module as before, then

$$x(\beta_{0l} + \beta_{1l}A + \cdots) = \phi(\lambda)x$$

where $\phi(\lambda) = \beta_0 + \beta_1\lambda + \cdots$. Hence, an endomorphism B commutes with every linear transformation belonging to $\mathfrak{A}$ if and only if $(\phi(\lambda)x)B = \phi(\lambda)(xB)$. Hence $\mathfrak{B}$ coincides with the set of $\Phi[\lambda]$-endomorphisms of $\mathfrak{R}$. We are therefore led again to adopt the module point of view, and to consider the problem of determining the set $\mathfrak{B}$ of $\mathfrak{o}$-endomorphisms of any finitely generated $\mathfrak{o}$-module, $\mathfrak{o}$ a principal ideal domain. It is evident that $\mathfrak{B}$ is a subring of the ring of endomorphisms of the group $\mathfrak{R}$.

As usual we denote the ring of the endomorphisms $\alpha_l: x \rightarrow \alpha x$, α in $\mathfrak{o}$, by $\mathfrak{o}_l$. This ring is a homomorphic image of $\mathfrak{o}$ and so it is commutative. It follows that $\mathfrak{o}_l \subseteq \mathfrak{B}$ the ring of $\mathfrak{o}$-endomorphisms. It is also clear that $\mathfrak{o}_l$ is contained in the center of $\mathfrak{B}$. It should be remarked that in the special case of the $\Phi[\lambda]$-module determined by a linear transformation A, the ring $\Phi[\lambda]_l$ is just the ring $\mathfrak{A} = \Phi_l[A]$.

We consider first the problem of determining $\mathfrak{B}$ in the special case of a cyclic $\mathfrak{o}$-module. It is not necessary to assume here that $\mathfrak{o}$ is a principal ideal domain, but only that $\mathfrak{o}$ is a commutative ring with an identity. We have the following

Theorem 17. *If $\mathfrak{o}$ is a commutative ring with an identity and $\mathfrak{R}$ is a cyclic $\mathfrak{o}$-module, then the only $\mathfrak{o}$-endomorphisms of $\mathfrak{R}$ are the mappings $x \rightarrow \alpha x$.*

Proof. Let $\mathfrak{R} = \{e\}$ and let B be an $\mathfrak{o}$-endomorphism of $\mathfrak{R}$. Suppose that $eB = \beta e$. Then if $x = \alpha e$, $xB = (\alpha e)B = \alpha(eB)$ $= \beta(\alpha e) = \beta x$. Thus $B = \beta_l$.

Corollary. *If A is a cyclic linear transformation in a vector space over a field, the only linear transformations that commute with A are the polynomials in A (with coefficients in Φ_l).*

EXERCISE

1. Show that the matrices that commute with

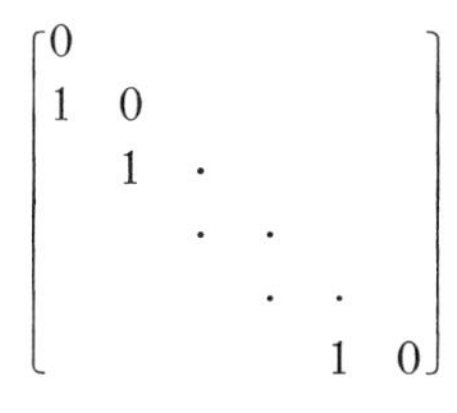

are given by

$$\begin{bmatrix} \alpha_0 & & & & 0 \\ \alpha_1 & \alpha_0 & & & \\ & \alpha_1 & \alpha_0 & & \\ \vdots & & \ddots & \ddots & \\ \alpha_{n-1} & & \cdots & \alpha_1 & \alpha_0 \end{bmatrix}$$

where the α_i are arbitrary in Φ.

16. Determination of the ring of $\mathfrak{o}$-endomorphisms of a finitely generated $\mathfrak{o}$-module, $\mathfrak{o}$ principal. We suppose now that $\mathfrak{R}$ is a finitely generated $\mathfrak{o}$-module, $\mathfrak{o}$ a principal ideal domain. We know that $\mathfrak{R} = \{f_1\} \oplus \{f_2\} \oplus \cdots \oplus \{f_t\}$ where, if the order ideal $\mathfrak{J}_{f_i} = (\delta_i)$, $i = 1, 2, \cdots, t$, then

$$(45) \qquad \delta_j = 0 \text{ if } j > u \text{ and } \delta_i \mid \delta_j \text{ if } i \leq j \text{ for all } i, j.$$

Let $B \varepsilon \mathfrak{B}$ the ring of $\mathfrak{o}$-endomorphisms of $\mathfrak{R}$, and suppose that $f_iB = g_i$, $i = 1, 2, \cdots, t$. Then if x is any element of $\mathfrak{R}$, $x = \Sigma\xi_i f_i$, ξ_i in $\mathfrak{o}$. Hence

$$xB = (\Sigma\xi_i f_i)B = \Sigma(\xi_i f_i)B = \Sigma\xi_i(f_iB) = \Sigma\xi_i g_i.$$

Thus B is completely determined by its effect on the generators f_i of $\mathfrak{R}$. We note next that since $\delta_i f_i = 0$, $\delta_i g_i = \delta_i(f_iB) = (\delta_i f_i)B = 0$. Hence if (ϵ_i) is the order ideal of g_i, then $\epsilon_i \mid \delta_i$.

Conversely, suppose that for each i, g_i is an element of $\mathfrak{R}$ whose order ideal (ϵ_i) satisfies the condition $\epsilon_i \mid \delta_i$. Define B to be the mapping $\Sigma\xi_i f_i \to \Sigma\xi_i g_i$. Then we assert that $B \varepsilon \mathfrak{B}$. We show first that B is single-valued. For suppose that $\Sigma\xi_i f_i = \Sigma\eta_i f_i$ are two representations of the same element. Then $\Sigma(\xi_i - \eta_i)f_i = 0$. Hence $\delta_i \mid (\xi_i - \eta_i)$. Consequently $\epsilon_i \mid (\xi_i - \eta_i)$, and this implies that $\Sigma(\xi_i - \eta_i)g_i = 0$, or $\Sigma\xi_i g_i = \Sigma\eta_i g_i$. This shows that the results obtained from the two representations are equal. The verification that B is an $\mathfrak{o}$-endomorphism is now immediate.

Our result is the following: There is a 1–1 correspondence between the elements $B \varepsilon \mathfrak{B}$ and the ordered sets $(g_1, g_2, \cdots, g_t)$ of elements g_i whose order ideals (ϵ_i) satisfy the condition $\epsilon_i \mid \delta_i$. We now set $g_i = \Sigma\beta_{ij}f_j$, $\beta_{ij} \varepsilon \mathfrak{o}$, and we associate with the ordered set $(g_1, g_2, \cdots, g_t)$ the matrix

$$(46) \qquad (\beta) = \begin{bmatrix} \beta_{11} & \beta_{12} & \cdots & \beta_{1t} \\ \beta_{21} & \beta_{22} & \cdots & \beta_{2t} \\ \cdot & \cdot & \cdots & \cdot \\ \beta_{t1} & \beta_{t2} & \cdots & \beta_{tt} \end{bmatrix}$$

in the ring $\mathfrak{o}_t$ of $t \times t$ matrices with elements in $\mathfrak{o}$. This matrix is not uniquely determined. For any β_{ij} may be replaced by a β_{ij}' such that $\beta_{ij}' \equiv \beta_{ij} (\text{mod } \delta_j)$. This is the only alteration that can be made without changing the g_i. Hence we may say that the elements of the j-th column of (β) are determined modulo δ_j. The condition $\epsilon_i \mid \delta_i$, or what is the same thing, $\delta_i g_i = 0$, is equivalent to the equations

$$\delta_i \beta_{ij} \equiv 0 \pmod{\delta_j}. \tag{47}$$

This, of course, means that there exist γ_{ij} such that $\delta_i \beta_{ij} = \gamma_{ij} \delta_j$. Hence (47) is equivalent to the following condition on the matrix (β) of (46). There exists a matrix (γ) such that

$$\begin{bmatrix} \delta_1 & & & \\ & \delta_2 & & \\ & & \ddots & \\ & & & \delta_t \end{bmatrix} \begin{bmatrix} \beta_{11} & \beta_{12} & \cdots & \beta_{1t} \\ \beta_{21} & \beta_{22} & \cdots & \beta_{2t} \\ \cdot & \cdot & \cdots & \cdot \\ \beta_{t1} & \beta_{t2} & \cdots & \beta_{tt} \end{bmatrix} = \begin{bmatrix} \gamma_{11} & \gamma_{12} & \cdots & \gamma_{1t} \\ \gamma_{21} & \gamma_{22} & \cdots & \gamma_{2t} \\ \cdot & \cdot & \cdots & \cdot \\ \gamma_{t1} & \gamma_{t2} & \cdots & \gamma_{tt} \end{bmatrix} \begin{bmatrix} \delta_1 & & & \\ & \delta_2 & & \\ & & \ddots & \\ & & & \delta_t \end{bmatrix}. \tag{48}$$

The totality $\mathfrak{M}$ of matrices (β) which satisfy this condition is a subring of the matrix ring $\mathfrak{o}_t$. The matrix (β) determines an element B of $\mathfrak{B}$ such that $f_i B = \Sigma \beta_{ij} f_j$. It is easy to verify that the correspondence $(\beta) \to B$ is a homomorphism between $\mathfrak{M}$ and $\mathfrak{B}$. Now the endomorphism B determined by (β) is 0 if and only if $\beta_{ij} \equiv 0 (\mathrm{mod}\ \delta_j)$. Hence the kernel of our homomorphism is the set $\mathfrak{N}$ of matrices (ν) in which ν_{ij} is a multiple $\mu_{ij}\delta_j$. Thus $B \in \mathfrak{N}$ if and only if there exists a (μ) in $\mathfrak{o}_t$ such that

$$(\nu) = (\mu)(\delta) \tag{49}$$

where $(\delta) = \mathrm{diag}\,\{\delta_1, \delta_2, \cdots, \delta_t\}$. The ring $\mathfrak{B}$ is isomorphic to the difference ring $\mathfrak{M}/\mathfrak{N}$.

Theorem 18. *Let* $\mathfrak{R} = \{f_1\} \oplus \{f_2\} \oplus \cdots \oplus \{f_t\}$ *where the order ideal of* f_i *is* (δ_i). *Then the ring* $\mathfrak{B}$ *of* $\mathfrak{o}$*-endomorphisms of* $\mathfrak{R}$ *is isomorphic to the difference ring* $\mathfrak{M}/\mathfrak{N}$ *where* $\mathfrak{M}$ *is the subring of*

$\mathfrak{o}_t$ *of matrices* (β) *for which a* (γ) *exists such that* (48) *holds and* $\mathfrak{N}$ *is the ideal of matrices* (ν) *for which a* (μ) *exists such that* (49) *holds.*

An explicit determination of the matrices of $\mathfrak{M}$ can be made if use is made of the conditions (45) on the δ's. We note the following cases of (47):

1. $i \geq j$. Here $\delta_i \equiv 0(\text{mod } \delta_j)$. Hence these β_{ij} are arbitrary.
2. $i \leq u, j > u$. Here (47) and (45) imply that these $\beta_{ij} = 0$.
3. $i, j > u$. Here $\delta_i = \delta_j = 0$ and the β's are arbitrary.
4. $i < j \leq u$. Let $\eta_{ij} = \delta_i^{-1}\delta_j$. Then (47) is equivalent to the condition $\beta_{ij} \equiv 0(\text{mod } \eta_{ij})$.

Thus (β) has the following form:

$$
(50) \qquad \left[\begin{array}{cccc|ccc}
\beta_{11} & \beta_{12} & \cdots & \beta_{1u} & & & \\
\beta_{21} & \beta_{22} & \cdots & \beta_{2u} & & & \\
\cdot & \cdot & \cdots & \cdot & & 0 & \\
\beta_{u1} & \beta_{u2} & \cdots & \beta_{uu} & & & \\
\hline
\beta_{u+1,1} & \beta_{u+1,2} & \cdots & \beta_{u+1,u} & \beta_{u+1,u+1} & \cdots & \beta_{u+1,t} \\
\cdot & \cdot & \cdots & \cdot & \cdot & \cdots & \cdot \\
\cdot & \cdot & \cdots & \cdot & \cdot & \cdots & \cdot \\
\beta_{t1} & \beta_{t2} & \cdots & \beta_{tu} & \beta_{t,u+1} & \cdots & \beta_{tt}
\end{array}\right]
$$

where all the β's are arbitrary except above the main diagonal in the upper left-hand block. Here $\beta_{ij} = \mu_{ij}\eta_{ij}$ where μ_{ij} is arbitrary and $\eta_{ij} = \delta_i^{-1}\delta_j$. The condition $\beta_{ij} \equiv 0(\text{mod } \delta_j)$ for these β's is equivalent to $\mu_{ij} \equiv 0(\text{mod } \delta_i)$.

17. The linear transformations which commute with a given linear transformation. We specialize $\mathfrak{N}$ to be the $\Phi[\lambda]$-module determined by the linear transformation A. Here every $\delta_i \neq 0$ so that $u = t$. The ring $\mathfrak{M}$ now consists of all the matrices (β) in which the β_{ij} are arbitrary if $i \geq j$ and $\beta_{ij} = \mu_{ij}\eta_{ij}$, $\eta_{ij} = \delta_i^{-1}\delta_j$ for $i < j$. Any β_{ij} may be replaced by β_{ij}' in the same coset $(\text{mod } \delta_j)$. Consequently μ_{ij} may be replaced by μ_{ij}' in the same coset $(\text{mod } \delta_i)$. Thus if $n_i = \deg \delta_i$ then we may suppose that

$$
\begin{aligned}
\deg \beta_{ij} &< n_j \quad \text{if} \quad i \geq j \\
\deg \mu_{ij} &< n_i \quad \text{if} \quad i < j.
\end{aligned}
$$

A matrix in $\mathfrak{M}$ that satisfies these conditions will be called a *normalized* matrix. It is clear that two normalized matrices determine the same B in the ring $\mathfrak{B}$ of $\mathfrak{o}$-endomorphisms if and only if they are identical. Hence the correspondence $(\beta) \rightarrow B$ determined before is 1–1 between the set $\mathfrak{U}$ of normalized matrices and the ring $\mathfrak{B}$.

We know that $\mathfrak{B} \supseteq \mathfrak{A} = \Phi_l[A] \supseteq \Phi_l$. It follows that $\mathfrak{B}$ is a subspace of the vector space $\mathfrak{L}$ over Φ. We wish now to compute the dimensionality of $\mathfrak{B}$ over Φ.

We note first that the normalized matrix corresponding to the scalar multiplication α_l is the scalar matrix $\alpha 1 = \text{diag}\,\{\alpha, \alpha, \cdots, \alpha\}$ of t rows and columns; for we have the relations $f_i \alpha_l = \alpha f_i$. Also it is clear that the set $\mathfrak{U}$ of normalized matrices is closed under addition and under multiplication by scalar matrices. It follows that $\mathfrak{U}$ may be regarded as a vector space over Φ. Here $\alpha(\beta)$ for (β) in $\mathfrak{U}$ is defined to be the matrix $\alpha 1(\beta)$. If B_i, $i = 1, 2$, is in $\mathfrak{B}$ and $B_i \rightarrow (\beta_i) \in \mathfrak{U}$, then clearly $B_1 + B_2 \rightarrow (\beta_1) + (\beta_2)$ and $\alpha B_1 \rightarrow \alpha(\beta_1)$. Since our correspondence is 1–1, this shows that $\mathfrak{B}$ over Φ is equivalent to $\mathfrak{U}$ over Φ. We now determine the dimensionality of $\mathfrak{U}$.

Let $\mathfrak{U}_{ij}$ denote the subspace of $\mathfrak{U}$ of normalized matrices in which $\beta_{kl} = 0$ for all $(k, l) \neq (i, j)$. It is easy to see that if $i \geq j$, $\dim \mathfrak{U}_{ij}$, the dimensionality of the space of polynomials of degree $< n_j$ is n_j. Similarly if $i < j$, $\dim \mathfrak{U}_{ij} = n_i$. Since $\mathfrak{U}$ is a direct sum of the subspaces $\mathfrak{U}_{ij}$,

$$\begin{aligned}\dim \mathfrak{U} &= \sum_{j=1}^{t} (t - j + 1)n_j + \sum_{i=1}^{t-1} (t - i)n_i \\ &= \sum_{j=1}^{t} (2t - 2j + 1)n_j.\end{aligned}$$

This proves the following

Theorem 19. *(Frobenius) Let* $(\alpha) \in \Phi_n$ *and let* $\delta_1(\lambda), \delta_2(\lambda), \cdots, \delta_t(\lambda)$ *be the invariant factors* $\neq 1$ *of* $\lambda 1 - (\alpha)$. *Then if the degree of* $\delta_i(\lambda)$ *is* n_i, *the maximum number of linearly independent matrices commutative with* (α) *is given by the formula*

$$N = \sum_{j=1}^{t} (2t - 2j + 1)n_j.$$

Clearly, if $t > 1$, then $N > \sum_{j=1}^{t} n_j = n > n_t$. Since the dimensionality of $\mathfrak{A} = \Phi_l[A]$ over Φ_l is n_t, this shows that, in this case, $\mathfrak{B} \supset \mathfrak{A}$. If we recall that $t = 1$ is the condition that A be cyclic, we obtain the following converse of the corollary to Theorem 17.

Corollary. *If A is a linear transformation which is not cyclic, then there exist linear transformations that commute with A which are not polynomials in A.*

Example. Let

$$(\alpha) = \begin{bmatrix} 1 & 0 & 0 \\ 0 & 0 & 1 \\ 0 & -1 & 2 \end{bmatrix}.$$

Here, if A is the corresponding linear transformation, then $\mathfrak{R} = \{f_1\} \oplus \{f_2\}$. The invariant factors are $\delta_1 = \lambda - 1$ and $\delta_2 = (\lambda - 1)^2$. The general form of a normalized matrix (β) is

$$\begin{bmatrix} \beta_{11} & \beta_{12}(\lambda - 1) \\ \beta_{21} & \beta_{22} + \beta_{22}'\lambda \end{bmatrix}.$$

Since $\lambda f_1 = f_1$ and $\lambda^2 f_2 = (2\lambda - 1)f_2 = -f_2 + 2(\lambda f_2)$,

$$f_1 B = \beta_{11} f_1 - \beta_{12} f_2 + \beta_{12}(\lambda f_2)$$

$$f_2 B = \beta_{21} f_1 + \beta_{22} f_2 + \beta_{22}'(\lambda f_2)$$

$$(\lambda f_2) B = \beta_{21} f_1 - \beta_{22}' f_2 + (\beta_{22} + 2\beta_{22}')(\lambda f_2)\cdot$$

It follows that the general form of a matrix commutative with (α) is

$$\begin{bmatrix} \beta_{11} & -\beta_{12} & \beta_{12} \\ \beta_{21} & \beta_{22} & \beta_{22}' \\ \beta_{21} & -\beta_{22}' & \beta_{22} + 2\beta_{22}' \end{bmatrix}.$$

EXERCISES

1. Let $\mathfrak{R}$ be a finite commutative group and suppose that $\mathfrak{R}$ is a direct sum of cyclic groups of orders $n_1, n_2, \cdots, n_t$ where $n_i \mid n_j$ for $i \leq j$. Prove that the number of elements in the ring of endomorphisms of $\mathfrak{R}$ is

$$N = \prod_{j=1}^{t} n_j^{2t-2j+1}.$$

2. Determine the matrices that commute with

$$\begin{bmatrix} 0 & 0 & 0 & 0 & 0 \\ 1 & 0 & 0 & 0 & 0 \\ 0 & 0 & 0 & 0 & 0 \\ 0 & 0 & 1 & 0 & 0 \\ 0 & 0 & 0 & 1 & 0 \end{bmatrix}.$$

3. Determine the matrices which commute with

$$\begin{bmatrix} 1 & 0 & 0 & 0 \\ 0 & 0 & 1 & 0 \\ 0 & 0 & 0 & 1 \\ 0 & 1 & -3 & 3 \end{bmatrix}.$$

18. The center of the ring $\mathfrak{B}$. We return to the general case where $\mathfrak{R}$ is a finitely generated $\mathfrak{o}$-module, $\mathfrak{o}$ a principal ideal domain. As before, let $\mathfrak{R} = \{f_1\} \oplus \{f_2\} \oplus \cdots \oplus \{f_t\}$ where the orders (δ_i) satisfy (45). We shall prove the following

Theorem 20. *The center of the ring of $\mathfrak{o}$-endomorphisms of $\mathfrak{R}$ consists of the scalar multiplications.*

Let $\mathfrak{B}$ be the ring of $\mathfrak{o}$-endomorphisms, $\mathfrak{C}$ its center and $\mathfrak{o}_l$ the ring of scalar multiplications $x \to \alpha x$. We have seen that $\mathfrak{o}_l \subseteq \mathfrak{C}$. Now let C be any element of $\mathfrak{C}$. Let E_k, $k = 1, 2, \cdots, t$, be the $\mathfrak{o}$-endomorphism such that $f_j E_k = \delta_{jk} f_k$, $j = 1, 2, \cdots, t$. By the considerations of section 16, such endomorphisms exist in $\mathfrak{B}$. Also there exists $\mathfrak{o}$-endomorphisms E_{tk} such that $f_j E_{tk} = \delta_{jt} f_k$. Since C commutes with these endomorphisms, we have the following equations

$$f_t C = (f_t E_t) C = (f_t C) E_t = \gamma f_t, \ \gamma \,\varepsilon\, \sigma$$

$$f_k C = (f_t E_{tk}) C = (f_t C) E_{tk} = (\gamma f_t) E_{tk} = \gamma (f_t E_{tk}) = \gamma f_k.$$

Thus C coincides with the mapping $x \to \gamma x$. Hence $\mathfrak{C} = \mathfrak{o}_l$.

If C is any endomorphism in $\mathfrak{R}$ commuting with every element of $\mathfrak{B}$, then, in particular, C commutes with every element of $\mathfrak{o}_l$. Hence $C \,\varepsilon\, \mathfrak{B}$. Thus C is in the center of $\mathfrak{B}$. The converse is clear. This remark enables us to state Theorem 20 in the following alternative form:

Theorem 20′. *The only endomorphisms of $\mathfrak{R}$ which commute with every $\mathfrak{o}$-endomorphism are the scalar multiplications.*

This specializes to the

Corollary 1. *If C is a linear transformation that commutes with every linear transformation which commutes with A, then C is a polynomial in A.*

This corollary enables us to determine the center of the complete ring $\mathfrak{L}$ of linear transformations. For let $A = 1$. Then the ring $\mathfrak{B}$ of linear transformations that commute with A is the complete ring $\mathfrak{L}$. Hence Corollary 1 states that the only linear transformations that commute with every linear transformation are the polynomials in 1. Since a linear transformation is expressible as a polynomial in 1 if and only if it is a scalar multiplication, this gives the important

Corollary 2. *The center of the ring of linear transformations of a vector space over a field* Φ *is the set* Φ_l *of scalar multiplications.*

A slightly more direct proof of this result will be given later (Chapter VIII, p. 229).

EXERCISES

1. Prove that the center of the ring of endomorphisms of a finite group consists of the endomorphisms $x \rightarrow mx$, m an integer.
2. Prove that a linear transformation A is cyclic if and only if the ring $\mathfrak{B}$ of linear transformations commutative with A is commutative.
3. Prove the following extension of Theorem 20′: The only endomorphisms of $\mathfrak{R}$ which commute with every idempotent $\mathfrak{o}$-endomorphism are the scalar multiplications.

Chapter IV

SETS OF LINEAR TRANSFORMATIONS

In this chapter we shall introduce some general concepts which are fundamental in the study of arbitrary sets of linear transformations. A deeper study of these notions belongs more properly to the so-called theory of representations of rings and is beyond the scope of the present volume. An introduction to these notions will serve to put into better perspective the results of the preceding chapter. We shall also be able to extend some of these results to sets of commutative linear transformations.

1. Invariant subspaces. For the most part we shall be concerned in this chapter with the general case of a vector space over a division ring Δ. Let $\mathfrak{R}$ be a finite dimensional vector space over Δ and let Ω be a set of linear transformations in $\mathfrak{R}$ over Δ. If $(e_1, e_2, \cdots, e_n)$ is a basis for $\mathfrak{R}$ and $A \varepsilon \Omega$, then $e_iA = \Sigma\alpha_{ij}e_j$ and (α) is the matrix of A relative to the given basis. The matrices (α) determined in this way by the $A \varepsilon \Omega$ constitute a set ω that we shall call the set of *matrices of* Ω *relative to* $(e_1, e_2, \cdots, e_n)$. If $(f_1, f_2, \cdots, f_n)$ is a second basis and $f_i = \Sigma\mu_{ij}e_j$, then the matrices of Ω relative to this new basis is the set $\{(\mu)(\alpha)(\mu)^{-1}\}$, (α) in ω. We denote this set as $(\mu)\omega(\mu)^{-1}$.

From the geometric point of view a fundamental problem in the study of a set of linear transformations is that of determining the invariant subspaces relative to this set. As in the case of a single linear transformation a subspace $\mathfrak{S}$ is called *invariant* under Ω if $\mathfrak{S}A \subseteq \mathfrak{S}$ for every $A \varepsilon \Omega$. If Ω consists of a single transformation A, then the cyclic subspaces $\{x\}$ are examples of invariant subspaces. Other simple examples are the following:

1. Ω consists of the linear transformation 0. Here any subspace is invariant since $\mathfrak{S}0 = 0 \subseteq \mathfrak{S}$ for any $\mathfrak{S}$.

2. $\Omega = \mathfrak{L}$ the complete set of linear transformations. Here the only invariant subspaces are the zero space 0 and the whole space $\mathfrak{R}$. Thus let $\mathfrak{S}$ be an invariant subspace $\neq 0$ and let y be a non-zero vector in $\mathfrak{S}$. Then if x is any vector in $\mathfrak{R}$, there exists a linear transformation A such that $yA = x$. Since $\mathfrak{S}$ is invariant $x = yA \varepsilon \mathfrak{S}A \subseteq \mathfrak{S}$. Hence $x \varepsilon \mathfrak{S}$ and since x is arbitrary $\mathfrak{S} = \mathfrak{R}$.

Obviously the whole space $\mathfrak{R}$ is an invariant subspace relative to any set Ω. Also the 0 space is invariant since $0A = 0$ for any linear transformation A. The second example above shows that there exist sets Ω for which these two "trivial" subspaces are the only invariant subspaces. Such a set is called an *irreducible* set. It is also convenient at times to say that $\mathfrak{R}$ is *irreducible relative to the set* Ω.

Reducibility, or the existence of a proper ($\neq 0$, $\neq \mathfrak{R}$) invariant subspace manifests itself as a simple condition on the sets of matrices of Ω. Suppose $\mathfrak{S}$ is a proper invariant subspace and let $(f_1, f_2, \cdots, f_n)$ be a basis for $\mathfrak{R}$ such that $(f_1, f_2, \cdots, f_r)$ is a basis for $\mathfrak{S}$. Since $\mathfrak{S}$ is invariant, $f_iA \varepsilon \mathfrak{S}$ for each $i = 1, 2, \cdots, r$ and each $A \varepsilon \Omega$. Hence the relations that give the matrix A relative to $(f_1, f_2, \cdots, f_n)$ are

$$(1)\qquad \begin{aligned} f_iA &= \sum_{j=1}^{r} \beta_{ij} f_j, \quad i = 1, 2, \cdots, r \\ f_kA &= \sum_{l=1}^{n} \beta_{kl} f_l, \quad k = r + 1, r + 2, \cdots, n. \end{aligned}$$

Hence the matrix (β) of A has the form

$$(2)\qquad \begin{bmatrix} \beta_{11} & \beta_{12} & \cdots & \beta_{1r} & 0 & 0 & \cdots & 0 \\ \beta_{21} & \beta_{22} & \cdots & \beta_{2r} & 0 & 0 & \cdots & 0 \\ \cdot & \cdot & \cdots & \cdot & \cdot & \cdot & \cdots & \cdot \\ \beta_{r1} & \beta_{r2} & \cdots & \beta_{rr} & 0 & 0 & \cdots & 0 \\ \beta_{r+1,1} & \beta_{r+1,2} & \cdots & \beta_{r+1,r} & \beta_{r+1,r+1} & \cdot & \cdots & \beta_{r+1,n} \\ \cdot & \cdot & \cdots & \cdot & \cdot & \cdot & \cdots & \cdot \\ \cdot & \cdot & \cdots & \cdot & \cdot & \cdot & \cdots & \cdot \\ \beta_{n1} & \beta_{n2} & \cdots & \beta_{nr} & \beta_{n,r+1} & \cdot & \cdots & \beta_{n,n} \end{bmatrix}.$$

A matrix that has an $r \times n - r$ block of 0's in its upper right-hand corner is said to have *reduced form*. Hence the existence of a proper invariant subspace implies the existence of a basis relative to which the matrices of Ω all have reduced form. Another way of putting this is that, if ω is the set of matrices of Ω relative to some basis $(e_1, e_2, \cdots, e_n)$, then the existence of a proper invariant subspace implies that there exists a non-singular matrix (μ) such that all the matrices of $(\mu)\omega(\mu)^{-1}$ have the same reduced form (that is, with the same r for all).

The converse is valid, too. For suppose that all of the matrices of $(\mu)\omega(\mu)^{-1}$ have the reduced form (2). Then, if $f_i = \Sigma\mu_{ij}e_j$, the matrices of Ω relative to $(f_1, f_2, \cdots, f_n)$ constitute the set $(\mu)\omega(\mu)^{-1}$. Because of the form of these matrices, relations (1) hold, and these show that the space $\mathfrak{S} = [f_1, f_2, \cdots, f_r]$ is invariant under Ω.

EXERCISES

1. Let Ω be a set of linear transformations, and let B be a linear transformation that commutes with every $A \in \Omega$. Show that if $\mathfrak{S}$ is an invariant subspace relative to Ω, then $\mathfrak{S}B$ is also invariant. Also, show that the subset $\mathfrak{N}$ of vectors $y \in \mathfrak{S}$ such that $yB = 0$ is invariant relative to Ω.

2. Prove that a subspace $\mathfrak{S}$ is invariant relative to a set Ω if and only if the following operator condition holds: The relation $EAE = EA$ holds for every projection E onto $\mathfrak{S}$ and every $A \in \Omega$.

2. Induced linear transformations. If $\mathfrak{S}$ is a subspace invariant under Ω, then the linear transformations $A \in \Omega$ induce transformations in $\mathfrak{S}$. It is evident that these transformations are linear. We shall now show that in a certain sense the $A \in \Omega$ also induce linear transformations in the factor space $\bar{\mathfrak{R}} = \mathfrak{R}/\mathfrak{S}$. We recall that a vector $\bar{x}$ of $\bar{\mathfrak{R}}$ is a coset consisting of the vectors of the form $x + y$ where x is fixed and y ranges over $\mathfrak{S}$. If $A \in \Omega$, $(x + y)A = xA + yA = xA + y'$ where $y' = yA \in \mathfrak{S}$. Hence the image of any vector in the coset $\bar{x}$ is a vector in the coset $\overline{xA}$ determined by xA. Thus the mapping $\bar{A}$ that associates with $\bar{x}$ the vector $\overline{xA}$ of $\bar{\mathfrak{R}}$ is single-valued. We call $\bar{A}$ the transformation induced by A in $\bar{\mathfrak{R}}$. This mapping is linear, since

$$(\bar{x}_1 + \bar{x}_2)\bar{A} = (\overline{x_1 + x_2})\bar{A} = (\overline{(x_1 + x_2)A}) = (\overline{x_1A + x_2A})$$
$$= (\overline{x_1A}) + (\overline{x_2A}) = \bar{x}_1\bar{A} + \bar{x}_2\bar{A}$$

and

$$(\alpha\bar{x})\bar{A} = \overline{(\alpha x)}\bar{A} = \overline{((\alpha x)A)} = \overline{(\alpha(xA))} = \alpha\overline{(xA)} = \alpha(\bar{x}\bar{A}).$$

When there is no risk of confusion, we shall simplify matters by denoting the induced transformation by A also.

On the other hand, it is sometimes necessary to distinguish carefully between the transformations A and $\bar{A}$. The precise relation between these can be made explicit in the following way:

Consider the mapping P: $x \rightarrow \bar{x} = x + \mathfrak{S}$ of $\mathfrak{R}$ onto $\bar{\mathfrak{R}}$. It is immediate from the definitions of the compositions in $\bar{\mathfrak{R}}$ that P is a linear transformation of $\mathfrak{R}$ onto $\bar{\mathfrak{R}}$. We shall refer to P as the *natural mapping* of $\mathfrak{R}$ onto $\bar{\mathfrak{R}}$. Now we have defined $\bar{A}$ by the rule that $\bar{x}\bar{A} = \overline{xA}$. Thus we have the relation $P\bar{A} = AP$ connecting A and $\bar{A}$.

Now suppose the basis $(f_1, f_2, \cdots, f_n)$ is chosen as in the preceding section so that $(f_1, f_2, \cdots, f_r)$ is a basis for $\mathfrak{S}$. Then it is clear from the first set of relations in (1) that the matrix

$$\begin{bmatrix} \beta_{11} & \beta_{12} & \cdots & \beta_{1r} \\ \beta_{21} & \beta_{22} & \cdots & \beta_{2r} \\ \cdot & \cdot & \cdots & \cdot \\ \beta_{r1} & \beta_{r2} & \cdots & \beta_{rr} \end{bmatrix}$$

is a matrix of the linear transformation induced by A in the invariant subspace $\mathfrak{S}$. Moreover, by the second set of equations in (1) we have

$$\bar{f}_k\bar{A} = \overline{(f_kA)} = \overline{\left(\sum_1^n \beta_{kl}f_l\right)} = \sum_1^n \beta_{kl}\bar{f}_l = \sum_{r+1}^n \beta_{kl}\bar{f}_l.$$

Now we know that the vectors $(\bar{f}_{r+1}, \bar{f}_{r+2}, \cdots, \bar{f}_n)$ form a basis for $\bar{\mathfrak{R}}$. Hence these relations show that the matrix of $\bar{A}$ relative to this basis is the lower diagonal block which appears in (2):

$$\begin{bmatrix} \beta_{r+1,r+1} & \cdots & \beta_{r+1,n} \\ \beta_{r+2,r+1} & \cdots & \beta_{r+2,n} \\ \cdot & \cdots & \cdot \\ \beta_{n,r+1} & \cdots & \beta_{nn} \end{bmatrix}.$$

We have now found interpretations for the two diagonal blocks which occur in (2). It is natural to ask for the meaning of the remaining block, appearing in the lower left-hand corner. To obtain the significance of this block we observe first that $[f_{r+1}, \cdots, f_n]$ is a complement $\mathfrak{U}$ of the space $\mathfrak{S}$. The decomposition $\mathfrak{R} = \mathfrak{S} \oplus \mathfrak{U}$ defines a projection E of $\mathfrak{R}$ onto $\mathfrak{S}$. Since $\mathfrak{S}$ is invariant relative to Ω, $EAE = EA$ holds for every $A \varepsilon \Omega$ (Ex. 2, p. 117). We consider now the linear transformation $AE - EA = AE - EAE$. This transformation sends $\mathfrak{R}$ into $\mathfrak{S}$. Moreover, if $y \varepsilon \mathfrak{S}$, then $y(AE - EA) = yA - yA = 0$. It follows from this that $AE - EA$ defines a linear transformation A_E of $\bar{\mathfrak{R}} = \mathfrak{R}/\mathfrak{S}$ into $\mathfrak{S}$. Thus we define

$$\bar{x}A_E = x(AE - EA),$$

and, by the remark that we have made, we see that A_E is single-valued. One verifies directly that A_E is a linear transformation of $\mathfrak{R}/\mathfrak{S}$ into $\mathfrak{S}$. Moreover, by definition we have the relation $AE - EA = PA_E$.

We shall now show that the matrix

$$\begin{bmatrix} \beta_{r+1,1} & \cdots & \beta_{r+1,r} \\ \cdot & \cdots & \cdot \\ \cdot & \cdots & \cdot \\ \beta_{n,1} & \cdots & \beta_{n,r} \end{bmatrix}$$

is the matrix of A_E relative to the bases $(\bar{f}_{r+1}, \cdots, \bar{f}_n)$, $(f_1, \cdots, f_r)$ for $\bar{\mathfrak{R}}$ and $\mathfrak{S}$. We have the relations

$$\begin{aligned} \bar{f}_k A_E &= f_k(AE - EA) = f_k AE \\ &= \sum_{i=1}^{r} \beta_{ki} f_i, \quad k = r + 1, \cdots, n, \end{aligned}$$

which prove our assertion.

EXERCISES

1. Prove that a relation such as $A + B = C$ or $AB = C$ for A, B, C in Ω implies a corresponding relation $\bar{A} + \bar{B} = \bar{C}$, $\bar{A}\bar{B} = \bar{C}$ for the induced linear transformation in $\bar{\mathfrak{R}}$.

2. Show that A is 1-1 in $\mathfrak{R}$ if and only if A is 1-1 in $\mathfrak{S}$ and $\bar{A}$ is 1-1 in $\bar{\mathfrak{R}} = \mathfrak{R}/\mathfrak{S}$.

3. Composition series. Let $\mathfrak{S}$ and $\mathfrak{U}$ be invariant subspaces such that $\mathfrak{R} \supseteq \mathfrak{U} \supseteq \mathfrak{S}$. Then $\bar{\mathfrak{U}} = \mathfrak{U}/\mathfrak{S}$ is a subspace of $\bar{\mathfrak{R}} = \mathfrak{R}/\mathfrak{S}$. If $\bar{u} \varepsilon \bar{\mathfrak{U}}$, $\bar{u} = u + \mathfrak{S}$ where $u \varepsilon \mathfrak{U}$. Hence for any A in Ω, $\bar{u}\bar{A} = \overline{uA} \varepsilon \bar{\mathfrak{U}}$. This shows that $\bar{\mathfrak{U}}$ is invariant relative to the set $\bar{\Omega}$ of the induced linear transformations $\bar{A}$ in $\bar{\mathfrak{R}}$.

We shall now show that the converse holds, namely, any invariant subspace of $\bar{\mathfrak{R}}$ has the form $\bar{\mathfrak{U}} = \mathfrak{U}/\mathfrak{S}$ where $\mathfrak{U}$ is an invariant subspace of $\mathfrak{R}$ containing $\mathfrak{S}$. Thus let $\bar{\mathfrak{U}}$ be a subspace of $\bar{\mathfrak{R}}$ invariant relative to $\bar{\Omega}$. Let $\mathfrak{U}$ be the totality of vectors contained in the cosets that belong to $\bar{\mathfrak{U}}$. If u_1 and $u_2 \varepsilon \mathfrak{U}$, $\bar{u}_1 = u_1 + \mathfrak{S}$ and $\bar{u}_2 = u_2 + \mathfrak{S}$ are in $\bar{\mathfrak{U}}$. Hence $\bar{u}_1 + \bar{u}_2 = (u_1 + u_2) + \mathfrak{S}$ is in $\bar{\mathfrak{U}}$. Hence $u_1 + u_2 \varepsilon \mathfrak{U}$. Similarly $\alpha u \varepsilon \mathfrak{U}$ for any α in Δ and any u in $\mathfrak{U}$. Since $(u + \mathfrak{S})\bar{A} = uA + \mathfrak{S} \varepsilon \bar{\mathfrak{U}}$, $uA \varepsilon \mathfrak{U}$. Thus $\mathfrak{U}$ is an invariant subspace of $\mathfrak{R}$. Clearly $\bar{\mathfrak{U}} = \mathfrak{U}/\mathfrak{S}$.

A sequence of invariant subspaces

$$0 \subset \mathfrak{S}_1 \subset \mathfrak{S}_2 \subset \cdots \subset \mathfrak{S}_t = \mathfrak{R} \tag{3}$$

is called a *composition series* for $\mathfrak{R}$ relative to Ω if each $\mathfrak{S}_i$ is irreducible over $\mathfrak{S}_{i-1}$ in the sense that there exists no invariant $\mathfrak{S}'$ such that $\mathfrak{S}_i \supset \mathfrak{S}' \supset \mathfrak{S}_{i-1}$. By what we have shown it is clear that $\mathfrak{S}_i$ is irreducible over $\mathfrak{S}_{i-1}$ if and only if $\mathfrak{S}_i/\mathfrak{S}_{i-1}$ is irreducible relative to the set of linear transformations induced by the $A \varepsilon \Omega$. The irreducible spaces

$$\mathfrak{S}_1, \mathfrak{S}_2/\mathfrak{S}_1, \cdots, \mathfrak{S}_t/\mathfrak{S}_{t-1} \tag{4}$$

are called the *composition factors* of the series (3).

We now choose a basis $(f_1, f_2, \cdots, f_{n_1})$ for $\mathfrak{S}_1$. This can be supplemented to a basis $(f_1, f_2, \cdots, f_{n_1+n_2})$ for $\mathfrak{S}_2$. Continuing in this way we obtain finally a basis $(f_1, f_2, \cdots, f_n)$ for $\mathfrak{R}$ such that $(f_1, f_2, \cdots, f_{n_1+\cdots+n_i})$ is a basis for $\mathfrak{S}_i$. Then if $A \varepsilon \Omega$

$$\begin{aligned}
f_iA &= \sum_1^{n_1} \beta_{ij} f_j, & i &= 1, 2, \cdots, n_1 \\
f_kA &= \sum_1^{n_1+n_2} \beta_{kl} f_l, & k &= n_1 + 1, \cdots, n_1 + n_2 \\
&\cdots\cdots\cdots\cdots \\
f_pA &= \sum_1^{n_1+\cdots+n_t} \beta_{pq} f_q, & p &= n_1 + \cdots + n_{t-1} + 1, \cdots, \\
&& &\quad n_1 + \cdots + n_t = n
\end{aligned} \tag{5}$$

Consequently the matrix of A relative to our basis has the form

$$(6) \qquad \begin{bmatrix} (\beta_1) & & & 0 \\ * & (\beta_2) & & \\ & & \ddots & \\ * & * & \cdots & (\beta_t) \end{bmatrix}$$

in which the "blocks" above the main diagonal are all 0. The cosets $(\bar{f}_{n_1+\cdots+n_{i-1}+1}, \bar{f}_{n_1+\cdots+n_{i-1}+2}, \cdots, \bar{f}_{n_1+\cdots+n_i})$ form a basis for the factor space $\mathfrak{S}_i/\mathfrak{S}_{i-1}$, and by (5) we have

$$\bar{f}_r\bar{A} = \sum_{n_1+\cdots+n_{i-1}+1}^{n_1+\cdots+n_i} \beta_{rs}\bar{f}_s$$

if $\bar{A}$ denotes the transformation induced by A in the factor space $\mathfrak{S}_i/\mathfrak{S}_{i-1}$. The matrix of $\bar{A}$ relative to this basis is therefore the diagonal block (β_i) that appears in (6). Our assumption that $\mathfrak{S}_i$ is irreducible over $\mathfrak{S}_{i-1}$ means that it is impossible to find a matrix (μ_i) such that all of the matrices $(\mu_i)(\beta_i)(\mu_i)^{-1}$ all have the same reduced form.

It is easy to prove the existence of a composition series for any set Ω. First, if $\mathfrak{R}$ is irreducible, then $0 \subset \mathfrak{R}$ is such a series. Otherwise let $\mathfrak{S}$ be a proper invariant subspace. If $\mathfrak{S}$ is irreducible relative to the set of induced linear transformations, we take $\mathfrak{S}_1 = \mathfrak{S}$. Otherwise we let $\mathfrak{S}'$ be a proper invariant subspace of $\mathfrak{S}$. Now $\dim \mathfrak{R} > \dim \mathfrak{S} > \dim \mathfrak{S}'$. Hence this process cannot be continued indefinitely. Eventually we get an irreducible invariant subspace $\mathfrak{S}_1 \neq 0$. We repeat this argument with $\bar{\mathfrak{R}} = \mathfrak{R}/\mathfrak{S}_1$ and the set $\bar{\Omega}$ of induced transformations. Then we see that, if $\bar{\mathfrak{R}} \neq 0$ (that is, $\mathfrak{R} \neq \mathfrak{S}_1$), $\bar{\mathfrak{R}}$ contains an irreducible invariant subspace $\bar{\mathfrak{S}}_2 \neq 0$. This space has the form $\mathfrak{S}_2/\mathfrak{S}_1$, and $\mathfrak{S}_2$ is invariant relative to Ω and is irreducible over $\mathfrak{S}_1$. We consider next $\mathfrak{R}/\mathfrak{S}_2$. If this space is $\neq 0$, we obtain in the same way a subspace $\mathfrak{S}_3$ invariant under Ω and irreducible over $\mathfrak{S}_2$. Since $\dim \mathfrak{S}_1 < \dim \mathfrak{S}_2 < \dim \mathfrak{S}_3 < \cdots$, this process, too, breaks off after, say, t steps, with the space $\mathfrak{R}$. Then $0 \subset \mathfrak{S}_1 \subset \mathfrak{S}_2 \subset \cdots \subset \mathfrak{S}_t = \mathfrak{R}$ is a composition series.

4. Decomposability of a set of linear transformations. We consider now a decomposition of the space $\mathfrak{R}$ as

$$\mathfrak{R} = \mathfrak{R}_1 \oplus \mathfrak{R}_2 \oplus \cdots \oplus \mathfrak{R}_s, \tag{7}$$

a direct sum of the subspaces $\mathfrak{R}_i$ that are assumed to be invariant relative to Ω. Such a decomposition is called *proper* if each $\mathfrak{R}_i \neq 0$ and $s > 1$. If $\mathfrak{R}$ has a proper decomposition, we say that $\mathfrak{R}$ is *decomposable relative to* Ω and that Ω is a *decomposable* set of linear transformations.

Clearly, a set of linear transformations which is irreducible is indecomposable. On the other hand, there exist reducible sets which are indecomposable. Hence decomposability is an essentially stronger condition than reducibility. We prove this assertion by citing the following

Example. Let Ω consist of the linear transformation A with matrix

$$\begin{bmatrix} 0 & & & & & \\ 1 & 0 & & & & \\ & 1 & \cdot & & & \\ & & \cdot & \cdot & & \\ & & & \cdot & \cdot & \\ & & & & 1 & 0 \end{bmatrix}$$

relative to the basis $(e_1, e_2, \cdots, e_n)$. Then $e_1A = 0$ and $e_iA = e_{i-1}$ if $i > 1$. Hence the subspaces $\mathfrak{S}_i = [e_1, e_2, \cdots, e_i]$, $i = 1, 2, \cdots, n$, are invariant. We shall show that these are the only non-zero invariant subspaces relative to A. For let $\mathfrak{S}$ be such a subspace. Let h be the smallest integer such that $\mathfrak{S} \subseteq \mathfrak{S}_h$. Then $\mathfrak{S}$ contains a vector

$$\gamma_1 e_1 + \gamma_2 e_2 + \cdots + \gamma_h e_h$$

with $\gamma_h \neq 0$. We may assume that $\gamma_h = 1$ so that

$$y = \gamma_1 e_1 + \cdots + \gamma_{h-1} e_{h-1} + e_h \in \mathfrak{S}.$$

Then the vectors

$$yA = \gamma_2 e_1 + \cdots + \gamma_{h-1} e_{h-2} + e_{h-1}$$

$$yA^2 = \gamma_3 e_1 + \cdots + \gamma_{h-1} e_{h-3} + e_{h-2}$$

$$\cdots\cdots\cdots\cdots\cdots\cdots$$

are in $\mathfrak{S}$. Evidently $e_1, e_2, \cdots, e_h$ are linearly dependent on these vectors. Hence $\mathfrak{S}_h \subseteq \mathfrak{S}$ and so $\mathfrak{S} = \mathfrak{S}_h$. Since $\mathfrak{S}_i \subset \mathfrak{S}_j$ if $i < j$, it is clear that no two of these spaces are independent. Hence $\mathfrak{R}$ cannot be written as a direct sum of these invariant subspaces.

If $\mathfrak{R}$ is a direct sum of the invariant subspaces $\mathfrak{R}_i$, $i = 1, 2, \cdots, s$, $s \geq 1$, we may choose a basis $(f_1, f_2, \cdots, f_n)$ for $\mathfrak{R}$ such that

$$(8) \qquad (f_{n_1+\cdots+n_{i-1}+1}, f_{n_1+\cdots+n_{i-1}+2}, \cdots, f_{n_1+\cdots+n_i})$$

is a basis for $\mathfrak{R}_i$. Since $\mathfrak{R}_i$ is invariant, any vector in (8) is transformed into a linear combination of these vectors by any $A \varepsilon \Omega$. Hence the matrix of A has the form

$$(9) \qquad \begin{bmatrix} (\beta_1) & & & \\ & (\beta_2) & & \\ & & \ddots & \\ & & & (\beta_s) \end{bmatrix}.$$

Here the diagonal block is the matrix of the transformations induced by A in $\mathfrak{R}_i$ relative to the basis (8). Conversely, if Ω is any set of linear transformations and there exists a basis $(f_1, f_2, \cdots, f_n)$ such that all the matrices relative to this basis have the form (9), then $\mathfrak{R}$ is a direct sum of the invariant subspaces

$$\mathfrak{R}_i = [f_{n_1+\cdots+n_{i-1}+1}, \cdots, f_{n_1+\cdots+n_i}].$$

Suppose now that E_i, $i = 1, 2, \cdots, s$, are the projections determined by our decomposition. We recall that, if

$$(10) \qquad x = x_1 + x_2 + \cdots + x_s$$

where $x_i \varepsilon \mathfrak{R}_i$, then E_i is the mapping $x \to x_i$. We have also seen (p. 60) that the following relations hold:

$$(11) \quad E_i^2 = E_i, \quad E_iE_j = 0, \quad i \neq j, \quad E_1 + E_2 + \cdots + E_s = 1.$$

The space $\mathfrak{R}_i = \mathfrak{R}E_i$. Hence each E_i is $\neq 0$. Now let $A \varepsilon \Omega$. Then

$$xA = x_1A + x_2A + \cdots + x_sA$$

and, since $\mathfrak{R}_i$ is invariant, $x_iA \varepsilon \mathfrak{R}_i$. Thus the component in $\mathfrak{R}_i$ of xA is x_iA, or,

$$(12) \qquad xAE_i = x_iA = xE_iA.$$

This shows that the projections E_i commute with every $A \varepsilon \Omega$. Conversely suppose the E_i are linear transformations $\neq 0$ which satisfy (11) and which commute with every $A \varepsilon \Omega$. Then we know that $\mathfrak{R} = \mathfrak{R}_1 \oplus \mathfrak{R}_2 \oplus \cdots \oplus \mathfrak{R}_s$ where $\mathfrak{R}_i = \mathfrak{R}E_i$. Moreover, if $A \varepsilon \Omega$ and $x_i \varepsilon \mathfrak{R}_i$, then $x_i = xE_i$ and $x_iA = xE_iA = (xA)E_i \varepsilon \mathfrak{R}_i$. Hence $\mathfrak{R}_i$ is invariant relative to Ω.

If $s > 1$ in our discussion, each $\mathfrak{R}_i$ is a proper subspace and each $E_i \neq 1$. Hence we see that, if Ω is a decomposable set of linear transformations, then there exist projections $\neq 0$, $\neq 1$ which commute with every A in Ω. Conversely, if E_1 is a projection $\neq 0$, $\neq 1$ which commutes with every A in Ω, then $E_2 = 1 - E_1$ has these properties also. Moreover, E_1 and E_2 are orthogonal. Hence $\mathfrak{R} = \mathfrak{R}E_1 \oplus \mathfrak{R}E_2$ and the $\mathfrak{R}E_i$ are proper invariant subspaces. This proves the following important criterion.

Theorem 1. *A set Ω of linear transformations is decomposable if and only if there exist projections $E \neq 0$, $\neq 1$ which commute with every transformation in Ω.*

EXERCISE

1. Show that, if $\mathfrak{R}$ is irreducible (indecomposable) relative to a subset of Ω, then it is irreducible (indecomposable) relative to Ω. Use this to prove that the set of linear transformations corresponding to the set of triangular matrices

$$\begin{bmatrix} \alpha_{11} & & & 0 \\ \alpha_{21} & \alpha_{22} & & \\ \vdots & & \ddots & \\ \alpha_{n1} & \alpha_{n2} & & \alpha_{nn} \end{bmatrix}$$

is an indecomposable set.

5. Complete reducibility. If $\mathfrak{S}_1$ and $\mathfrak{S}_2$ are invariant subspaces under Ω, then so is $\mathfrak{S}_1 \cap \mathfrak{S}_2$ and $\mathfrak{S}_1 + \mathfrak{S}_2$. Hence the totality of invariant subspaces is a sublattice L_Ω of the complete lattice L of subspaces of $\mathfrak{R}$. It is natural to apply lattice-theoretic ideas in the study of the set L_Ω and this is, in fact, what we have done in the foregoing discussion. Thus the statement that Ω is irreducible amounts to saying that L_Ω contains just two elements. Also the statement that Ω is decomposable may be formulated as a property of the lattice L_Ω. Of the properties of L we singled out in Chapter I, it is clear that the chain conditions and the Dedekind law are preserved in passing to the sublattice

L_Ω. The complementation property of L, however, may or may not be valid in L_Ω. It certainly holds if Ω consists of the identity 1 only, for then $L_\Omega = L$. Also, if Ω is irreducible, then L_Ω is complemented, and there are other less trivial examples which will be encountered later.

If L_Ω is complemented, we say that Ω is a *completely reducible set* of linear transformations. This, of course, means that, if $\mathfrak{S}$ is *any* invariant subspace relative to Ω, then there exists a second invariant subspace $\mathfrak{S}'$ such that $\mathfrak{R} = \mathfrak{S} \oplus \mathfrak{S}'$. This condition is admittedly rather elusive since it applies to every invariant subspace $\mathfrak{S}$. It is therefore remarkable that this (possibly infinite) set of conditions can be replaced by the single condition given in the following

Theorem 2. *A set Ω of linear transformations is completely reducible if and only if $\mathfrak{R}$ can be expressed as a direct sum of subspaces $\mathfrak{R}_i$ that are invariant and irreducible relative to Ω.*

Sufficiency. Let $\mathfrak{R} = \mathfrak{R}_1 \oplus \mathfrak{R}_2 \oplus \cdots \oplus \mathfrak{R}_s$ where the $\mathfrak{R}_i$ are irreducible invariant subspaces. If $\mathfrak{S}$ is any invariant subspace, either $\mathfrak{S} = \mathfrak{R}$ or there exists an $\mathfrak{R}_i$, say $\mathfrak{R}_1$, such that $\mathfrak{R}_1 \not\subseteq \mathfrak{S}$. Then we set $\mathfrak{S}_1 = \mathfrak{S} + \mathfrak{R}_1$. Now $\mathfrak{S} \cap \mathfrak{R}_1 \,\varepsilon\, L_\Omega$ and since $\mathfrak{S} \cap \mathfrak{R}_1$ is contained in the irreducible invariant subspace $\mathfrak{R}_1$, either $\mathfrak{S} \cap \mathfrak{R}_1 = \mathfrak{R}_1$ or $\mathfrak{S} \cap \mathfrak{R}_1 = 0$. Since $\mathfrak{S} \cap \mathfrak{R}_1 = \mathfrak{R}_1$ is equivalent to $\mathfrak{S} \supseteq \mathfrak{R}_1$, we must have $\mathfrak{S} \cap \mathfrak{R}_1 = 0$. Hence $\mathfrak{S}_1 = \mathfrak{S} + \mathfrak{R}_1 = \mathfrak{S} \oplus \mathfrak{R}_1$. We now repeat the argument with $\mathfrak{S}_1$ in place of $\mathfrak{S}$. Then either $\mathfrak{S}_1 = \mathfrak{R}$, in which case $\mathfrak{R}_1$ is a complement of $\mathfrak{S}$, or there exists an $\mathfrak{R}_i$, say $\mathfrak{R}_2$, such that $\mathfrak{S}_2 = \mathfrak{S}_1 + \mathfrak{R}_2 = \mathfrak{S}_1 \oplus \mathfrak{R}_2$. Then $\mathfrak{S}_2 = \mathfrak{S} \oplus \mathfrak{R}_1 \oplus \mathfrak{R}_2$. Eventually we obtain for a suitable choice of the notation that $\mathfrak{R} = \mathfrak{S} \oplus \mathfrak{R}_1 \oplus \mathfrak{R}_2 \oplus \cdots \oplus \mathfrak{R}_h$. Then $\mathfrak{S}' = \mathfrak{R}_1 \oplus \mathfrak{R}_2 \oplus \cdots \oplus \mathfrak{R}_h$ is a complement of $\mathfrak{S}$ in $\mathfrak{R}$.

Necessity. Suppose that $\mathfrak{R}_1$ is an irreducible invariant subspace of $\mathfrak{R}$. Then either $\mathfrak{R} = \mathfrak{R}_1$ is irreducible or $\mathfrak{R} = \mathfrak{R}_1 \oplus \mathfrak{R}_1'$ where $\mathfrak{R}_1'$ is an invariant subspace $\neq 0$. Next let $\mathfrak{R}_2$ be an irreducible invariant subspace of $\mathfrak{R}_1'$. If $\mathfrak{R}_1' = \mathfrak{R}_2$, we have $\mathfrak{R} = \mathfrak{R}_1 \oplus \mathfrak{R}_2$, $\mathfrak{R}_i$ irreducible invariant as required. Otherwise, $\mathfrak{R}_1 + \mathfrak{R}_2$ has a complement $\mathfrak{R}_2'$, $\mathfrak{R}_2'$ invariant. Then $\mathfrak{R} = \mathfrak{R}_1 \oplus \mathfrak{R}_2 \oplus \mathfrak{R}_2'$. We repeat the argument with $\mathfrak{R}_2'$. This leads finally to $\mathfrak{R} = \mathfrak{R}_1 \oplus \mathfrak{R}_2 \oplus \cdots \oplus \mathfrak{R}_s$ where the $\mathfrak{R}_i$ are invariant and irreducible.

EXERCISES

1. Use the argument in the first part of Theorem 2 to show that, if $\mathfrak{R}$ is a sum (not necessarily direct) of the irreducible invariant subspaces $\mathfrak{R}_i$, then $\mathfrak{R}$ is completely reducible relative to Ω.

2. Show that, if ω is any set of diagonal matrices, then the corresponding set Ω of linear transformations is completely reducible.

3. Let $G = \{S_1, S_2, \cdots, S_m\}$ be a finite group of linear transformations, that is, a subgroup of the group of 1–1 linear transformations of $\mathfrak{R}$ over Δ. Assume that the order m of G is not divisible by the characteristic of Δ. Show that, if $\mathfrak{S}$ is an invariant subspace relative to G and E is a projection onto $\mathfrak{S}$, then $E_0 = \frac{1}{m}\left(\sum_{i=1}^{m} S_i E S_i^{-1}\right)$ is a projection on $\mathfrak{S}$ which commutes with the S_i. Hence prove the important theorem:

Any finite group of linear transformations whose order is not divisible by the characteristic of the division ring is completely reducible.

4. Let Ω be an arbitrary set of linear transformations and let E be any projection such that $EA = EAE$ holds for every A in Ω. As in § 2 define the linear transformation A_E of $\bar{\mathfrak{R}} = \mathfrak{R}/\mathfrak{R}E$ into $\mathfrak{S} = \mathfrak{R}E$ by $\bar{x}A_E = x(AE - EA)$. Prove that, if there exists a linear transformation D of $\bar{\mathfrak{R}}$ into $\mathfrak{S}$ such that $A_E = \bar{A}D - DA$ holds for all A, then $\mathfrak{S}$ has a complementary invariant subspace.

5. Prove that if Ω is completely reducible, and $\mathfrak{S}$ is an invariant subspace, then the set of linear transformations induced in $\mathfrak{S}$ is completely reducible.

***6. Relation to the theory of operator groups and the theory of modules.** The theory of sets of linear transformations that we are considering here can be regarded as a specialization of the theory of groups with operators (M-groups).* A reader who is familiar with the latter concept will observe that we are dealing here with the additive group $\mathfrak{R}$ considered as a group with operator set $M = \Delta_l \cup \Omega$, Δ_l the set of scalar multiplications. The concept of M-subgroup evidently coincides with that of invariant subspace relative to the set of linear transformations Ω. Hence the concepts of reducibility, decomposability, composition series coincide with these concepts for the M-group $\mathfrak{R}$.

The theory of M-groups also suggests the introduction of the following concept of homomorphism between invariant subspaces or factor spaces relative to the set Ω. A mapping θ is said to be an *Ω-linear transformation* of one such space into a second one if θ is a linear transformation and $(xA)\theta = (x\theta)A$ holds for all the induced transformations $A \,\varepsilon\, \Omega$. Similarly we say that two sub-

* See Volume I, Chapters V and VI.

spaces $\mathfrak{R}_i$, $i = 1, 2$, are *Ω-isomorphic* or *equivalent* if there exists a 1–1 Ω-linear transformation θ of $\mathfrak{R}_1$ onto $\mathfrak{R}_2$. If $(e_1, e_2, \cdots, e_r)$ is a basis for $\mathfrak{R}_1$, then $(e_1\theta, e_2\theta, \cdots, e_r\theta)$ is a basis for $\mathfrak{R}_2$, and it is immediately verifiable that every A has the same matrix relative to $(e_1, \cdots, e_r)$ and to $(e_1\theta, \cdots, e_r\theta)$. It follows that for arbitrary choice of bases in $\mathfrak{R}_1$ and $\mathfrak{R}_2$ the matrices (α_1) and (α_2) of any $A \varepsilon \Omega$ are related by $(\alpha_2) = (\mu)(\alpha_1)(\mu)^{-1}$ where the non-singular matrix (μ) is independent of A.

We can now state the following two fundamental theorems which are taken from the theory of M-groups.

Jordan-Hölder theorem. *If* $0 \subset \mathfrak{R}_1 \subset \mathfrak{R}_2 \subset \cdots \subset \mathfrak{R}_t = \mathfrak{R}$ *and* $0 \subset \mathfrak{S}_1 \subset \mathfrak{S}_2 \subset \cdots \subset \mathfrak{S}_u = \mathfrak{R}$ *are two composition series for a set* Ω *of linear transformations, then* $t = u$ *and the factors of composition* $\mathfrak{R}_i/\mathfrak{R}_{i-1}$, $\mathfrak{S}_j/\mathfrak{S}_{j-1}$ *can be put into* 1–1 *correspondence in such a way that corresponding pairs are Ω-isomorphic.*

Krull-Schmidt theorem. *If* $\mathfrak{R} = \mathfrak{R}_1 \oplus \mathfrak{R}_2 \oplus \cdots \oplus \mathfrak{R}_h = \mathfrak{S}_1 \oplus \mathfrak{S}_2 \oplus \cdots \oplus \mathfrak{S}_k$ *are two decompositions of* $\mathfrak{R}$ *into non-zero invariant and indecomposable subspaces relative to* Ω, *then* $h = k$ *and, if the* $\mathfrak{S}_i$ *are suitably ordered,* $\mathfrak{R}_i$ *and* $\mathfrak{S}_i$ *are Ω-isomorphic.*

We refer the reader to Volume I, Chapter V, for the proofs of these theorems.

We can also absorb the present theory into the theory of modules. This comes about from the fact that $\mathfrak{R}$ is commutative. Hence the endomorphisms of $\mathfrak{R}$ form a ring and the set $M = \Delta_l \cup \Omega$ generates a subring $\mathfrak{o}(M)$ of the ring of endomorphisms of $\mathfrak{R}$. We call $\mathfrak{o} = \mathfrak{o}(M)$ the *enveloping ring* of M. Now if $\mathfrak{S}$ is a subgroup of the additive group $\mathfrak{R}$, then the set of endomorphisms of $\mathfrak{R}$ that map $\mathfrak{S}$ into itself is a subring of the ring of endomorphisms. Hence if $\mathfrak{S}$ is an Ω-subspace, this subring contains M and consequently it also contains $\mathfrak{o}(M)$. Thus we see that any Ω-subspace of $\mathfrak{R}$ is an $\mathfrak{o}(M)$-subgroup. The converse is, of course, clear. In a similar manner we see that, if θ is an Ω-linear transformation, then θ is an $\mathfrak{o}(M)$-homomorphism, and generally speaking nothing is changed in shifting from the set M to its enveloping ring. We have seen that in dealing with a ring of endomorphisms it is often convenient to regard the underlying group as a module.

It is usually difficult to obtain the structure of the enveloping ring $\mathfrak{o}(M)$ and, when this is the case, there is no particular gain in shifting from the point of view of sets of linear transformations to that of modules. If $\Delta = \Phi$ is a field, the problem of determining the structure of $\mathfrak{o}$ is easier than in the general case. Here we note also that Φ_l is contained in the ring of linear transformations and $\mathfrak{o}$ can be regarded as an algebra over Φ. Consequently in this case one speaks of the *enveloping algebra* of the set Ω.

7. Reducibility, decomposability, complete reducibility for a single linear transformation. We adopt again the original point of view and we consider in this section the special case in which Ω consists of a single linear transformation and $\Delta = \Phi$ is a field.

If f is a non-zero vector in $\mathfrak{R}$, then the cyclic subspace $\{f\}$ is an invariant subspace $\neq 0$. Let $\mu_f(\lambda)$ be the order of f and suppose that $\mu_f(\lambda) = \pi(\lambda)\nu(\lambda)$ where $\pi(\lambda)$ is irreducible and has positive degree and leading coefficient $= 1$. Then $g = f\nu(A)$ has order $\pi(\lambda)$. Thus $\mathfrak{R}$ contains a vector whose order is a prime. Suppose now that A is irreducible. In this case we have $\mathfrak{R} = \{g\}$ where $\mu_g(\lambda) = \pi(\lambda)$ a prime. Conversely suppose that $\mathfrak{R}$ has this form and let $\mathfrak{S}$ be a subspace $\neq 0$ invariant under A. If h is a vector in $\mathfrak{S}$, the order $\mu_h(\lambda)$ is a factor of $\pi(\lambda)$. Hence if $h \neq 0$, $\mu_h(\lambda) = \pi(\lambda)$. Hence $\dim \{h\} = \deg \pi(\lambda) = \dim \{g\}$. It follows that $\{h\} = \{g\}$. We have therefore proved the following

Theorem 3. *A linear transformation A in $\mathfrak{R}$ over Φ is irreducible if and only if it is cyclic and has prime minimum polynomial.*

We consider next the question of decomposability of a single linear transformation. We know that $\mathfrak{R} = \{f_1\} \oplus \{f_2\} \oplus \cdots \oplus \{f_t\}$. Hence a necessary condition for indecomposability is that $t = 1$, that is, $\mathfrak{R} = \{f\}$ is cyclic. We have also seen that, if the minimum polynomial of a cyclic linear transformation can be expressed as $\mu_1(\lambda)\mu_2(\lambda)$ where $(\mu_1(\lambda), \mu_2(\lambda)) = 1$, then $\mathfrak{R} = \{g_1\} \oplus \{g_2\}$ where $\mu_{g_i}(\lambda) = \mu_i(\lambda)$. Hence a cyclic linear transformation is decomposable unless its minimum polynomial has the form $\pi(\lambda)^k$, $\pi(\lambda)$ prime. Conversely, these conditions are sufficient; for suppose that A is cyclic with $\pi(\lambda)^k$ as its minimum polynomial. Assume that $\mathfrak{R} = \mathfrak{R}_1 \oplus \mathfrak{R}_2$ where $\mathfrak{R}_i$ is invariant under A. Let $\mu_i(\lambda)$ be the minimum polynomial of the transformation induced by A in $\mathfrak{R}_i$. Then since $x_i\pi(A)^k = 0$ for all x_i in $\mathfrak{R}_i$, $\mu_i(\lambda) \mid \pi(\lambda)^k$. Hence $\mu_i(\lambda) = \pi(\lambda)^{k_i}$, $k_i \leq k$. Now $x_i\pi(A)^{k_i} = 0$ for all x_i in $\mathfrak{R}_i$, and this implies that, if $\bar{k} = \max(k_1, k_2)$, then $\pi(A)^{\bar{k}} = 0$. Thus $\bar{k} = k$ and so we may suppose that $k_1 = k$. This means that the minimum polynomial of A in $\mathfrak{R}_1$ is $\pi(\lambda)^k$. Consequently there exists a vector f_1 in $\mathfrak{R}_1$ whose order is $\pi(\lambda)^k$. Then $\dim \{f_1\} = \deg \pi(\lambda)^k = \dim \mathfrak{R}$. Hence $\{f_1\} = \mathfrak{R}$ and since $\{f_1\} \subseteq \mathfrak{R}_1$, $\mathfrak{R}_1 = \mathfrak{R}$. Thus the decomposition $\mathfrak{R} = \mathfrak{R}_1 \oplus \mathfrak{R}_2$ is not proper.

Theorem 4. *A linear transformation A in $\mathfrak{R}$ over Φ is indecomposable if and only if it is cyclic and its minimum polynomial is a power of a prime.*

We suppose next that A is completely reducible. Then $\mathfrak{R} = \mathfrak{R}_1 \oplus \mathfrak{R}_2 \oplus \cdots \oplus \mathfrak{R}_s$ where the $\mathfrak{R}_i$ are invariant and irreducible. By Theorem 3 the minimum polynomial of A in $\mathfrak{R}_i$ is a prime $\pi_i(\lambda)$. It follows that the minimum polynomial $\mu(\lambda)$ of A in $\mathfrak{R}$ is the least common multiple of the $\pi_i(\lambda)$. Hence $\mu(\lambda)$ is a product of distinct primes. Conversely, let A be a linear transformation whose minimum polynomial $\mu(\lambda)$ is a product of distinct primes. Then each invariant factor has this form too. Hence the elementary divisor ideals are of the form $(\pi_i(\lambda))$, $\pi_i(\lambda)$ a prime. Now, we know that this implies that $\mathfrak{R}$ is a direct sum of cyclic subspaces $\mathfrak{R}_i = \{g_i\}$ where the order $\mu_{g_i}(\lambda) = \pi_i(\lambda)$. By Theorem 3 each $\mathfrak{R}_i$ is irreducible. Hence $\mathfrak{R}$ is completely reducible. This proves the following

Theorem 5. *A linear transformation A in $\mathfrak{R}$ over Φ is completely reducible if and only if its minimum polynomial is a product of distinct primes.*

EXERCISES

1. Let A be cyclic with minimum polynomial $\mu(\lambda)$. Show that, if $\mathfrak{S}$ is invariant under A, then $\mathfrak{S}$ is a cyclic subspace. Show that the invariant subspaces of $\mathfrak{R}$ can be put into 1–1 correspondence with the factors having leading coefficient 1 of $\mu(\lambda)$.
2. Let Φ be infinite and let $\mathfrak{R} = \{f_1\} \oplus \{f_2\}$ be a decomposition of $\mathfrak{R}$ relative to A such that $\mu_{f_1}(\lambda) = \mu_{f_2}(\lambda) = \pi(\lambda)$ a prime. Show that $\mathfrak{R}$ has an infinite number of subspaces invariant under A.
3. Prove that, if Φ is infinite, then the number of invariant subspaces relative to A is finite if and only if A is cyclic.

8. The primary components of a space relative to a linear transformation. The decomposition of a vector space into indecomposable subspaces relative to a linear transformation is not uniquely determined. On the other hand, as we shall show in this section, there is a decomposition which is not as refined as the one into indecomposable components but which has the important advantage of uniqueness.

Let $\mu(\lambda)$ be the minimum polynomial of A and let

$$\mu(\lambda) = \pi_1(\lambda)^{k_1}\pi_2(\lambda)^{k_2} \cdots \pi_s(\lambda)^{k_s} \tag{13}$$

be the factorization of $\mu(\lambda)$ into powers of distinct irreducible polynomials having leading coefficients equal 1. Let $\mathfrak{R}_i$ be the subspace of vectors $x_i \in \mathfrak{R}$ such that

$$x_i\pi_i(A)^{k_i} = 0. \tag{14}$$

Then we assert that the spaces $\mathfrak{R}_i$ are invariant relative to A and that

$$\mathfrak{R} = \mathfrak{R}_1 \oplus \mathfrak{R}_2 \oplus \cdots \oplus \mathfrak{R}_s. \tag{15}$$

The invariance of the $\mathfrak{R}_i$ is an immediate consequence of the fact that $\pi_i(A)^{k_i}$ commutes with A (cf. Ex. 1, p. 117). The proof of (15) can be obtained by using the results of the preceding section; however, we shall give an independent discussion that has some points of interest of its own.

We note first that the polynomials

$$\begin{aligned}(16)\quad \mu_i(\lambda) &= \mu(\lambda) \,/\, \pi_i(\lambda)^{k_i} = \\ &\pi_1(\lambda)^{k_1} \cdots \pi_{i-1}(\lambda)^{k_{i-1}}\pi_{i+1}(\lambda)^{k_{i+1}} \cdots \pi_s(\lambda)^{k_s}, \quad i = 1, 2, \cdots, s\end{aligned}$$

are relatively prime. Hence there exist polynomials $\phi_1(\lambda)$, $\phi_2(\lambda), \cdots, \phi_s(\lambda)$ such that

$$\phi_1(\lambda)\mu_1(\lambda) + \phi_2(\lambda)\mu_2(\lambda) + \cdots + \phi_s(\lambda)\mu_s(\lambda) = 1. \tag{17}$$

We can substitute A in this relation and obtain

$$\phi_1(A)\mu_1(A) + \phi_2(A)\mu_2(A) + \cdots + \phi_s(A)\mu_s(A) = 1. \tag{18}$$

Since $\mu_i(\lambda)\mu_j(\lambda)$ is divisible by $\mu(\lambda)$ if $i \neq j$,

$$\phi_i(A)\mu_i(A)\phi_j(A)\mu_j(A) = 0.$$

We set $E_i = \phi_i(A)\mu_i(A)$. Then this relation reads

$$E_iE_j = 0, \quad i \neq j \tag{19}$$

and (18) becomes

$$E_1 + E_2 + \cdots + E_s = 1. \tag{20}$$

If we multiply this relation by E_i and use (19), we obtain

$$E_i^2 = E_i. \tag{21}$$

Thus the E_i are orthogonal projections with sum 1 and therefore

$$\mathfrak{R} = \mathfrak{R}E_1 \oplus \mathfrak{R}E_2 \oplus \cdots \oplus \mathfrak{R}E_s.$$

Since the E_i are polynomials in A, they commute with A; hence the spaces $\mathfrak{R}E_i$ are invariant relative to A. We shall now show that $\mathfrak{R}E_i$ coincides with the space $\mathfrak{R}_i$ of vectors x_i such that $x_i\pi_i(A)^{k_i} = 0$. First let $y_i \varepsilon \mathfrak{R}E_i$. Then $y_i = uE_i = u\phi_i(A)\mu_i(A)$. Since $\mu_i(\lambda)\pi_i(\lambda)^{k_i} = \mu(\lambda)$, this gives

$$y_i\pi_i(A)^{k_i} = u\phi_i(A)\mu_i(A)\pi_i(A)^{k_i} = u\phi_i(A)\mu(A) = 0.$$

Hence $y_i \varepsilon \mathfrak{R}_i$. Conversely, let

$$x = x_1 + x_2 + \cdots + x_s,$$

where $x_j \varepsilon \mathfrak{R}E_j$, be a vector in $\mathfrak{R}_i$. Then

$$0 = x\pi_i(A)^{k_i} = x_1\pi_i(A)^{k_i} + x_2\pi_i(A)^{k_i} + \cdots + x_s\pi_i(A)^{k_i}$$

and since $x_j\pi_i(A)^{k_i} \varepsilon \mathfrak{R}E_j$ each

$$x_j\pi_i(A)^{k_i} = 0.$$

On the other hand, $x_j\pi_j(A)^{k_j} = 0$, and since $\pi_i(\lambda)$, $\pi_j(\lambda)$ are distinct if $i \neq j$, this implies that $x_j = 0$ for $j \neq i$. Hence $x = x_i \varepsilon \mathfrak{R}E_i$. This completes the proof that $\mathfrak{R}_i = \mathfrak{R}E_i$.

It is clear that the minimum polynomial of A acting in $\mathfrak{R}_i$ is of the form $\pi_i(\lambda)^{l_i}$, $l_i \leq k_i$. Moreover, if $x = x_1 + x_2 + \cdots + x_s$ where $x_i \varepsilon \mathfrak{R}_i$, and $\mu^*(\lambda) = \pi_1(\lambda)^{l_1}\pi_2(\lambda)^{l_2} \cdots \pi_s(\lambda)^{l_s}$, then $x_i\mu^*(A) = 0$ for all i. Hence $x\mu^*(A) = 0$ and $\mu^*(A) = 0$. Consequently $\mu(\lambda) \mid \mu^*(\lambda)$. Evidently this implies that $\mu(\lambda) = \mu^*(\lambda)$ and that $l_i = k_i$ for $i = 1, 2, \cdots, s$. This completes the proof of the following

Theorem 6. *Let $\mu(\lambda)$ be the minimum polynomial of the linear transformation A in $\mathfrak{R}$ over Φ and let* (13) *be the factorization of $\mu(\lambda)$ into prime powers. Then, if $\mathfrak{R}_i$ is defined to be the subspace of vectors x_i such that $x_i\pi_i(A)^{k_i} = 0$,*

$$\mathfrak{R} = \mathfrak{R}_1 \oplus \mathfrak{R}_2 \oplus \cdots \oplus \mathfrak{R}_s$$

and the minimum polynomial of the induced transformation in $\mathfrak{R}_i$ is $\pi_i(\lambda)^{k_i}$.

We shall call the spaces $\mathfrak{R}_i$ the *primary components* of $\mathfrak{R}$ relative to A. The projections E_i determined by the decomposition (15) will be called the *principal idempotent elements* of A.

We specialize now by assuming that Φ is algebraically closed. In this case the $\pi_i(\lambda)$ are of first degree, say, $\pi_i(\lambda) = \lambda - \rho_i$. We set

$$N_i = (A - \rho_i 1)E_i$$

so that

$$A = AE_1 + AE_2 + \cdots + AE_s = (\rho_1 E_1 + N_1) + (\rho_2 E_2 + N_2) + \cdots + (\rho_s E_s + N_s). \tag{22}$$

Evidently

$$N_i N_j = 0, \quad E_i N_j = 0 = N_j E_i \tag{23}$$

if $i \neq j$ and

$$E_i N_i = N_i = N_i E_i. \tag{24}$$

Also $N_i^{k_i} = (A - \rho_i 1)^{k_i} E_i = E_i (A - \rho_i 1)^{k_i}$ and if x is any vector then $xE_i \in \mathfrak{R}_i$. Furthermore $x_i(A - \rho_i 1)^{k_i} = 0$. This proves that

$$N_i^{k_i} = 0. \tag{25}$$

The nilpotent linear transformations N_i will be called the *principal nilpotent* elements of A. Like the E_i these linear transformations are polynomials in A.

EXERCISE

1. Let Φ be algebraically closed of characteristic 0 and suppose that $E_i, N_i, i = 1, 2, \cdots, s$, are the principal idempotent and principal nilpotent elements of A and that $A = \Sigma(\rho_i E_i + N_i)$. Show that, if $\phi(\lambda)$ is a polynomial, then

$$\phi(A) = \Sigma\left[\phi(\rho_i)E_i + \frac{\phi'(\rho_i)}{1!}N_i + \frac{\phi''(\rho_i)}{2!}N_i^2 + \cdots + \frac{\phi^{(k_i-1)}(\rho_i)}{(k_i-1)!}N_i^{k_i-1}\right].$$

9. Sets of commutative linear transformations. We suppose now that Ω is a set of commutative linear transformations in $\mathfrak{R}$ over Φ. Assume first that Ω is indecomposable. Then we shall show that the minimum polynomial of any $A \varepsilon \Omega$ is a prime power. For otherwise we can obtain a proper decomposition of $\mathfrak{R}$ as $\mathfrak{R}_1 \oplus \mathfrak{R}_2 \oplus \cdots \oplus \mathfrak{R}_s$ where the $\mathfrak{R}_i$ are the primary components relative to A. Now it is clear that if B is a linear transformation that commutes with A then $\mathfrak{R}_i B \subseteq \mathfrak{R}_i$. Hence the $\mathfrak{R}_i$ are invariant relative to Ω, and this contradicts the indecomposability of Ω.

Assume next that $\mathfrak{R}$ is irreducible. Then we assert that the minimum polynomial $\mu(\lambda)$ of every A is irreducible. For if $\pi(\lambda)$ is an irreducible factor of $\mu(\lambda)$, then the space $\mathfrak{S}$ of vectors y such that $y\pi(A) = 0$ is $\neq 0$. Evidently $\mathfrak{S}$ is invariant relative to Ω. Hence $\mathfrak{S} = \mathfrak{R}$ and $\pi(\lambda) = \mu(\lambda)$.

Now let Ω be an arbitrary commutative set of linear transformations. We first decompose $\mathfrak{R}$ as

$$\mathfrak{R} = \mathfrak{R}^{(1)} \oplus \mathfrak{R}^{(2)} \oplus \cdots \oplus \mathfrak{R}^{(h)}$$

where the $\mathfrak{R}^{(i)}$ are indecomposable. It is clear that the transformations induced by the $A \varepsilon \Omega$ in any invariant subspace and in any factor space commute. This holds in particular for the $\mathfrak{R}^{(i)}$. Our result in the indecomposable case, therefore, shows that the minimum polynomial of any A acting in $\mathfrak{R}^{(i)}$ is a prime power $\pi_i(\lambda)^{k_i}$. We now choose a composition series

$$0 \subset \mathfrak{R}_1{}^{(i)} \subset \mathfrak{R}_2{}^{(i)} \subset \cdots \subset \mathfrak{R}_{t_i}{}^{(i)} = \mathfrak{R}^{(i)} \tag{26}$$

for each $\mathfrak{R}^{(i)}$. Then each factor space $\mathfrak{R}_j{}^{(i)}/\mathfrak{R}_{j-1}{}^{(i)}$ is irreducible relative to the induced transformations. Hence the minimum polynomial of the transformation $\bar{A}$ induced by A in $\mathfrak{R}_j{}^{(i)}/\mathfrak{R}_{j-1}{}^{(i)}$ is irreducible. On the other hand, $\pi_i(\bar{A})^{k_i} = 0$. Hence the minimum polynomial of $\bar{A}$ is $\pi_i(\lambda)$.

We now choose a basis for $\mathfrak{R}^{(i)}$ corresponding to (26), that is, the first group of vectors is a basis for $\mathfrak{R}_1{}^{(i)}$, the second group supplement these to give a basis for $\mathfrak{R}_2{}^{(i)}$, etc. The bases thus determined for the different $\mathfrak{R}^{(i)}$ constitute a basis for $\mathfrak{R}$. With respect to this basis the matrix of A has the block form

$$(27) \qquad \begin{bmatrix} (\alpha_1) & & & \\ & (\alpha_2) & & \\ & & \ddots & \\ & & & (\alpha_h) \end{bmatrix}$$

where the block (α_i) that goes with $\mathfrak{R}^{(i)}$ has the form

$$(28) \qquad \begin{bmatrix} (\beta_1) & & & 0 \\ & (\beta_2) & & \\ & & \ddots & \\ * & & & (\beta_{t_i}) \end{bmatrix}.$$

The matrix (β_j) is a matrix of the induced linear transformation $\bar{A}$. Hence its minimum polynomial is $\pi_i(\lambda)$.

Suppose now that Φ is algebraically closed. Then $\pi_i(\lambda) = \lambda - \rho_i$. Hence $\bar{A} = \rho_i 1$ is a scalar multiplication in $\mathfrak{R}_j{}^{(i)}/\mathfrak{R}_{j-1}{}^{(i)}$. Since any subspace is invariant relative to a scalar multiplication, it follows that $\mathfrak{R}_j{}^{(i)}/\mathfrak{R}_{j-1}{}^{(i)}$ is one dimensional. Hence the matrices (β_j) are one-rowed. Our final result can therefore be stated as the following theorem on matrices:

Theorem 7. *Let Φ be an algebraically closed field and let ω be a set of commutative matrices belonging to Φ_n. Then there exists a non-singular matrix (μ) in Φ_n such that*

$$(\mu)(\alpha)(\mu)^{-1} = \begin{bmatrix} (\alpha_1) & & & \\ & (\alpha_2) & & \\ & & \ddots & \\ & & & (\alpha_n) \end{bmatrix}$$

where

$$(\alpha_i) = \begin{bmatrix} \rho_i & & & 0 \\ & \rho_i & & \\ & & \ddots & \\ * & & & \rho_i \end{bmatrix}$$

for all (α).

EXERCISES

1. Let Φ be the field of complex numbers and let

$$(\alpha) = \begin{bmatrix} 1 & 0 & 0 \\ 0 & 0 & -1 \\ 0 & 1 & 2 \end{bmatrix} \qquad (\beta) = \begin{bmatrix} 0 & 1 & 1 \\ -1 & 1 & -1 \\ 1 & 1 & 3 \end{bmatrix}.$$

Verify that $(\alpha)(\beta) = (\beta)(\alpha)$ and illustrate Theorem 7 with this pair of matrices.

2. (Ingraham) Let A be a matrix in the block form

$$\begin{bmatrix} A_{11} & A_{12} & \cdots & A_{1m} \\ A_{21} & A_{22} & \cdots & A_{2m} \\ \cdot & \cdot & \cdots & \cdot \\ A_{m1} & A_{m2} & \cdots & A_{mm} \end{bmatrix}$$

where the A_{ij} are $r \times r$ matrices with elements in an algebraically closed field. Assume that the A_{ij} commute with each other and define

$$\det_R A = \Sigma \pm A_{1i_1}A_{2i_2} \cdots A_{mi_m}$$

where the sum is taken over all permutations $i_1, i_2, \cdots, i_m$ of $1, 2, \cdots, m$ and the sign is $+$ or $-$ according as the permutation is even or odd. Use Theorem 7 to prove the following transitivity property of determinants

$$\det(\det_R A) = \det A.$$

(This holds for arbitrary base fields. Compare Sec. 9, Chapter VII.)

3. (Schur) Show that the maximum number of linearly independent commutative matrices which can be chosen in Φ_n is $\left[\frac{n^2}{4}\right] + 1$, where, in general, $[\alpha]$ denotes the greatest integer in the real number α.

Chapter V

BILINEAR FORMS

This chapter is devoted to the study of certain types of functions, called bilinear, which are defined for pairs of vectors (x, y') where x is in a left vector space $\mathfrak{R}$ and y' is in a right vector space $\mathfrak{R}'$. The values of $g(x, y')$ are assumed to belong to Δ, and the functions of one variable $g_x(y') = g(x, y')$ and $g_{y'}(x) = g(x, y')$ obtained by fixing the other variable are linear. Of particular interest are the non-degenerate bilinear forms. These determine 1–1 linear transformations of $\mathfrak{R}'$ onto the space of linear functions on $\mathfrak{R}$. Consequently, if A is a linear transformation in $\mathfrak{R}$, there is a natural way of associating with it a transposed linear transformation in $\mathfrak{R}'$.

If the division ring Δ possesses an anti-automorphism, then any left vector space $\mathfrak{R}$ over Δ can also be regarded as a right vector space over Δ. Hence in this case one has the possibility of defining bilinear forms connecting the space with itself. Such forms are called scalar products. Their study is equivalent to the study of a certain type of equivalence for matrices called cogredience. The most important types of scalar products are the hermitian, symmetric and alternate scalar products. We shall obtain canonical matrices for such forms, and, in certain special cases that are of interest in elementary geometry, complete solutions of the cogredience problem will be given. We shall also prove Witt's theorem for hermitian forms and apply it to define the concept of signature for such forms over an arbitrary division ring of characteristic $\neq 2$.

1. Bilinear forms. If $\mathfrak{R}$ is a (left) vector space over Δ, we have defined a linear function as a mapping $x \to f(x) \in \Delta$ such that

$$f(x+y) = f(x) + f(y), \quad f(\alpha x) = \alpha f(x). \tag{1}$$

These mappings form a right vector space $\mathfrak{R}^*$ relative to the compositions

$$(f+g)(x) = f(x) + g(x), \quad (f\alpha)(x) = f(x)\alpha. \tag{2}$$

As we have seen, these definitions imply that, for each fixed vector $x \in \mathfrak{R}$, the mapping $f \to x(f) \equiv f(x)$ is a linear function in $\mathfrak{R}^*$. Because of this symmetry it is natural to regard $f(x) = x(f)$ as a function of the pair $x \in \mathfrak{R}, f \in \mathfrak{R}^*$. We therefore denote the value $f(x)$ as $s(x, f)$, and the above equations now read

$$s(x+y, f) = s(x, f) + s(y, f), \quad s(\alpha x, f) = \alpha s(x, f) \tag{3}$$

$$s(x, f+g) = s(x, f) + s(x, g), \quad s(x, f\alpha) = s(x, f)\alpha. \tag{4}$$

We shall now generalize this situation by assuming that $\mathfrak{R}'$ is any right vector space over Δ. Then a function $g(x, y')$ defined for all pairs (x, y'), x in $\mathfrak{R}$, y' in $\mathfrak{R}'$, with values $g(x, y')$ in Δ is called a *bilinear form* if

$$g(x_1 + x_2, y') = g(x_1, y') + g(x_2, y'), \tag{5}$$
$$g(\alpha x, y') = \alpha g(x, y'),$$

$$g(x, y_1' + y_2') = g(x, y_1') + g(x, y_2'), \tag{6}$$
$$g(x, y'\alpha) = g(x, y')\alpha.$$

Clearly $s(x, f)$ is a bilinear form for the space $\mathfrak{R}$ and its conjugate space $\mathfrak{R}^*$.

On the other hand, as we proceed to show, the conjugate space $\mathfrak{R}^*$ can be used to give an alternative definition of a bilinear form. First, let $g(x, y')$ be a bilinear form connecting the left vector space $\mathfrak{R}$ and the right vector space $\mathfrak{R}'$. We fix the vector y' and regard $g(x, y')$ as a function of x. Accordingly we write $g(x, y') = g_{y'}(x)$. Then, by (5), $g_{y'}(x)$ is linear, that is, it belongs to the conjugate space $\mathfrak{R}^*$. Now let y' vary and consider the mapping

$y' \to g_{y'} \varepsilon \mathfrak{R}^*$. We denote this mapping as R and we observe that, by (6), R is linear, since

$$g_{y_1'+y_2'}(x) = g(x, y_1' + y_2') = g(x, y_1') + g(x, y_2')$$

$$= g_{y_1'}(x) + g_{y_2'}(x),$$

$$g_{y'\alpha}(x) = g(x, y'\alpha) = g(x, y')\alpha = g_{y'}(x)\alpha.$$

In a similar manner, if we fix x, then the function $g_x(y') = g(x, y')$ is linear in y'. Hence g_x is in the conjugate space $(\mathfrak{R}')^*$ of $\mathfrak{R}'$. The mapping $L: x \to g_x$ is a linear transformation of $\mathfrak{R}$ into $(\mathfrak{R}')^*$.

Conversely, suppose that we are given a linear transformation $R: y' \to g_{y'}(x)$ of the right vector space $\mathfrak{R}'$ into the space $\mathfrak{R}^*$ of linear functions on $\mathfrak{R}$. Then we can regard $g_{y'}(x)$ as $g(x, y')$, a function of (x, y'), x in $\mathfrak{R}$, y' in $\mathfrak{R}'$, and we can verify that this function is a bilinear form. Thus we see that an equivalent definition for the concept of a bilinear form is that of a linear transformation of the right vector space $\mathfrak{R}'$ into the conjugate $\mathfrak{R}^*$ of $\mathfrak{R}$. Similarly, we could also say that a bilinear form is a linear transformation of $\mathfrak{R}$ into the conjugate space $(\mathfrak{R}')^*$ of $\mathfrak{R}'$. The original definition, however, has the advantage of symmetry over the present formulations, and it will be given preference in the sequel.

EXERCISE

1. Show that, if $g(x, y')$ is a bilinear form and A is a linear transformation in $\mathfrak{R}$, then $g(xA, y')$ is a bilinear form.

2. Matrices of a bilinear form. Suppose now that $\mathfrak{R}$ and $\mathfrak{R}'$ are finite dimensional and that $(e_1, e_2, \cdots, e_n)$, $(f_1', f_2', \cdots, f_{n'}')$, respectively, are bases for these spaces. We shall call the matrix

$$\begin{bmatrix} g(e_1,f_1') & g(e_1,f_2') & \cdots & g(e_1,f_{n'}') \\ g(e_2,f_1') & g(e_2,f_2') & \cdots & g(e_2,f_{n'}') \\ \cdot & \cdot & \cdots & \cdot \\ g(e_n,f_1') & g(e_n,f_2') & \cdots & g(e_n,f_{n'}') \end{bmatrix} \tag{7}$$

the matrix of the bilinear form g relative to the given bases. The form is completely determined by this matrix; for if x and y'

are arbitrary in $\mathfrak{R}$ and $\mathfrak{R}'$, we can write $x = \sum_1^n \xi_i e_i$, $y' = \sum_1^{n'} f_j' \eta_j$, and we obtain

$$g(x, y') = g(\Sigma \xi_i e_i, \Sigma f_j' \eta_j) = \Sigma \xi_i g(e_i, f_j') \eta_j.$$

Hence $g(x, y')$ is known from the representations of x and y' and from the entries $g(e_i, f_j')$ of (7). Also it is clear that, if (β_{ij}) is any $n \times n'$ matrix with elements in Δ, then there exists a bilinear form $h(x, y')$ which has this matrix as its matrix relative to $(e_1, e_2, \cdots, e_n)$, $(f_1', f_2', \cdots, f_{n'}')$; for we can define

$$g(x, y') = \Sigma \xi_i \beta_{ij} \eta_j$$

and it is easy to verify that this function is bilinear. Since $g(e_i, f_j') = \beta_{ij}$ the matrix of g is the given one.

We consider now the effect on the matrix (β) of $g(x, y')$ of changes of bases in the two spaces. Let $(u_1, u_2, \cdots, u_n)$ be a second basis in $\mathfrak{R}$, $u_i = \Sigma \mu_{ij} e_j$ where $(\mu) \varepsilon L(\Delta, n)$, and let $(v_1', v_2', \cdots, v_{n'}')$ be a second basis in $\mathfrak{R}'$, $v_k' = \Sigma f_l' \nu_{lk}$, $(\nu) \varepsilon L(\Delta, n')$. Then

$$g(u_i, v_k') = g(\Sigma \mu_{ij} e_j, \Sigma f_l' \nu_{lk}) = \Sigma \mu_{ij} \beta_{jl} \nu_{lk}$$

so that the new matrix is $(\mu)(\beta)(\nu)$ a matrix equivalent to (β). We remark that in the commutative case this relation is usually encountered in a slightly different form. Here one generally considers all vector spaces as left vector spaces. Then the change of basis in $\mathfrak{R}'$ reads $v_k' = \Sigma \nu_{kl} f_l'$, and the new matrix of the bilinear form is $(\mu)(\beta)(\nu)'$ where $(\nu)'$ is the transpose of the matrix $(\nu) = (\nu_{kl})$.

We return now to the general case of an arbitrary Δ. We have proved in Chapter II (p. 45) in connection with the theory of linear transformations that, if (β) is any rectangular matrix with elements in Δ, then there exist non-singular square matrices (μ) and (ν) such that

$$(\mu)(\beta)(\nu) = \text{diag}\,\{1, \cdots, 1, 0, \cdots, 0\}. \tag{8}$$

This matrix result yields the following fundamental theorem on bilinear forms.

Theorem 1. *If $g(x, y')$ is a bilinear form connecting the vector spaces $\mathfrak{R}$ and $\mathfrak{R}'$, then there exist bases $(u_1, u_2, \cdots, u_n)$, $(v_1', v_2', \cdots, v_{n'}')$ for these spaces such that*

$$(9)\qquad \begin{aligned} g(u_i, v_j') &= \delta_{ij} \quad \text{if} \quad i, j = 1, 2, \cdots, r \\ g(u_i, v_j') &= 0 \quad \text{if} \quad i > r \quad \text{or} \quad j > r. \end{aligned}$$

Evidently the number r is the rank (row or column) of the matrix (β) of $g(x, y')$. An abstract characterization of this number will be given in the next section.

3. Non-degenerate forms. If $g(x, y')$ is a bilinear form, the totality of vectors $z \, \varepsilon \, \mathfrak{R}$ such that $g(z, y') = 0$ for all $y' \, \varepsilon \, \mathfrak{R}'$ is a subspace $\mathfrak{N}$ which we shall call the *left radical* of g. Similarly we define the right radical $\mathfrak{N}'$ as the subspace of $\mathfrak{R}'$ of vectors z' such that $g(x, z') = 0$ for all x. It is clear from the definition that the left (right) radical is just the null space of the linear transformation L (R) of $\mathfrak{R}$ $(\mathfrak{R}')$ into the space of linear functions on $\mathfrak{R}'$ $(\mathfrak{R})$. If $(f_1', f_2', \cdots, f_{n'}')$ is a basis for $\mathfrak{R}'$ and $z \, \varepsilon \, \mathfrak{N}$, then $g(z, f_j') = 0$ for $j = 1, 2, \cdots, n'$. Moreover the n' equations $g(z, f_j') = 0$ imply that $g(z, x') = 0$ for all x'. Hence these conditions are also sufficient that $z \, \varepsilon \, \mathfrak{N}$. Now let $(e_1, e_2, \cdots, e_n)$ be a basis for $\mathfrak{R}$ and write $z = \sum_1^n \zeta_i e_i$. Then

$$(10)\quad 0 = g(z, f_j') = g(\Sigma \zeta_i e_i, f_j') = \sum_{i=i}^{n} \zeta_i \beta_{ij}, \quad j = 1, 2, \cdots, n',$$

are the conditions that $z = \Sigma \zeta_i e_i \, \varepsilon \, \mathfrak{N}$. It follows that dim $\mathfrak{N}$ is the maximum number of linearly independent solutions of (10). Hence dim $\mathfrak{N} = n - r$ where r is the rank of (β). Similarly we see that dim $\mathfrak{N}' = n' - r$. We remark also that, if we use the normalized bases given in Theorem 1, then we see that $\mathfrak{N} = [u_{r+1}, u_{r+2}, \cdots, u_n]$ and $\mathfrak{N}' = [v_{r+1}', v_{r+2}', \cdots, v_{n'}']$. For it is clear that the u_j with $j > r$ are in $\mathfrak{N}$. Moreover, if $\Sigma \gamma_i u_i \, \varepsilon \, \mathfrak{N}$, then, in particular, for $k \leq r$,

$$0 = (\Sigma \gamma_i u_i, v_k') = \gamma_k.$$

Hence the vector has the form $\sum_{r+1}^{n} \gamma_j u_j$, and therefore it belongs to $[u_{r+1}, u_{r+2}, \cdots, u_n]$. This proves the assertion about $\mathfrak{N}$. A similar argument applies to $\mathfrak{N}'$.

We know that $\dim \mathfrak{R}L + \dim \mathfrak{N} = n$ and $\dim \mathfrak{R}'R + \dim \mathfrak{N}' = n'$. It follows that $\dim \mathfrak{R}L = r = \dim \mathfrak{R}'R$.

The bilinear form g is said to be *non-degenerate* if $\mathfrak{N} = 0$ and $\mathfrak{N}' = 0$. Thus g is non-degenerate if the only vector z in $\mathfrak{R}$ such that $g(z, y') = 0$ for all y' is $z = 0$ and the only vector z' in $\mathfrak{R}'$ such that $g(x, z') = 0$ for all x is $z' = 0$. If g is non-degenerate, then the mappings L and R are 1–1. Also since $\dim \mathfrak{N} = 0 = \dim \mathfrak{N}'$, then $n = r = n'$. Thus we see that, if $\mathfrak{R}$ and $\mathfrak{R}'$ are connected by a non-degenerate bilinear form, then these spaces have the same dimensionality n. We recall also that a space and its space of linear functions have the same dimensionality. Hence the mappings L and R are 1–1 linear transformations of vector spaces into vector spaces of the same dimension. It follows that these are mappings *onto* the corresponding spaces. The relations $n = r = n'$ are also sufficient that the bilinear form be non-degenerate; for they clearly imply that $\dim \mathfrak{N} = n - r = 0$ and $\dim \mathfrak{N}' = n' - r = 0$. Our results can be summarized in the following

Theorem 2. *Necessary and sufficient conditions that a bilinear form $g(x, y')$ connecting $\mathfrak{R}$ and $\mathfrak{R}'$ be non-degenerate are*: 1) *$\mathfrak{R}$ and $\mathfrak{R}'$ have the same dimensionality*; 2) *the matrix of the form relative to any pair of bases is non-singular. If g is non-degenerate and $f_1(y')$ is a linear function on $\mathfrak{R}'$, then there exists one and only one vector x in $\mathfrak{R}$ such that $g(x, y') = f_1(y')$ holds for all y'. Also if $f_2(x)$ is any linear function on $\mathfrak{R}$, then there is a unique vector y' in $\mathfrak{R}'$ such that $g(x, y') = f_2(x)$ holds for all x.*

If two vector spaces $\mathfrak{R}$ and $\mathfrak{R}'$ are connected by a non-degenerate bilinear form g, then we shall say that these spaces are *dual* relative to g. Suppose that this is the case and let $(e_1, e_2, \cdots, e_n)$ be a given basis in $\mathfrak{R}$. We choose a basis $(f_1', f_2', \cdots, f_n')$ for $\mathfrak{R}'$, and we obtain the non-singular matrix (β) where $\beta_{ij} = g(e_i, f_j')$. Now let (ν) be any non-singular matrix in Δ_n and let $v_k' = \Sigma f_i'\nu_{ik}$ be the corresponding new basis in $\mathfrak{R}'$. Then we have seen that the matrix of $g(x, y')$ relative to the pair of bases $(e_1, e_2, \cdots, e_n)$, $(v_1', v_2', \cdots, v_n')$ is $(\beta)(\nu)$. Thus there exists a uniquely determined basis $(e_1', e_2', \cdots, e_n')$ for $\mathfrak{R}'$ such that the matrix of

g relative to $(e_1, e_2, \cdots, e_n)$, $(e_1', e_2', \cdots, e_n')$ is the identity matrix, that is,

$$g(e_i, e_j') = \delta_{ij}, \quad i, j = 1, 2, \cdots, n. \tag{11}$$

The e_i' are obtained by taking $(\nu) = (\beta)^{-1}$. We call the bases $(e_1, e_2, \cdots, e_n)$, $(e_1', e_2', \cdots, e_n')$ *complementary* bases for the dual spaces. Our argument shows that any basis in $\mathfrak{R}$ ($\mathfrak{R}'$) determines a unique complementary basis in $\mathfrak{R}'$ ($\mathfrak{R}$).

Complementary bases have been used in our study of the notion of incidence between subspaces of $\mathfrak{R}$ and of its conjugate space $\mathfrak{R}^*$. These spaces are dual relative to the fundamental bilinear form $s(x, f) \equiv f(x)$; for if $f(x) = 0$ for all x, then $f = 0$ by definition, and if $x \neq 0$, then there exists a linear function f such that $f(x) \neq 0$. We can carry over to the general case the results on incident spaces. Thus let $\mathfrak{R}$ and $\mathfrak{R}'$ be dual relative to $g(x, y')$. Then if $\mathfrak{S}$ is a subspace of $\mathfrak{R}$, we define the space $j(\mathfrak{S})$ *incident* to $\mathfrak{S}$ to be the subspace of $\mathfrak{R}'$ of vectors y' such that $g(x, y') = 0$ for all $x \in \mathfrak{S}$. In a similar manner we define $j(\mathfrak{S}')$ for a subspace $\mathfrak{S}'$ of $\mathfrak{R}'$. We can prove as in the special case of $\mathfrak{R}$ and $\mathfrak{R}^*$ (p. 55) that $\dim j(\mathfrak{S}) = n - \dim \mathfrak{S}$ and $\dim j(\mathfrak{S}') = n - \dim \mathfrak{S}'$. These relations, together with the obvious relations $j(j(\mathfrak{S})) \supseteq \mathfrak{S}$, $j(j(\mathfrak{S}')) \supseteq \mathfrak{S}'$, imply that

$$j(j(\mathfrak{S})) = \mathfrak{S}, \quad j(j(\mathfrak{S}')) = \mathfrak{S}'.$$

Also, as in the special case of the space of linear functions we can prove that the mapping $\mathfrak{S} \to j(\mathfrak{S})$ is an anti-automorphism of the lattice of subspaces of $\mathfrak{R}$ onto the lattice of subspaces of $\mathfrak{R}'$. The inverse of this anti-automorphism is, of course, the mapping $\mathfrak{S}' \to j(\mathfrak{S}')$. We leave it as an exercise to the reader to fill in the details of this discussion.

EXERCISE

1. Let $\mathfrak{R}$ and $\mathfrak{R}'$ be connected by the bilinear form $g(x, y')$ and let $\mathfrak{N}$ and $\mathfrak{N}'$ be the radicals determined by this form. If $x + \mathfrak{N} \in \mathfrak{R}/\mathfrak{N}$ and $y' + \mathfrak{N}' \in \mathfrak{R}'/\mathfrak{N}'$, set $g(x + \mathfrak{N}, y' + \mathfrak{N}') = g(x, y')$. Show that this defines a non-degenerate bilinear form for the factor spaces $\mathfrak{R}/\mathfrak{N}$, $\mathfrak{R}'/\mathfrak{N}'$.

4. Transpose of a linear transformation relative to a pair of bilinear forms. Let $\mathfrak{R}_i$, $i = 1, 2$, be a left vector space and let

$\mathfrak{R}_i'$ be dual to $\mathfrak{R}_i$ relative to the bilinear form $g_i(x_i, y_i')$. If A is a linear transformation of $\mathfrak{R}_1$ into $\mathfrak{R}_2$, then we know how to define the transpose A^* as a linear transformation of the conjugate space $\mathfrak{R}_2^*$ into the conjugate space $\mathfrak{R}_1^*$. If $f \varepsilon \mathfrak{R}_2^*$, then fA^* is, by definition, the resultant Af. We can now make use of the equivalences R_i of $\mathfrak{R}_i'$ onto $\mathfrak{R}_i^*$ to define a linear transformation of $\mathfrak{R}_2'$ into $\mathfrak{R}_1'$, namely, we set

$$A' = R_2 A^* R_1^{-1}. \tag{12}$$

Clearly this product can be defined since we have the following scheme of linear mappings:

$$\begin{aligned} R_2: &\quad \mathfrak{R}_2' \to \mathfrak{R}_2^* \\ A^*: &\quad \mathfrak{R}_2^* \to \mathfrak{R}_1^* \\ R_1^{-1}: &\quad \mathfrak{R}_1^* \to \mathfrak{R}_1'. \end{aligned}$$

Also it is clear that A' is a linear transformation of $\mathfrak{R}_2'$ into $\mathfrak{R}_1'$. We shall call this transformation the *transpose* of A (*relative to the bilinear forms* g_1 *and* g_2).

We determine next the form of A'. Let $y_2' \varepsilon \mathfrak{R}_2'$. Then $y_2'R_2$ is the linear function $g_2(x_2, y_2')$ on the space $\mathfrak{R}_2$. Moreover, $y_2'R_2A^*$ is the linear function f_1 on $\mathfrak{R}_1$ such that

$$f_1(x_1) = g_2(x_1A, y_2'). \tag{13}$$

Finally $y_2'R_2A^*R_1^{-1}$ is the vector y_1' of $\mathfrak{R}_1'$ such that

$$f_1(x_1) = g_2(x_1A, y_2') = g_1(x_1, y_1') \tag{14}$$

holds for all x_1 in $\mathfrak{R}_1$. The mapping A' sends y_2' into y_1' and, according to the above equation, $y_1' = y_2'A'$ is the uniquely determined vector of $\mathfrak{R}_1'$ such that

$$g_1(x_1, y_2'A') = g_2(x_1A, y_2') \tag{15}$$

holds for all x_1 in $\mathfrak{R}_1$.

Since the notion of duality is a symmetric one, we can interchange the roles of $\mathfrak{R}_i$ and $\mathfrak{R}_i'$ in the foregoing discussion. Thus if A' is a linear transformation of $\mathfrak{R}_2'$ into $\mathfrak{R}_1'$, then we define its transpose (relative to the given forms) to be the mapping $A'' =$

$L_1A'^*L_2^{-1}$ where L_1 is the fundamental equivalence of $\mathfrak{R}_1$ onto $\mathfrak{R}_1'^*$, A'^* is the transpose of A', and L_2 is the equivalence of $\mathfrak{R}_2$ onto $\mathfrak{R}_2'^*$. In analogy with (15) we can verify that x_1A'' is the uniquely determined vector such that

$$g_2(x_1A'', y_2') = g_1(x_1, y_2'A') \tag{16}$$

holds for all y_2' in $\mathfrak{R}_2'$. Now it is clear from (15) and (16) that, if A' is the transpose of A, then A is the transpose of A'. Also if A'' is the transpose of A', then A' is the transpose of A''. Thus the two correspondences $A \to A'$, $A' \to A''$ are inverses of each other. Hence $A \to A'$ is a 1–1 mapping of the set $\mathfrak{L}(\mathfrak{R}_1, \mathfrak{R}_2)$ of linear transformations $\mathfrak{R}_1$ into $\mathfrak{R}_2$ onto the set $\mathfrak{L}(\mathfrak{R}_2', \mathfrak{R}_1')$ of linear transformations of $\mathfrak{R}_2'$ into $\mathfrak{R}_1'$. Of course, this can also be seen directly from the definition of the transpose and the properties of the mapping $A \to A'$.

The algebraic properties of the mapping $A \to A'$ can also be deduced from the previously established properties of the mapping $A \to A^*$. We state these without proofs: 1) If B is a second linear transformation of $\mathfrak{R}_1$ into $\mathfrak{R}_2$, then

$$(A + B)' = A' + B'. \tag{17}$$

2) If $\mathfrak{R}_3$ and $\mathfrak{R}_3'$ are dual relative to the non-degenerate form $g_3(x_3, y_3')$ and C is a linear transformation of $\mathfrak{R}_2$ into $\mathfrak{R}_3$, then we denote the transpose of C relative to g_2, g_3 by C'. Also we denote the transpose of AC relative to g_1, g_3 by $(AC)'$. Then we have the relation

$$(AC)' = C'A'. \tag{18}$$

3) As usual if we deal with a single vector space $\mathfrak{R}$ and its dual relative to $g(x, y')$, then 1) and 2) show that the mapping $A \to A'$, the transpose of A relative to g, is an anti-isomorphism of the ring of linear transformations in $\mathfrak{R}$ onto the ring of linear transformations in $\mathfrak{R}'$.

Finally, we wish to determine the relation between the matrices of a linear transformation and its transpose. Again let A be a linear transformation of $\mathfrak{R}_1$ into $\mathfrak{R}_2$ and let $\mathfrak{R}_1'$ and $\mathfrak{R}_2'$ be the duals relative to the forms g_1 and g_2 respectively. Let $(e_1, e_2, \cdots, e_{n_1})$ be a basis for $\mathfrak{R}_1$; $(e_1', e_2', \cdots, e_{n_1}')$, the complementary basis in $\mathfrak{R}_1'$. Similarly let $(f_1, f_2, \cdots, f_{n_2})$ be a basis for $\mathfrak{R}_2$ and $(f_1', f_2', \cdots, f_{n_2}')$ the complementary basis. Suppose that

$$e_iA = \sum_{k=1}^{n_2} \alpha_{ik} f_k, \qquad i = 1, 2, \cdots, n_1$$

$$f_l'A' = \sum_{j=1}^{n_1} e_j' \alpha_{jl}', \qquad l = 1, 2, \cdots, n_2.$$

Then the conditions

$$g_2(e_iA, f_l') = g_1(e_i, f_l'A')$$

yield the relations

$$\alpha_{il} = \alpha_{il}'$$

for the matrices. Thus as in the special case of spaces and the dual spaces of linear functions, we have the rule: *If complementary bases are used in the dual spaces, then the matrix of a transformation A and of its transpose A' are equal.*

EXERCISE

1. Supply proofs of statements 1), 2) and 3) above.

5. Another relation between linear transformations and bilinear forms. We assume again that the spaces $\mathfrak{R}_i$ and $\mathfrak{R}_i'$, $i = 1, 2$, are dual relative to g_i. Let u' and v be fixed vectors in $\mathfrak{R}_1'$ and $\mathfrak{R}_2$ respectively. Then if x ranges over $\mathfrak{R}_1$, the mapping

$$x \rightarrow g_1(x, u')v \tag{19}$$

is a linear transformation of $\mathfrak{R}_1$ into $\mathfrak{R}_2$. This is clear from the properties of the bilinear form. Since our transformation is determined by the pair u', v, we shall denote it as the "product" $u' \times v$. More generally if $u_1', u_2', \cdots, u_m'$ are in $\mathfrak{R}_1'$ and $v_1, v_2, \cdots, v_m$ are in $\mathfrak{R}_2$, then the mapping $u_1' \times v_1 + u_2' \times v_2 + \cdots + u_m' \times v_m$ is the linear transformation

$$x \rightarrow g_1(x, u_1')v_1 + g_1(x, u_2')v_2 + \cdots + g_1(x, u_m')v_m.$$

We show next that any linear transformation of $\mathfrak{R}_1$ into $\mathfrak{R}_2$ has this form. For let $A \in \mathfrak{L}(\mathfrak{R}_1, \mathfrak{R}_2)$ and let $(v_1, v_2, \cdots, v_r)$ be a basis for the rank space $\mathfrak{R}_1A$. Then if x is any vector in $\mathfrak{R}_1$, xA

can be written in one and only one way in the form

$$xA = \phi_1 v_1 + \phi_2 v_2 + \cdots + \phi_r v_r \tag{20}$$

where the coefficients ϕ_i are in Δ. Since these coefficients are uniquely determined by x, we may consider them as functions of x, and accordingly we write $\phi_i = \phi_i(x)$. Thus (20) can be rewritten as

$$xA = \phi_1(x)v_1 + \phi_2(x)v_2 + \cdots + \phi_r(x)v_r. \tag{20'}$$

If y is a second vector in $\mathfrak{R}_1$, then

$$\Sigma\phi_i(x + y)v_i = (x + y)A = xA + yA = \Sigma\phi_i(x)v_i + \Sigma\phi_i(y)v_i.$$

Hence $\phi_i(x + y) = \phi_i(x) + \phi_i(y)$. Also by (20′), if $\alpha \in \Delta$, then

$$(\alpha x)A = \alpha(xA) = \alpha\phi_1(x)v_1 + \alpha\phi_2(x)v_2 + \cdots + \alpha\phi_r(x)v_r,$$

and this implies that $\phi_i(\alpha x) = \alpha\phi_i(x)$. Hence the functions $\phi_i(x)$ are linear. Since $g_1(x, y')$ is a non-degenerate form, it follows that there exist vectors u_i' in $\mathfrak{R}_1'$ such that $\phi_i(x) = g_1(x, u_i')$ holds for all x. Thus

$$xA = g_1(x, u_1')v_1 + g_1(x, u_2')v_2 + \cdots + g_1(x, u_r')v_r$$

and $A = u_1' \times v_1 + u_2' \times v_2 + \cdots + u_r' \times v_r$ as required.

The pairs of vectors $(u_1', v_1), (u_2', v_2), \cdots, (u_m', v_m)$ also define a linear transformation A' of $\mathfrak{R}_2'$ into $\mathfrak{R}_1'$. If $x' \in \mathfrak{R}_2'$, then we define

$$x'A' = u_1'g_2(v_1, x') + u_2'g_2(v_2, x') + \cdots + u_m'g_2(v_m, x').$$

We shall denote this transformation as

$$u_1' \times' v_1 + u_2' \times' v_2 + \cdots + u_m' \times' v_m.$$

As is indicated by our notation the transformations A and A' are transposes relative to the bilinear forms g_1 and g_2; for if x is any vector in $\mathfrak{R}_1$ and x' is any vector in $\mathfrak{R}_2'$, then

$$g_1(x, x'A') = g_1(x, \sum_i u_i'g_2(v_i, x')) = \sum_i g_1(x, u_i')g_2(v_i, x')$$

and

$$g_2(xA, x') = g_2\left(\sum_i g_1(x, u_i')v_i, x'\right) = \sum_i g_1(x, u_i')g_2(v_i, x').$$

Hence $g_1(x, x'A') = g_2(xA, x')$, and this proves our assertion.

We consider again the representation of the linear transformation A as $\sum_i u_i' \times v_i$. It is clear from the definition of the product $u' \times v$ that

$$(u_1' + u_2') \times v = u_1' \times v + u_2' \times v, \tag{21}$$

$$u' \times (v_1 + v_2) = u' \times v_1 + u' \times v_2$$

$$u'\alpha \times v = u' \times \alpha v. \tag{22}$$

Hence if we express the u_i' as $u_i' = \Sigma e_j'\beta_{ji}$, then $\Sigma u_i' \times v_i = \Sigma e_j'\beta_{ji} \times v_i = \Sigma e_j' \times \beta_{ji}v_i = \Sigma e_j' \times w_j$ where $w_j = \Sigma\beta_{ji}v_i$. Thus if we can write the u_i', $i = 1, 2, \cdots, m$, in terms of fewer than m vectors e_j', then we can obtain expressions for $A = \Sigma u_i' \times v_i$ as a sum of fewer than m products $e_j' \times w_j$. Similarly if we can write the v_i as $v_i = \Sigma\alpha_{ij}f_j$ where the number of f's is less than m, then we can obtain an expression for A as a sum of less than m products.

Now let $A = \sum_1^r u_i' \times v_i$ be an expression for A for which r is minimal. Then our remarks show that the sets $(u_1', u_2', \cdots, u_r')$ and $(v_1, v_2, \cdots, v_r)$ are linearly independent. Now there exist vectors $(u_1, u_2, \cdots, u_r)$ in $\mathfrak{R}_1$ such that

$$g_1(u_i, u_j') = \delta_{ij}, \quad i, j = 1, 2, \cdots, r.$$

Then

$$u_iA = \sum_j g_1(u_i, u_j')v_j = v_i$$

and $[v_1, v_2, \cdots, v_r] \subseteq \mathfrak{R}_1A$. On the other hand, for any x

$$xA = \sum_j g_1(x, u_j')v_j \,\varepsilon\, [v_1, v_2, \cdots, v_r].$$

Hence $\mathfrak{R}_1A = [v_1, v_2, \cdots, v_r]$, and r is the rank of A. If we use the form of the transpose determined above, we see that the rank space $\mathfrak{R}_2'A' = [u_1', u_2', \cdots, u_r']$. We remark that this gives still another proof of the theorem that rank A = rank A'.

6. Scalar products. We know that, if $\mathfrak{R}$ is a left vector space over a field Φ, then $\mathfrak{R}$ can also be regarded as a right vector space over Φ. Hence we have the possibility of defining bilinear forms connecting $\mathfrak{R}$ with itself and of regarding $\mathfrak{R}$ as a dual of itself. These considerations can also be applied to the division ring case, provided that Δ possesses an anti-automorphism.

Thus suppose that Δ is a division ring in which there is defined an anti-automorphism $\alpha \to \bar{\alpha}$. The defining properties of such a mapping are that it is 1–1 of Δ onto itself and that it satisfies the conditions

$$\overline{\alpha + \beta} = \bar{\alpha} + \bar{\beta}, \quad \overline{\alpha\beta} = \bar{\beta}\bar{\alpha}. \tag{23}$$

Thus, for example, Δ can be taken to be the division ring of Hamilton's quaternions and $\alpha \to \bar{\alpha}$ the mapping of a quaternion onto its conjugate. We remark also that, if $\Delta = \Phi$ is a field, then an anti-automorphism is just an automorphism in Φ. Hence in this case $\alpha \to \bar{\alpha}$ can be taken to be any automorphism of Φ. In particular, we can take $\bar{\alpha} = \alpha$, that is, the mapping is the identity automorphism. This is the classical case, and it will receive special emphasis in our discussion.

Let $\mathfrak{R}$ be a (left) vector space over Δ. By using the given anti-automorphism $\alpha \to \bar{\alpha}$ it is easy to turn $\mathfrak{R}$ into a right vector space over Δ. We have merely to set $x\bar{\alpha} = \alpha x$, or, in other words, if $\alpha \to \alpha^A$ is the inverse of $\alpha \to \bar{\alpha}$, then $x\alpha = \alpha^A x$. One verifies that this definition of right multiplication by scalars satisfies the basic requirement (cf. p. 6). Hence $\mathfrak{R}$ becomes, in this way, a right vector space over Δ. The two dimensionalities, left and right, of $\mathfrak{R}$ over Δ are equal. In fact, it is clear that, if $(e_1, e_2, \cdots, e_n)$ is a left (right) basis for $\mathfrak{R}$, then it is also a right (left) basis; for, if $x = \Sigma\xi_i e_i$, then $x = \Sigma e_i\bar{\xi}_i$. Moreover, if $\Sigma e_i\delta_i = 0$, then $\Sigma\delta_i^A e_i = 0$, $\delta_i^A = 0$ and $\delta_i = 0$. We remark also that any left subspace of $\mathfrak{R}$ is a right subspace and conversely.

We can now consider bilinear forms connecting the left vector space $\mathfrak{R}$ and the right vector space $\mathfrak{R}$. Such forms will be called *scalar products*. Thus by definition a scalar product is a function $g(x, y)$ defined for pairs of vectors (x, y) belonging to $\mathfrak{R}$, having values in Δ, and satisfying the equations

$$g(x_1 + x_2, y) = g(x_1, y) + g(x_2, y), \quad g(\alpha x, y) = \alpha g(x, y) \tag{24}$$

$$g(x, y_1 + y_2) = g(x, y_1) + g(x, y_2), \quad g(x, \alpha y) = g(x, y)\bar{\alpha}; \tag{25}$$

for since $\alpha y = y\bar{\alpha}$, the second condition in (25) is equivalent to the second part of (6).

If $(e_1, e_2, \cdots, e_n)$ is a basis for $\mathfrak{R}$ over Δ, the matrix (β), $\beta_{ij} = g(e_i, e_j)$ is called *the matrix of the scalar product* relative to this basis. Evidently this concept is a specialization of our former one of the matrix of a bilinear form determined by bases in the spaces connected by the form. As in the general case, the matrix and the basis completely determine the function $g(x, y)$; for if $x = \Sigma\xi_i e_i$ and $y = \Sigma\eta_i e_i$, then by (24) and (25)

$$g(x, y) = \sum_{i,j} \xi_i \beta_{ij} \bar{\eta}_j. \tag{26}$$

Conversely, if $(e_1, e_2, \cdots, e_n)$ and (β) are given, then the equation can be used to define a scalar product in $\mathfrak{R}$.

We consider now the effect of a change of basis on the matrix of a scalar product. Let $(f_1, f_2, \cdots, f_n)$ be a second basis of $\mathfrak{R}$ and suppose that $f_i = \Sigma\mu_{ij}e_j$. Then if we regard these as right vector spaces, the relation between them is $f_i = \Sigma e_j \nu_{ji}$ where $\nu_{ji} = \bar{\mu}_{ij}$. Thus the matrix connecting the right bases is $(\nu) = (\bar{\mu})'$, and this matrix is called the *conjugate transpose* of the matrix (μ). Now we have seen (p. 139) that the new matrix of $g(x, y)$ is $(\gamma) = (\mu)(\beta)(\nu)$. Hence we have the relation

$$(\gamma) = (\mu)(\beta)(\bar{\mu})' \tag{27}$$

where (μ) is the matrix that gives the change of left bases. Two matrices (β) and (γ) connected as in (27) by a non-singular matrix (μ) are said to be *cogredient* (relative to the given anti-automorphism). It is easy to see that this relation is an equivalence. The result we have established is that any two matrices of a scalar product are cogredient. It is also clear that, if (β) is the matrix of a scalar product, then any matrix cogredient to (β) is also a matrix of the scalar product.

If $\mathfrak{S}$ is a subspace of $\mathfrak{R}$, then it is evident that the contraction of the function $g(x, y)$ to the pairs of vectors in $\mathfrak{S}$ is a scalar product for $\mathfrak{S}$. It is now natural to introduce the following notion of equivalence between subspaces of $\mathfrak{R}$. We say that the subspaces $\mathfrak{S}_1$ and $\mathfrak{S}_2$ are *g-equivalent* if there exists a 1–1 linear transformation U of $\mathfrak{S}_1$ onto $\mathfrak{S}_2$ such that $g(x_1, y_1) = g(x_1U, y_1U)$ holds for all x_1, y_1 in $\mathfrak{S}_1$. If $(e_1, e_2, \cdots, e_r)$ is a basis for $\mathfrak{S}_1$, then $(e_1U, e_2U, \cdots, e_rU)$ is a basis for $\mathfrak{S}_2$, and $g(e_i, e_j) = g(e_iU, e_jU)$. Hence the matrices of the contractions of g relative to these bases are identical. It follows that, if arbitrary bases are chosen in g-equivalent subspaces, then the matrices determined by these bases are cogredient. Conversely, suppose that $\mathfrak{S}_1$ and $\mathfrak{S}_2$ are subspaces such that the matrices of the contractions of g to these spaces are cogredient. Then by an appropriate choice of basis in $\mathfrak{S}_2$ we can suppose that we have a basis $(e_1, e_2, \cdots, e_r)$ for $\mathfrak{S}_1$ and a basis $(f_1, f_2, \cdots, f_r)$ for $\mathfrak{S}_2$ such that $g(e_i, e_j) = g(f_i, f_j)$. Now let $x_1 = \Sigma\xi_i e_i$, $y_1 = \Sigma\eta_j e_j$ be any two vectors in $\mathfrak{S}_1$. We have $\sum_{i,j} \xi_i g(e_i, e_j)\bar{\eta}_j = \sum_{i,j} \xi_i g(f_i, f_j)\bar{\eta}_j$, and this gives $g(x_1, y_1) = g(\Sigma\xi_i f_i, \Sigma\eta_j f_j)$. Now the mapping $x_1 = \Sigma\xi_i e_i \to \Sigma\xi_i f_i$ is a 1–1 linear transformation of $\mathfrak{S}_1$ onto $\mathfrak{S}_2$. Hence the foregoing relation shows that $\mathfrak{S}_1$ and $\mathfrak{S}_2$ are g-equivalent.

7. Hermitian scalar products. We shall assume now the anti-automorphism $\alpha \to \bar{\alpha}$ is *involutorial* in the sense that $\bar{\bar{\alpha}} = \alpha$ for all α. We consider first the theory of scalar products which are *hermitian* in the sense that

$$g(y, x) = \overline{g(x, y)} \tag{28}$$

for all x and y. If $\Delta = \Phi$ is a field and $\bar{\alpha} \equiv \alpha$ (this is allowed), then the hermitian condition reads

$$g(y, x) = g(x, y). \tag{29}$$

A scalar product satisfying this condition is said to be *symmetric*.

If $(e_1, e_2, \cdots, e_n)$ is a basis for $\mathfrak{R}$ over Δ, then the condition (28) implies in particular that

$$\beta_{ij} = g(e_i, e_j) = \overline{g(e_j, e_i)} = \bar{\beta}_{ji}.$$

Hence a necessary condition that $g(x, y)$ be hermitian is that the matrices (β) of the scalar product be *hermitian* in the sense that

$$(\bar{\beta})' = (\beta). \tag{30}$$

This condition is also sufficient; for, if it holds and $x = \Sigma\xi_i e_i$ and $y = \Sigma\eta_i e_i$, then

$$g(y, x) = \Sigma\eta_i\beta_{ij}\bar{\xi}_j = \Sigma\eta_i\bar{\beta}_{ji}\bar{\xi}_j$$

and

$$\overline{g(x, y)} = \overline{\Sigma\xi_i\beta_{ij}\bar{\eta}_j} = \Sigma\eta_j\bar{\beta}_{ij}\bar{\xi}_i.$$

Hence $g(y, x) = \overline{g(x, y)}$. In particular $g(x, y)$ is symmetric if and only if its matrices are symmetric $((\beta)' = (\beta))$.

If $g(x, y)$ is hermitian, $g(u, v) = 0$ for the particular vectors u, v implies $g(v, u) = \overline{g(u, v)} = 0$. If this holds, then we say that the vectors u and v are *orthogonal* (relative to g). Our remark shows that this relation is a symmetric one. If $\mathfrak{S}$ is a subspace of $\mathfrak{R}$, we define the *orthogonal complement* $\mathfrak{S}^{\perp}$ to be the space of vectors v that are orthogonal to every vector $u \in \mathfrak{S}$. This coincides with our earlier definition of $j(\mathfrak{S})$. It should be noted that, except in important special cases, some of which will be considered in the next chapter, $\mathfrak{S}^{\perp}$ is not in general a complement of $\mathfrak{S}$ in the lattice of subspaces.

The subspace $\mathfrak{R}^{\perp}$ of vectors z that are orthogonal to every $x \in \mathfrak{R}$ will, as before, be called *the radical* of $g(x, y)$. The scalar product is *non-degenerate* if its radical $\mathfrak{R}^{\perp} = 0$. In this case we know that the mapping $\mathfrak{S} \rightarrow \mathfrak{S}^{\perp}$ is an anti-automorphism of the lattice of subspaces onto itself. Also we know that $\dim \mathfrak{S}^{\perp} = n - \dim \mathfrak{S}$. Hence $\dim \mathfrak{S}^{\perp\perp} = \dim \mathfrak{S}$. Since it is clear that $\mathfrak{S}^{\perp\perp} \supseteq \mathfrak{S}$, this relation gives $\mathfrak{S}^{\perp\perp} = \mathfrak{S}$. Thus the mapping $\mathfrak{S} \rightarrow \mathfrak{S}^{\perp}$ determined by a non-degenerate hermitian scalar product is an involutorial mapping in the lattice of subspaces.

A subspace $\mathfrak{S}$ will be called *isotropic* if $\mathfrak{S} \cap \mathfrak{S}^{\perp} \neq 0$. This evidently implies that $\mathfrak{S}$ contains a non-zero vector u which is *isotropic* in the sense that $g(u, u) = 0$. A subspace $\mathfrak{S}$ will be called *totally isotropic* if $\mathfrak{S} \subseteq \mathfrak{S}^{\perp}$.

If g is non-degenerate and $\mathfrak{S}$ is a non-isotropic subspace, then $\mathfrak{S} \cap \mathfrak{S}^{\perp} = 0$ and $\dim \mathfrak{S}^{\perp} = n - \dim \mathfrak{S}$. Hence in this case we have the decomposition $\mathfrak{R} = \mathfrak{S} \oplus \mathfrak{S}^{\perp}$.

EXERCISES

1. Show that the existence of a hermitian scalar product $\neq 0$ implies that the anti-automorphism $\alpha \rightarrow \bar{\alpha}$ is involutorial. (Hence the latter condition is superfluous in the above discussion.)

2. Show that the existence of a scalar product $\neq 0$ which satisfies $g(x, y) = g(y, x)$ implies that $\bar{\alpha} \equiv \alpha$ and hence that Δ is a field.

3. Show that, if (α) and (β) are hermitian, then so are $(\alpha) \pm (\beta)$, $(\alpha)^r$, $(\alpha)(\beta) + (\beta)(\alpha)$.

4. A scalar product $h(x, y)$ is called *skew-hermitian* if $h(y, x) = -\overline{h(x, y)}$. Let μ be an element of Δ that satifies $\bar{\mu} = -\mu \neq 0$ and let $h(x, y)$ be skew-hermitian. Prove that the mapping $\xi \rightarrow \xi^* \equiv \mu^{-1}\bar{\xi}\mu$ is an involutorial anti-automorphism in Δ and that $g(x, y) = h(x, y)\mu$ is hermitian relative to this anti-automorphism.

8. Matrices of hermitian scalar products. In our discussion of hermitian scalar products we shall find it convenient to treat separately the symmetric forms over fields of characteristic 2. We therefore assume in this section that, if g is symmetric, then the characteristic of $\Delta = \Phi$ is not two.

We consider first the function $g(x, x)$ of a single vector x determined by the scalar product. If $g(x, y)$ is symmetric, the associated function $g(x, x)$ is called *the quadratic form* determined by the symmetric form. In general, we shall say that the element β of Δ is *represented* by the scalar product if there exists a vector $u \neq 0$ such that $g(u, u) = \beta$. Since

$$g(u, u) = \overline{g(u, u)}, \quad \beta = \bar{\beta}.$$

Hence the elements represented by the scalar product are invariant under the anti-automorphism. An important step in proving the main result on hermitian scalar products is the following

Lemma. *If $g(x, y)$ is a hermitian scalar product $\neq 0$, then there exist non-zero elements which are represented by the scalar product. (In other words, if $\mathfrak{R}$ is not totally isotropic, then it contains a non-isotropic vector.)*

Proof. If the conclusion is false, $g(u, u) = 0$ for all u. Hence

$$g(x, y) + g(y, x) = g(x + y, x + y) - g(x, x) - g(y, y) = 0$$

for all x, y. Since $g(y, x) = \overline{g(x, y)}$ this gives $\overline{g(x, y)} = -g(x, y)$. Now since $g(x, y) \neq 0$, there exist vectors u, v such that $\rho = g(u, v) \neq 0$. If we replace u by $\rho^{-1}u$ and change the notation, then we can suppose that $g(u, v) = 1$. Then for any α in Δ, $\overline{g(\alpha u, v)} = -g(\alpha u, v)$ and $\bar{\alpha} = -\alpha$. Since $\bar{1} = 1$, this implies that the characteristic is two and that $\bar{\alpha} = \alpha$. Hence our anti-automorphism is the identity mapping and $\Delta = \Phi$ is commutative. This case is ruled out by assumption.

The argument we have just used is not a constructive one. However, in the special case in which $g(x, y)$ is symmetric, we can easily give such a method to find a vector u such that $g(u, u) \neq 0$. Thus let $(e_1, e_2, \cdots, e_n)$ be a basis for $\mathfrak{R}$ over Φ. Then $g(e_i, e_j) \neq 0$ for some pair e_i, e_j. If $i = j$, we can take $u = e_i$.

Otherwise, we can assume that $g(e_i, e_i) = 0 = g(e_j, e_j)$ and $g(e_i, e_j) \neq 0$ for $i \neq j$. Then $g(e_i + e_j, \ e_i + e_j) = g(e_i, e_i) + g(e_i, e_j) + g(e_j, e_i) + g(e_j, e_j) = g(e_i, e_j) + g(e_j, e_i) = 2g(e_i, e_j) \neq 0$.

We now return to the general case and we shall prove the following

Theorem 3. *If $g(x, y)$ is a hermitian scalar product, then there exists a basis $(u_1, u_2, \cdots, u_r, z_1, z_2, \cdots, z_{n-r})$ such that*

$$g(u_i, u_i) = \beta_i \neq 0, \quad i = 1, 2, \cdots, r, \tag{31}$$

and all other products are 0.

Proof. The result is trivial if $g = 0$; for then any basis serves as a set of z's. If $g \neq 0$, we can take u_1 to be any vector such that $g(u_1, u_1) = \beta_1 \neq 0$. Such vectors exist by the Lemma. Now suppose that we have already determined linearly independent vectors $(u_1, u_2, \cdots, u_k)$ such that $g(u_i, u_i) = \beta_i \neq 0$ and $g(u_i, u_j) = 0$ if $i \neq j$. We introduce the mapping E_k defined by

$$x \to \sum_1^k g(x, u_i)\beta_i^{-1}u_i. \tag{32}$$

Clearly E_k is linear and maps $\mathfrak{R}$ into the space $\mathfrak{S}_k = [u_1, u_2, \cdots, u_k]$. Also $u_jE_k = \sum_i g(u_j, u_i)\beta_i^{-1}u_i = u_j$. Hence E_k is the identity mapping in $\mathfrak{S}_k$. It follows that $E_k^2 = E_k$ so that, if we set $F_k = 1 - E_k$, then $\mathfrak{R} = \mathfrak{S}_k \oplus \mathfrak{R}F_k$. Also we have

$$\begin{aligned} g(xE_k, u_j) &= g\left(\sum_{i=1}^k g(x, u_i)\beta_i^{-1}u_i, u_j\right) \\ &= \sum_{i=1}^k g(x, u_i)\beta_i^{-1}g(u_i, u_j) \\ &= g(x, u_j). \end{aligned}$$

Hence $g(xF_k, u_j) = g(x(1 - E_k), u_j) = g(x, u_j) - g(xE_k, u_j) = 0$. Thus the vectors in $\mathfrak{R}F_k$ are orthogonal to every vector in $\mathfrak{S}_k$, that is, $\mathfrak{R}F_k \subseteq \mathfrak{S}_k^{\perp}$. We consider the scalar product $g(x, y)$ in $\mathfrak{R}F_k$. If this is 0 in $\mathfrak{R}F_k$, we choose a basis $(z_1, z_2, \cdots, z_m)$ for $\mathfrak{R}F_k$. Since $\mathfrak{R} = \mathfrak{S}_k \oplus \mathfrak{R}F_k$, $(u_1, u_2, \cdots, u_k, z_1, z_2, \cdots, z_m)$ is a basis for $\mathfrak{R}$. Since the z's are orthogonal to the u's and to the z's, this is a basis of the required type. On the other hand, if $g \neq 0$ in $\mathfrak{R}F_k$, then we can find a vector u_{k+1} in this space such that $g(u_{k+1}, u_{k+1}) = \beta_{k+1} \neq 0$. Then $(u_1, u_2, \cdots, u_{k+1})$ is a linearly independent set and, since u_{k+1} is orthogonal to the other u_i, the new set satisfies the same conditions as $(u_1, u_2, \cdots, u_k)$. The process can then be repeated.

The method that we have given for finding a normalized basis is in essence due to Lagrange. It should be noted that β_1 can be taken to be any element that is represented by the scalar product. Also we note that any u_i can be replaced by $\gamma_i u_i$, $\gamma_i \neq 0$. This replaces β_i by $\beta_i' = g(\gamma_i u_i, \gamma_i u_i) = \gamma_i \beta_i \bar{\gamma}_i$. In carrying out the above process it is necessary to have a basis for $\mathfrak{R}F_k$. This can be done by supplementing $(u_1, u_2, \cdots, u_k)$ to a basis $(u_1, u_2, \cdots, u_k, v_1, v_2, \cdots, v_{n-k})$ for $\mathfrak{R}$. Then the $v_i F_k$ form a basis for $\mathfrak{R}F_k$.

EXERCISES

1. If Φ is the field of rational numbers, find a diagonal matrix (β) cogredient to

$$(\alpha) = \begin{bmatrix} -2 & 3 & 5 \\ 3 & 1 & -1 \\ 5 & -1 & 4 \end{bmatrix}.$$

Also find a non-singular matrix (μ) such that $(\mu)(\alpha)(\mu)' = (\beta)$.

2. Show that the space $[z_1, z_2, \cdots, z_{n-r}]$ given in Theorem 3 is the radical of $g(x, y)$.

3. Suppose that Φ has characteristic $\neq 2$ and that $g(x, y)$ is a symmetric non-degenerate scalar product in $\mathfrak{R}$ over Φ. Prove that, if g represents 0, then g is *universal* in the sense that it represents every element of Φ.

9. Symmetric and hermitian scalar products over special division rings. The foregoing discussion reduces the problem of cogredience for hermitian matrices to that of finding conditions for cogredience of diagonal matrices. No general solution of this problem is known. The known results are all special in that they make use of special assumptions on the nature of Δ and of the anti-automorphism. For example, a complete solution of the cogredience problem is known for symmetric matrices over the field of rational numbers. The theory for this case, which is due principally to Minkowski and Hasse, is arithmetic in nature and will not be discussed here. Instead, our plan for the remainder of this chapter is as follows: In this section and in § 10 we consider the special cases of the cogredience problem which should be familiar to every student of mathematics. In the last two sections we consider questions which are of interest primarily to specialists in algebra. Thus in § 11 we consider again the general theory of hermitian forms and we discuss Witt's important generalization of the notion of signature. In the final section (§ 12) we consider the theory of symmetric scalar products over a field of characteristic 2.

We specialize first the result of the preceding section to symmetric matrices and symmetric scalar products ($\Delta = \Phi$ a field, $\bar{\alpha} \equiv \alpha$). We assume Φ has characteristic $\neq 2$. Suppose that $g(x, y)$ is a symmetric scalar product and let $(u_1, u_2, \cdots, u_n)$ be a basis such that

$$g(u_i, u_j) = \delta_{ij}\beta_i, \quad \beta_i \neq 0, \quad i = 1, 2, \cdots, r.$$

Then r is the rank of the matrix diag $\{\beta_1, \beta_2, \cdots, \beta_n\}$ of $g(x, y)$, and r is the common rank of all the matrices of this scalar product. As we have seen, we may replace u_i by $v_i = \gamma_i u_i$, $\gamma_i \neq 0$. Then the v's form another basis and the matrix determined by this basis is

$$\text{diag}\,\{\beta_1', \beta_2', \cdots, \beta_n'\}$$

where $\beta_i' = \gamma_i^2\beta_i$. Thus we see that any β_i can be replaced by $\beta_i' = \gamma_i^2\beta_i$, $\gamma_i \neq 0$. Suppose now that Φ is a field in which every element is a square. As usual, if $\gamma^2 = \beta$, we write $\gamma = \beta^{1/2}$. Then if $\beta_i \neq 0$, let $\gamma_i = \beta_i^{-1/2} \equiv (\beta_i^{1/2})^{-1}$. This choice of γ_i replaces β_i by $\beta_i' = 1$. Hence our symmetric scalar product has a matrix of the form

$$\text{diag}\,\{1, 1, \cdots, 1, 0, \cdots, 0\}. \tag{33}$$

This form is applicable in particular if Φ is an algebraically closed field.

If (β) is any symmetric matrix, then (β) can be used to define a symmetric scalar product in $\mathfrak{R}$ over Φ. The matrices of this scalar product constitute the cogredience class determined by (β). Thus we see that, if Φ is algebraically closed of characteristic $\neq 2$, then any symmetric matrix (β) in Φ_n is cogredient to a matrix of the form (33). Clearly (33) is completely determined by the rank of (β). This implies the following

Theorem 4. *If Φ is algebraically closed of characteristic $\neq 2$, then any two symmetric matrices in Φ_n are cogredient if and only if they have the same rank.*

We assume next that Φ is the field of real numbers. Let the vectors u_i be arranged so that

$$\beta_i > 0, i = 1, 2, \cdots, p; \quad \beta_j < 0, j = p + 1, \cdots, r.$$

Then we can extract a square root of β_i, $i \leq p$ and of $-\beta_j$, $r \geq j > p$. Accordingly, let $\gamma_i = \beta_i^{-1/2}$, $\gamma_j = (-\beta_j)^{-1/2}$. Using these γ_i as above, we obtain a matrix

$$\{\overbrace{1, \cdots, 1}^{p}, \overbrace{-1, \cdots, -1}^{r-p}, 0, \cdots, 0\}^* \tag{34}$$

for $g(x, y)$. It follows from this that any real symmetric matrix is cogredient to a matrix of the form (34). We shall show next that two matrices of the form (34) cannot be cogredient in Φ_n, Φ the real field, unless they are identical. This will follow from

Theorem 5 (Sylvester). *If the diagonal matrices*

$$\{\beta_1, \beta_2, \cdots, \beta_n\}, \quad \{\beta_1', \beta_2', \cdots, \beta_n'\}$$

are cogredient in Φ_n, Φ the field of real numbers, then the number p of positive β_i is the same as the number p' of positive β_i'.

Proof. The given matrices may be taken to be matrices of the same symmetric scalar product in $\mathfrak{R}$ over Φ. We may suppose that the first p β_i are > 0 and the first p' β_i' are > 0 and that in both matrices the first r elements are $\neq 0$ and the last $n - r$ are 0. Let $(u_1, u_2, \cdots, u_n)$ be a basis relative to which the matrix of $g(x, y)$ is $\{\beta_1, \beta_2, \cdots, \beta_n\}$ and $(v_1, v_2, \cdots, v_n)$ a basis for which the matrix is $\{\beta_1', \beta_2', \cdots, \beta_n'\}$. It is easy to see (Ex. 2, p. 154) that the radical

$$\mathfrak{R}^\perp = [u_{r+1}, u_{r+2}, \cdots, u_n] = [v_{r+1}, v_{r+2}, \cdots, v_n].$$

* In the remainder of this chapter we drop the symbol "diag" in the notation for diagonal matrices.

We now introduce the following subspaces:

$$\mathfrak{R}_+ = [u_1, u_2, \cdots, u_p]$$

$$\mathfrak{S}_+ = [v_1, v_2, \cdots, v_{p'}]$$

$$\mathfrak{R}_- = [u_{p+1}, u_{p+2}, \cdots, u_r]$$

$$\mathfrak{S}_- = [v_{p'+1}, v_{p'+2}, \cdots, v_r].$$

Let $y \in \mathfrak{R}_+ + \mathfrak{R}^\perp$. Then $y = \sum_1^p \eta_i u_i + \sum_{r+1}^n \eta_j u_j$ and

$$g(y, y) = \sum_1^p \eta_i^2 \beta_i.$$

Since $\beta_i > 0$ for $i = 1, 2, \cdots, p$, $g(y, y) \geq 0$ and $g(y, y) = 0$ only if all the $\eta_i = 0$, $i \leq p$. Thus if $y \in \mathfrak{R}_+ + \mathfrak{R}^\perp$, $g(y, y) \geq 0$ and $g(y, y) = 0$ only if $y \in \mathfrak{R}^\perp$. A similar result holds for $\mathfrak{S}_+ + \mathfrak{R}^\perp$. On the other hand, if $y \in \mathfrak{R}_- + \mathfrak{R}^\perp$ or to $\mathfrak{S}_- + \mathfrak{R}^\perp$, then a similar argument shows that $g(y, y) \leq 0$ and that $g(y, y) = 0$ only if $y \in \mathfrak{R}^\perp$. Now let $y \in (\mathfrak{R}_+ + \mathfrak{R}^\perp) \cap (\mathfrak{S}_- + \mathfrak{R}^\perp)$. Then $g(y, y) \geq 0$ and $g(y, y) \leq 0$. Hence $g(y, y) = 0$ and $y \in \mathfrak{R}^\perp$. This establishes the following relation:

$$(\mathfrak{R}_+ + \mathfrak{R}^\perp) \cap (\mathfrak{S}_- + \mathfrak{R}^\perp) = \mathfrak{R}^\perp. \tag{35}$$

We now make use of the general dimensionality relation (Ex. 3, p. 28):

$$\dim (\mathfrak{S}_1 \cap \mathfrak{S}_2) = \dim \mathfrak{S}_1 + \dim \mathfrak{S}_2 - \dim (\mathfrak{S}_1 + \mathfrak{S}_2)$$
$$\geq \dim \mathfrak{S}_1 + \dim \mathfrak{S}_2 - n.$$

Applying this to (35) we obtain

$$n - r \geq p + (n - r) + (r - p') + (n - r) - n.$$

Hence $p - p' \leq 0$. Similarly $p' \leq p$ and so $p = p'$.

Sylvester's theorem shows that the number of positive elements in any diagonal matrix that is cogredient to a given real symmetric matrix (β) is an invariant. Likewise the difference $2p - r$ between the number of positive elements and the number of negative elements in any diagonal form is an invariant. We call this number the *signature* (*inertial index*) of the matrix. The main theorem on real symmetric matrices may now be stated as

Theorem 6. *Two real symmetric matrices are cogredient if and only if they have the same rank and the same signature.*

We assume next that Φ is the field of complex numbers and that $\alpha \to \bar{\alpha}$ is the mapping of a complex number into its complex conjugate. Let $g(x, y)$ be a hermitian scalar product in $\mathfrak{R}$ over Φ associated with the usual mapping $\alpha \to \bar{\alpha}$. For any u, $\overline{g(u, u)} = g(u, u)$ is real. In particular, if $(u_1, u_2, \cdots, u_n)$ is a basis such that (31) holds, then the elements β_i are real. If we replace u_i by $v_i = \gamma_i u_i$, $\gamma_i \neq 0$, then β_i is replaced by $\gamma_i \bar{\gamma}_i \beta_i = |\gamma_i|^2 \beta_i$. It follows in this case, too, that $\mathfrak{R}$ has a basis relative to which the matrix has the form (34).

Sylvester's theorem holds in the present case also. It asserts that, if Φ is the complex field and $\{\beta_1, \beta_2, \cdots, \beta_n\}$ and $\{\beta_1', \beta_2', \cdots, \beta_n'\}$ are cogredient in the sense that there exists a matrix (μ) in $L(\Phi, n)$ such that

$$\{\beta_1', \beta_2', \cdots, \beta_n'\} = (\mu)\{\beta_1, \beta_2, \cdots, \beta_n\}(\bar{\mu})',$$

then the number of positive β_i is the same as the number of positive β_i'. The proof is exactly the same as before. The essential point is that, if

$$y = \sum_{i}^{p} \eta_i u_i + \sum_{r+1}^{n} \eta_j u_j$$

and $\beta_i > 0$ for $i \leq p$ and $\beta_j = 0$ for $j > r$, then

$$g(y, y) = \Sigma |\eta_i|^2 \beta_i \geq 0.$$

As in the real case we have the criterion that two hermitian matrices are cogredient relative to the anti-automorphism $\alpha \to \bar{\alpha}$ if and only if they have the same rank and the same signature. Here again we define the *signature* to be the difference between the number of positive elements and the number of negative elements in a diagonal matrix cogredient to the given hermitian matrix.

We consider finally the theory of quaternionic hermitian forms. Let Δ be the division ring of Hamilton's quaternions and let $\alpha \to \bar{\alpha}$ be the mapping of a quaternion $\alpha = \alpha_0 + \alpha_1 i + \alpha_2 j + \alpha_3 k$ into its conjugate

$$\bar{\alpha} = \alpha_0 - \alpha_1 i - \alpha_2 j - \alpha_3 k.$$

If $g(x, y)$ is hermitian in $\mathfrak{R}$ over Δ, $\overline{g(u, u)} = g(u, u)$ is real. Also $\alpha\bar{\alpha} = \alpha_0^2 + \alpha_1^2 + \alpha_2^2 + \alpha_3^2 \geq 0$ and equality holds only if $\alpha = 0$. These remarks indicate that this case is essentially the same as the complex hermitian case. Matrices of the form (34) serve as canonical forms under cogredience. Sylvester's theorem and the final result that equality of ranks and of signatures are necessary and sufficient conditions for cogredience hold.

EXERCISES

1. Prove that the number of cogredience classes of real symmetric matrices of n rows and columns is $(n + 1)(n + 2)/2$.
2. Find a matrix of the form (34) cogredient to the real symmetric matrix

$$\begin{bmatrix} 1 & 3 & -5 \\ 3 & -1 & 0 \\ -5 & 0 & 2 \end{bmatrix}.$$

3. Find the matrix of the form (34) that is cogredient to the hermitian quaternion matrix

$$\begin{bmatrix} -1 & 1 + i - 2j & -2i + k \\ 1 - i + 2j & 0 & 4j \\ 2i - k & -4 & 2 \end{bmatrix}.$$

4. Prove that, if $(\alpha), (\beta), \cdots, (\nu)$ are a) real symmetric, or b) complex hermitian, or c) quaternionic hermitian, then

$$(\alpha)^2 + (\beta)^2 + \cdots + (\nu)^2 = 0$$

can hold only if $(\alpha) = (\beta) = \cdots = (\nu) = 0$.

10. Alternate scalar products. A scalar product $g(x, y)$ is called *skew symmetric* if $g(x, y) = -g(y, x)$ for all x and y. In this case $g(x, x) = -g(x, x)$. If the characteristic is not two, then this implies that $g(x, x) = 0$. If the characteristic is two, it may still be true that $g(x, x) = 0$ for all x. We shall now call a skew symmetric scalar product *alternate* if $g(x, x) = 0$ for all x, and we consider the problem of finding suitable canonical bases for this type of scalar product. The proof of the lemma on p. 152 shows that, if $g(x, y)$ is alternate, then it is skew symmetric and that, if $g \neq 0$ is skew-symmetric, then $\bar{\alpha} \equiv \alpha$ and $\Delta = \Phi$ is commutative.

Suppose that $g(x, y)$ is alternate and not identically 0. Then we can find a pair of vectors u, v such that $g(u, v) \neq 0$. By replacing v by a suitable multiple v_1 of v we then obtain $u_1 = u$ and v_1 such that $g(u_1, v_1) = 1$. Then $g(v_1, u_1) = -1$, $g(u_1, u_1) = 0 = g(v_1, v_1)$. It follows that u_1 and v_1 are linearly independent. Now suppose that we have already found k pairs of vectors $u_1,v_1, u_2,v_2, \cdots, u_k,v_k$ that are linearly independent and satisfy

$$g(u_i, v_i) = 1, \quad g(v_i, u_i) = -1 \tag{36}$$

with all other products 0. Let E_k denote the linear transformation

$$x \rightarrow \sum_1^k g(x, v_i)u_i - \sum_1^k g(x, u_i)v_i. \tag{37}$$

As in the proof of Theorem 3 we see that E_k is a projection on the space $\mathfrak{S}_k = [u_1,v_1, u_2,v_2, \cdots, u_k,v_k]$. Hence if $F_k = 1 - E_k$, $\mathfrak{R} = \mathfrak{S}_k \oplus \mathfrak{R}F_k$. Also we can verify that $g(xE_k, u_i) = g(x, u_i)$ and $g(xE_k, v_i) = g(x, v_i)$. Hence $g(xF_k, u_i) = 0 = g(xF_k, v_i)$ and $\mathfrak{R}F_k \subseteq \mathfrak{S}_k^{\perp}$.* Now either g is 0 in $\mathfrak{R}F_k$ or we can choose a pair of vectors u_{k+1}, v_{k+1} in this space such that

$$g(u_{k+1}, v_{k+1}) = 1 = -g(v_{k+1}, u_{k+1}).$$

Since $\mathfrak{S}_k \cap \mathfrak{R}F_k = 0$, $(u_1,v_1, u_2,v_2, \cdots, u_{k+1},v_{k+1})$ is an independent set and since the last two vectors are orthogonal to the preceding ones the set of $2(k + 1)$ vectors satisfies the same conditions as the set $(u_1,v_1, u_2,v_2, \cdots, u_k,v_k)$. Eventually we either span the whole space or obtain a space $\mathfrak{R}F_r$ in which g is 0. In the latter case we choose any basis $(z_{2r+1}, \cdots, z_n)$ for this space.

* $\mathfrak{S}_k^{\perp}$ is the orthogonal complement of $\mathfrak{S}_k$, defined as for hermitian scalar products.

Evidently the matrix of $g(x, y)$ relative to the basis

$$(u_1,v_1,\ u_2,v_2,\ \cdots,\ u_r,v_r,\ z_1,z_2,\ \cdots,\ z_{n-2r})$$

is

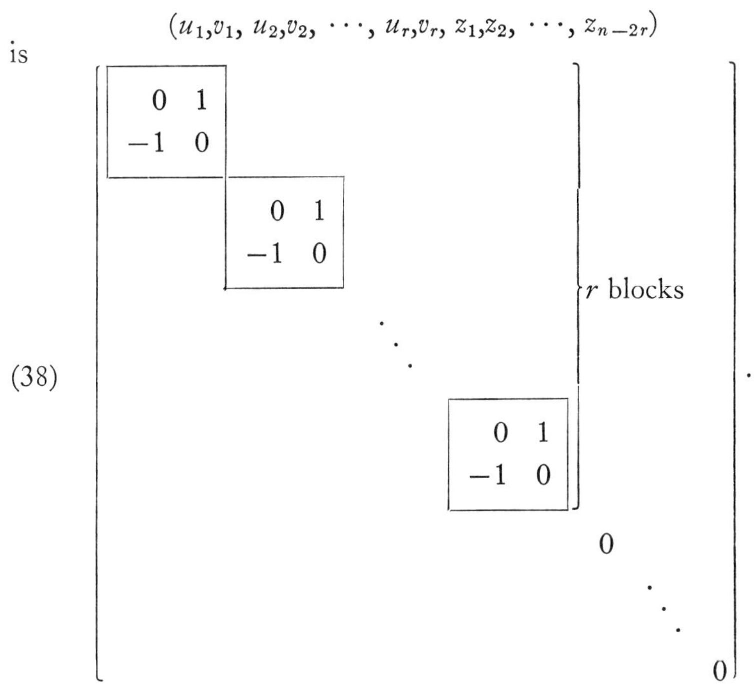

This proves the following

Theorem 7. *If $g(x, y)$ is an alternate scalar product, there exists a basis for $\mathfrak{R}$ relative to which the matrix has the form* (38).

If $g(x, y) \neq 0$ is alternate, the anti-automorphism is the identity and the matrices (β) of $g(x, y)$ are *alternate* in the sense that $(\beta)' = -(\beta)$ and $\beta_{ii} = 0$ for $i = 1, 2, \cdots, n$. These conditions are also sufficient; for if $x = \Sigma\xi_i e_i$, then

$$g(x, x) = \sum_{i,j} \beta_{ij}\xi_i\xi_j = \sum_i \beta_{ii}\xi_i^2 + \sum_{i<j} (\beta_{ij} + \beta_{ji})\xi_i\xi_j = 0.$$

Hence $g(x, y)$ is alternate. The following results are now easy consequences of Theorem 7.

Corollary 1. *The rank of an alternate matrix with elements in a field is even.*

Corollary 2. *Two alternate matrices in Φ_n are cogredient if and only if they have the same rank.*

EXERCISE

1. Prove that

$$(\alpha) = \begin{bmatrix} 0 & 2 & -1 & 3 \\ -2 & 0 & 4 & -2 \\ 1 & -4 & 0 & 1 \\ -3 & 2 & -1 & 0 \end{bmatrix}, \quad (\beta) = \begin{bmatrix} 0 & 1 & 0 & 0 \\ -1 & 0 & 0 & 0 \\ 0 & 0 & 0 & 1 \\ 0 & 0 & -1 & 0 \end{bmatrix}$$

are cogredient in Φ_4, Φ the field of rational numbers. Also find a (μ) in $L(\Phi, 4)$ such that $(\beta) = (\mu)(\alpha)(\mu)'$.

***11. Witt's theorem.** We now take up again the general theory of hermitian scalar products over a division ring. In the present discussion we shall assume that the basic anti-automorphism $\alpha \to \bar{\alpha}$ satisfies the following solvability condition.

Axiom S. *The equation $\xi + \bar{\xi} = \beta$ has a solution for every hermitian element β of the division ring.*

This axiom is automatically satisfied if the characteristic of Δ is not two; for, in this case, we may take $\xi = \frac{1}{2}\beta$. It is also satisfied in the characteristic two case if there exists an element γ of the center Γ of Δ such that $\bar{\gamma} \neq \gamma$. Then $\delta = \gamma + \bar{\gamma} \neq 0$ is in Γ since the anti-automorphism maps Γ into itself. Hence if $\xi = \beta\gamma\delta^{-1}$, then

$$\xi + \bar{\xi} = \beta\gamma\delta^{-1} + \beta\bar{\gamma}\delta^{-1} = \beta\delta^{-1}(\gamma + \bar{\gamma}) = \beta.$$

We remark finally that our axiom rules out the case $\Delta = \Phi$, a field of characteristic two, $\bar{\alpha} \equiv \alpha$. This is clear since in this case $\xi + \bar{\xi} = 0$ while β need not be 0.

Assume that $\mathfrak{R}$ is a vector space over Δ and that $g(x, y)$ is a non-degenerate hermitian scalar product relative to the anti-automorphism $\alpha \to \bar{\alpha}$. We recall that, if $\mathfrak{S}$ is a subspace of $\mathfrak{R}$ that is not isotropic in the sense that $\mathfrak{S} \cap \mathfrak{S}^\perp = 0$, then we have the decomposition $\mathfrak{R} = \mathfrak{S} \oplus \mathfrak{S}^\perp$. The basic result of the theory which we shall develop is the following theorem.

Witt's theorem. *If $\mathfrak{S}_1$ and $\mathfrak{S}_2$ are non-isotropic and g-equivalent, then $\mathfrak{S}_1^\perp$ and $\mathfrak{S}_2^\perp$ are g-equivalent.**

* Witt proved this result for symmetric scalar products over a field of characteristic $\neq 2$ (*Journal für Math.*, Vol. 176 (1937)). The extension to division rings of characteristic $\neq 2$ is due to Pall (*Bulletin Amer. Math. Soc.*, Vol. 51 (1945)). In the present discussion we assume only the foregoing Axiom S. Cf. also Kaplansky, *Forms in infinite dimensional spaces*, Anais Acad. Brasil Ci. 22 (1950), pp. 1–7. Witt's theorem does not hold for symmetric scalar products over a field of characteristic 2 (see the next section).

Proof. It suffices to prove the result for $\dim \mathfrak{S}_1 = 1$; for, if it is known in this case, then we can use induction on $\dim \mathfrak{S}_1$ as follows. We choose a vector u_1 in $\mathfrak{S}_1$ such that $[u_1]$ is not isotropic (Lemma to Theorem 3). Then $\mathfrak{S}_1 = [u_1] \oplus \mathfrak{U}_1$ where $\mathfrak{U}_1 \subseteq [u_1]^\perp$. Using the equivalence of $\mathfrak{S}_1$ and $\mathfrak{S}_2$, we can write $\mathfrak{S}_2 = [u_2] \oplus \mathfrak{U}_2$ where $\mathfrak{U}_2 \subseteq [u_2]^\perp$, $[u_1]$ and $[u_2]$ are equivalent and $\mathfrak{U}_1$ and $\mathfrak{U}_2$ are equivalent. Then $[u_1]^\perp = \mathfrak{U}_1 \oplus \mathfrak{S}_1{}^\perp$ and $[u_2]^\perp = \mathfrak{U}_2 \oplus \mathfrak{S}_2{}^\perp$ are equivalent under a transformation U. Hence $[u_2]^\perp = \mathfrak{U}_2 \oplus \mathfrak{S}_2{}^\perp = \mathfrak{U}_1 U \oplus \mathfrak{S}_1{}^\perp U$. Now $\mathfrak{U}_2$ and $\mathfrak{U}_1 U$ are equivalent and $\mathfrak{U}_1 U$ and $\mathfrak{S}_1{}^\perp U$ are orthogonal. Since $\dim \mathfrak{U}_1 U < \dim \mathfrak{S}_1$ and $\mathfrak{U}_1$ is not isotropic, we can assume that $\mathfrak{S}_1{}^\perp U$ and $\mathfrak{S}_2{}^\perp$ are equivalent. This implies that $\mathfrak{S}_1{}^\perp$ and $\mathfrak{S}_2{}^\perp$ are equivalent.

We note next that the case: $\mathfrak{S}_1 = [u_1]$, $\dim \mathfrak{R}$ arbitrary, can be deduced from the special case: $\mathfrak{S}_1 = [u_1]$, $\dim \mathfrak{R} = 2$. Thus, suppose we know the result in the special case. Let $\mathfrak{S}_2 = [u_2]$ be equivalent to $[u_1]$. We have the decompositions $\mathfrak{R} = [u_1] \oplus [u_1]^\perp = [u_2] \oplus [u_2]^\perp$ and $g(u_1, u_1) = g(u_2, u_2)$. If $[u_1, u_2]$ is one-dimensional, $[u_1]^\perp = [u_2]^\perp$, and the theorem holds. Hence assume that $\dim [u_1, u_2] = 2$. Consider first the case in which this space is not isotropic. Here

$$
\begin{aligned}
\mathfrak{R} &= [u_1, u_2] \oplus [u_1, u_2]^\perp \\
&= [u_1] \oplus ([u_1, u_2] \cap [u_1]^\perp) \oplus [u_1, u_2]^\perp \qquad (39) \\
&= [u_2] \oplus ([u_1, u_2] \cap [u_2]^\perp) \oplus [u_1, u_2]^\perp.
\end{aligned}
$$

Then $[u_1, u_2] \cap [u_1]^\perp$ and $[u_1, u_2] \cap [u_2]^\perp$ are equivalent by the case $\dim \mathfrak{R} = 2$ (applied to $\mathfrak{R} = [u_1, u_2]$). Hence by (39)

$$[u_1]^\perp = ([u_1, u_2] \cap [u_1]^\perp) \oplus [u_1, u_2]^\perp$$

and

$$[u_2]^\perp = ([u_1, u_2] \cap [u_2]^\perp) \oplus [u_1, u_2]^\perp$$

are g-equivalent. Next consider the case in which $[u_1, u_2]$ is isotropic. Here we have a vector $w \neq 0$ in this space such that $g(w, u_1) = 0 = g(w, u_2)$. We can find a t in $\mathfrak{R}$ such that $g(w, t) \neq 0$. Then the matrix of g in $[u_1, u_2, t]$ relative to the basis (w, u_1, t) is

$$\begin{bmatrix} 0 & 0 & g(w, t) \\ 0 & g(u_1, u_1) & * \\ g(t, w) & * & * \end{bmatrix}.$$

It is evident that the row vectors of this matrix are linearly independent; hence the matrix is non-singular and $[u_1, u_2, t]$ is not isotropic. Hence

$$\begin{aligned}\mathfrak{R} &= [u_1, u_2, t] \oplus [u_1, u_2, t]^\perp \\ &= [u_1] \oplus ([u_1, u_2, t] \cap [u_1]^\perp) \oplus [u_1, u_2, t]^\perp \\ &= [u_2] \oplus ([u_1, u_2, t] \cap [u_2]^\perp) \oplus [u_1, u_2, t]^\perp.\end{aligned}$$

Thus, it suffices to show that $[u_1, u_2, t] \cap [u_1]^\perp$ and $[u_1, u_2, t] \cap [u_2]^\perp$ are equivalent. Now these are non-isotropic two-dimensional spaces that contain an isotropic one-dimensional subspace $[w]$. In a space of this type we can always select a basis relative to which the matrix is $\begin{bmatrix} 0 & 1 \\ 1 & 0 \end{bmatrix}\cdot$ To see this, choose q so that $g(w, q) = 1 = g(q, w)$ and set $z = q + \lambda w$. Then $g(z, w) = 1 = g(w, z)$ also while $g(z, z) = g(q, q) + \lambda + \bar{\lambda}$. By Axiom S we can choose λ so that $g(z, z) = 0$. This gives a matrix of the required form. It follows now that $[u_1, u_2, t] \cap [u_1]^\perp$ and $[u_1, u_2, t] \cap [u_2]^\perp$ are equivalent. This completes the reduction to the case: $\dim \mathfrak{R} = 2$, $\dim \mathfrak{S}_1 = 1$.

We consider finally this special case. The result we wish to prove here is equivalent to the statement that, if the non-singular diagonal hermitian matrices $\{\alpha, \beta_1\}$ and $\{\alpha, \beta_2\}$ are cogredient, then the elements β_1 and β_2 are cogredient. Let (μ) be a non-singular matrix such that

$$\begin{bmatrix} \mu_{11} & \mu_{12} \\ \mu_{21} & \mu_{22} \end{bmatrix} \begin{bmatrix} \alpha & 0 \\ 0 & \beta_1 \end{bmatrix} \begin{bmatrix} \bar{\mu}_{11} & \bar{\mu}_{21} \\ \bar{\mu}_{12} & \bar{\mu}_{22} \end{bmatrix} = \begin{bmatrix} \alpha & 0 \\ 0 & \beta_2 \end{bmatrix}. \tag{40}$$

If Δ is commutative, we take determinants and we obtain $\beta_2 = \mu\beta_1\bar{\mu}$ where $\mu = \det(\mu_{ij})$, which is the desired result. In the general case we shall work directly with the conditions that are given by the matrix relation. These are

$$\mu_{11}\alpha\bar{\mu}_{11} + \mu_{12}\beta_1\bar{\mu}_{12} = \alpha \tag{41}$$

$$\mu_{11}\alpha\bar{\mu}_{21} + \mu_{12}\beta_1\bar{\mu}_{22} = 0 = \mu_{21}\alpha\bar{\mu}_{11} + \mu_{22}\beta_1\bar{\mu}_{12} \tag{42}$$

$$\mu_{21}\alpha\bar{\mu}_{21} + \mu_{22}\beta_1\bar{\mu}_{22} = \beta_2. \tag{43}$$

If $\mu_{11} = 0$, $\mu_{22} = 0$ by (42) since $\mu_{12} \neq 0$. Hence $\alpha = \mu_{12}\beta_1\bar{\mu}_{12}$

and $\beta_2 = \mu_{21}\alpha\bar{\mu}_{21}$. These relations imply that β_1 and β_2 are cogredient.

Assume, therefore, that $\mu_{11} \neq 0$. Then by (42)

$$\alpha\bar{\mu}_{21} = -\mu_{11}{}^{-1}\mu_{12}\beta_1\bar{\mu}_{22}$$

$$\mu_{21} = -\mu_{22}\beta_1\bar{\mu}_{12}\bar{\mu}_{11}{}^{-1}\alpha^{-1}$$

so that by (43)

$$\mu_{22}(\beta_1\bar{\mu}_{12}\bar{\mu}_{11}{}^{-1}\alpha^{-1}\mu_{11}{}^{-1}\mu_{12}\beta_1 + \beta_1)\bar{\mu}_{22} = \beta_2.$$

Thus β_2 is cogredient to $\beta_1\bar{\mu}_{12}\bar{\mu}_{11}{}^{-1}\alpha^{-1}\mu_{11}{}^{-1}\mu_{12}\beta_1 + \beta_1$. We wish to show that the latter element is cogredient to β_1. For this purpose we try to solve

$$(44)\quad (1 + \beta_1\bar{\mu}_{12}\xi\mu_{12})\beta_1(1 + \bar{\mu}_{12}\bar{\xi}\mu_{12}\beta_1) = \beta_1\bar{\mu}_{12}\bar{\mu}_{11}{}^{-1}\alpha^{-1}\mu_{11}{}^{-1}\mu_{12}\beta_1 + \beta_1.$$

This will be satisfied if

$$(45)\qquad \xi + \bar{\xi} + \xi\mu_{12}\beta_1\bar{\mu}_{12}\bar{\xi} = (\mu_{11}\alpha\bar{\mu}_{11})^{-1}.$$

Replacing $\mu_{12}\beta_1\bar{\mu}_{12}$ according to (41) we obtain

$$(46)\qquad \xi + \bar{\xi} + \xi(\alpha - \mu_{11}\alpha\bar{\mu}_{11})\bar{\xi} = (\mu_{11}\alpha\bar{\mu}_{11})^{-1}.$$

If $\mu_{11} = 1$, (46) reduces to $\xi + \bar{\xi} = (\mu_{11}\alpha\bar{\mu}_{11})^{-1}$, which is solvable by Axiom S. If $\mu_{11} \neq 1$, we make the substitution $\xi = \eta^{-1}$ and multiply on the left by η and on the right by $\bar{\eta}$ to obtain

$$(47)\qquad \eta + \bar{\eta} + (\alpha - \mu_{11}\alpha\bar{\mu}_{11}) = \eta(\mu_{11}\alpha\bar{\mu}_{11})^{-1}\bar{\eta}.$$

Next we substitute $\eta = \zeta + \mu_{11}\alpha\bar{\mu}_{11}$ and obtain

$$(48)\qquad \zeta(\mu_{11}\alpha\bar{\mu}_{11})^{-1}\bar{\zeta} = \alpha.$$

Now (48) is satisfied by $\zeta = -\alpha\bar{\mu}_{11}$; hence (47) is satisfied by $\eta = (\mu_{11} - 1)\alpha\bar{\mu}_{11}$ which is not 0. Then $\xi = \eta^{-1}$ satisfies (46). Thus (44) holds and β_1 and β_2 are cogredient. This completes the proof.

A 1–1 linear transformation U of $\mathfrak{R}$ onto itself is said to be *g-unitary* if $g(xU, yU) = g(x, y)$ holds for every pair of vectors x, y in $\mathfrak{R}$. Evidently this condition is equivalent to the requirement that $UU' = 1$ where U' is the transpose of U relative to the scalar product. Now suppose again that $\mathfrak{S}_1$ and $\mathfrak{S}_2$ are non-isotropic spaces which are g-equivalent and let M be a g-equivalence of $\mathfrak{S}_1$ onto $\mathfrak{S}_2$. By Witt's theorem we can find an equivalence N of $\mathfrak{S}_1^{\perp}$ onto $\mathfrak{S}_2^{\perp}$. Since $\mathfrak{R} = \mathfrak{S}_1 \oplus \mathfrak{S}_1^{\perp}$, any vector x can be written in one and only one way as $u + v$ where $u \varepsilon \mathfrak{S}_1$ and $v \varepsilon \mathfrak{S}_1^{\perp}$. Then the mapping $U: x \rightarrow uM + vN$ is a 1–1 linear transformation of $\mathfrak{R}$ onto itself. Moreover, making use of the fact that $uM \varepsilon \mathfrak{S}_2$ and $vN \varepsilon \mathfrak{S}_2^{\perp}$, we can verify directly that U is g-unitary. Evidently U coincides with the given equivalence M on the subspace $\mathfrak{S}_1$. Thus we see that Witt's theorem implies that any g-equivalence between non-isotropic subspaces can be extended to a g-unitary transformation. We shall now show that this result holds also for isotropic subspaces of $\mathfrak{R}$.

Consider first an arbitrary subspace $\mathfrak{S}$ of $\mathfrak{R}$. If x is any vector in $\mathfrak{R}$, then the mapping $y \rightarrow g_x(y) \equiv g(y, x)$ of $\mathfrak{S}$ into Δ is a linear function. It is easy to see that the linear functions thus obtained fill up the conjugate space $\mathfrak{S}^*$ of $\mathfrak{S}$ (Ex. 2, p. 56). Hence if $(y_1, y_2, \cdots, y_m)$ is any basis for $\mathfrak{S}$, then we can find a vector v_1 such that

$$g(y_1, v_1) = 1, \quad g(y_i, v_1) = 0 \quad \text{for} \quad i > 1. \tag{49}$$

Assume now that $\mathfrak{S}$ is isotropic and that $(y_1, y_2, \cdots, y_\nu)$ is a basis for the radical of $\mathfrak{S}$. Then we can choose the vector v_1 so that in addition to (49) we have $g(v_1, v_1) = 0$. This can be seen by an argument used in the proof of Witt's theorem. Thus if v_1 is not orthogonal to itself, then we can replace this vector by $v_1 + \lambda y_1$ and choose λ so that $\lambda + \bar{\lambda} + g(v_1, v_1) = 0$. We then denote this new vector as v_1.

Now the space $[y_1, v_1]$ is a two-dimensional non-isotropic subspace of $\mathfrak{R}$. Hence $\mathfrak{R} = [y_1, v_1] \oplus [y_1, v_1]^{\perp}$. Also it is clear that $[y_2, \cdots, y_m] \subseteq [y_1, v_1]^{\perp}$ and that $[y_2, \cdots, y_\nu]$ is the radical of $[y_2, \cdots, y_m]$. Hence we can use induction on ν to prove the existence of a set of vectors $(v_1, v_2, \cdots, v_\nu)$ such that

$$\begin{aligned} g(y_j, v_j) &= 1, \quad j = 1, 2, \cdots, \nu \\ g(y_i, v_j) &= 0 \quad \text{otherwise} \\ g(v_j, v_k) &= 0 \quad j, k = 1, 2, \cdots, \nu. \end{aligned} \tag{50}$$

It is immediate that the vectors $(v_1, v_2, \cdots, v_\nu)$ are linearly independent and that the space $\mathfrak{B} = [v_1, v_2, \cdots, v_\nu]$ is totally isotropic and satisfies $\mathfrak{V} \cap \mathfrak{S} = 0$. The matrix of g in $\mathfrak{S} + \mathfrak{V}$ relative to the basis $(y_1, y_2, \cdots, y_m, v_1, v_2, \cdots, v_\nu)$ is

$$\left[\begin{array}{c|c|c} 0 & 0 & \begin{matrix} 1 & & \\ & \ddots & \\ & & 1 \end{matrix} \\ \hline 0 & B & 0 \\ \hline \begin{matrix} 1 & & \\ & \ddots & \\ & & 1 \end{matrix} & 0 & 0 \end{array}\right] \tag{51}$$

where B is the matrix relative to $(y_{\nu+1}, \cdots, y_m)$. Since $[y_1, \cdots, y_\nu]$ is the radical of $\mathfrak{S}$, B is non-singular. Hence (51) is non-singular and so $\mathfrak{S} + \mathfrak{V}$ is not isotropic.

Now let U be an equivalence of $\mathfrak{S}$ (onto $\mathfrak{S}U$). Evidently $[y_1U, \cdots, y_\nu U]$ is the radical of $\mathfrak{S}U$. Hence we can find a set of vectors $(\bar{v}_1, \bar{v}_2, \cdots, \bar{v}_\nu)$ such that

$$\begin{aligned} g(y_jU, \bar{v}_j) &= 1, \quad j = 1, 2, \cdots, \nu \\ g(y_iU, \bar{v}_j) &= 0 \quad \text{otherwise} \\ g(\bar{v}_j, \bar{v}_k) &= 0, \quad j, k = 1, 2, \cdots, \nu. \end{aligned}$$

Then it is clear that the linear transformation that sends v_j into $\bar{v}_j$ and that coincides with U on $\mathfrak{S}$ is an equivalence of $\mathfrak{S} + \mathfrak{V}$. Since $\mathfrak{S} + \mathfrak{V}$ is not isotropic, this mapping can be extended to a g-unitary transformation. Hence we have proved the following

Theorem 8. *Any g-equivalence of a subspace of $\mathfrak{R}$ can be extended to a g-unitary transformation in $\mathfrak{R}$.**

A hermitian scalar product is called *totally regular* if $g(x, x) \neq 0$ for every $x \neq 0$ in $\mathfrak{R}$. This is equivalent to saying that every non-zero subspace of $\mathfrak{R}$ is not isotropic. Hence if g is

* Cf. Dieudonné, *Sur les Groupes Classiques*, p. 18.

totally regular and $\mathfrak{S}$ is any subspace, then $\mathfrak{R} = \mathfrak{S} \oplus \mathfrak{S}^{\perp}$. We shall now show that the problem of cogredience of hermitian matrices can be reduced to the case in which the associated scalar products are totally regular.

Let $\mathfrak{S}$ be a totally isotropic subspace of $\mathfrak{R}$ which is maximal in the sense that it cannot be imbedded in a larger subspace of this type. Write $\mathfrak{S} = [y_1, \cdots, y_\nu]$ where the y_i are linearly independent. As before we determine a totally isotropic space $\mathfrak{V} = [v_1, \cdots, v_\nu]$ such that $\mathfrak{X} = \mathfrak{S} + \mathfrak{V} = \mathfrak{S} \oplus \mathfrak{V}$ and such that the matrix of g relative to the basis $(y_1, \cdots, y_\nu, v_1, \cdots, v_\nu)$ of $\mathfrak{X}$ is

$$\left[\begin{array}{c|c} 0 & \begin{matrix} 1 & & \\ & \ddots & \\ & & 1 \end{matrix} \\ \hline \begin{matrix} 1 & & \\ & \ddots & \\ & & 1 \end{matrix} & 0 \end{array}\right]. \tag{52}$$

Since $\mathfrak{S}$ is a maximal totally isotropic subspace and $\mathfrak{R} = \mathfrak{S} \oplus \mathfrak{V} \oplus \mathfrak{X}^{\perp}$, g is totally regular in $\mathfrak{X}^{\perp}$. We can choose a basis for $\mathfrak{R}$ so that the matrix of g has the form

$$\left[\begin{array}{c|c|c} 0 & \begin{matrix} 1 & & \\ & \ddots & \\ & & 1 \end{matrix} & 0 \\ \hline \begin{matrix} 1 & & \\ & \ddots & \\ & & 1 \end{matrix} & 0 & 0 \\ \hline 0 & 0 & B \end{array}\right] \tag{53}$$

where B is a matrix of g in $\mathfrak{X}^{\perp}$. Thus any hermitian matrix is cogredient to one of the form (53) in which B is totally regular (that is, the associated scalar product is totally regular).

Conversely suppose that we have any matrix of g of the form (53) where B is totally regular. Let $(y_1, \cdots, y_\nu, v_1, \cdots, v_\nu, z_1, \cdots, z_{n-2\nu})$ be a basis such that the matrix of g relative to this basis is the given matrix (53). Then $\mathfrak{S} = [y_1, \cdots, y_\nu]$ is totally isotropic. Moreover, $\mathfrak{S}$ is maximal totally isotropic; for, if there is a larger totally isotropic subspace containing $\mathfrak{S}$, then it contains a non-zero vector of the form $v + z$ where $v \varepsilon [v_1, \cdots, v_\nu]$ and $z \varepsilon [z_1, \cdots, z_{n-2\nu}]$. Then $g(v + z, v + z) = g(z, z) = 0$ so that $z = 0$. If $v = \Sigma\gamma_i v_i$, $g(v, y_i) = \gamma_i$. Hence also $v = 0$ contrary to $v + z \neq 0$.

We are now in a position to show that two matrices of the form (53) in which the B's are totally regular are cogredient if and only if the matrices B are cogredient. The "if" part is, of course, trivial, and the foregoing discussion shows that the "only if" part is equivalent to the statement that the spaces $\mathfrak{X}^{\perp}$ determined as above by a maximal totally isotropic subspace $\mathfrak{S}$ are g-equivalent.

We observe first that any two maximal totally isotropic subspaces have the same dimensionality. Thus let $\mathfrak{S}_1$ and $\mathfrak{S}_2$ be of this type and assume that $\dim \mathfrak{S}_1 \geq \dim \mathfrak{S}_2$. Then we can find a subspace $\mathfrak{U}_1$ of $\mathfrak{S}_1$ such that $\dim \mathfrak{U}_1 = \dim \mathfrak{S}_2$. Since $\mathfrak{U}_1$ and $\mathfrak{S}_2$ are totally isotropic, any 1–1 linear transformation of $\mathfrak{U}_1$ onto $\mathfrak{S}_2$ is a g-equivalence; hence it can be extended to a unitary transformation U. Then $\mathfrak{S}_1 U$ is totally isotropic and contains $\mathfrak{U}_1 U = \mathfrak{S}_2$. By the maximality of $\mathfrak{S}_2$ this implies that $\mathfrak{U}_1 = \mathfrak{S}_1$ and that $\dim \mathfrak{S}_1 = \dim \mathfrak{S}_2$.

Now let $\mathfrak{S}_i = [y_1^{(i)}, \cdots, y_\nu^{(i)}]$, $i = 1, 2$, and determine $\mathfrak{B}_i = [v_1^{(i)}, \cdots, v_\nu^{(i)}]$ as above so that the matrix of g in $\mathfrak{X}_i = \mathfrak{S}_i + \mathfrak{B}_i$ is (52). Then $\mathfrak{X}_1$ and $\mathfrak{X}_2$ are g-equivalent; hence also $\mathfrak{X}_1^{\perp}$ and $\mathfrak{X}_2^{\perp}$ are g-equivalent. We have therefore established the following

Theorem 9. *Any non-singular hermitian matrix is cogredient to one of the form* (53) *in which B is totally regular*; *two matrices of the form* (53) *in which B is totally regular are cogredient if and only if the submatrices B are cogredient.*

The number of rows and columns of the matrix B is, of course, an invariant of the cogredience class. This number is the dimensionality of the spaces $\mathfrak{X}^{\perp}$ and is also $n - 2\nu$ where ν is the maximum dimensionality of totally isotropic subspaces relative to g. We shall call this non-negative integer the *Witt signature* of g or of the associated matrices. It can be seen (Ex. 1 below) that for symmetric matrices over the real field or for hermitian matrices over the complex field or over real quaternions the Witt signature is the absolute value of the ordinary signature which we defined before.

EXERCISES

1. Prove the foregoing statement on signatures.
2. If g is an alternate non-degenerate scalar product, a 1–1 linear transformation S of $\mathfrak{R}$ such that $g(xS, yS) = g(x, y)$ is called a *symplectic* transformation. Prove the following analogue of Theorem 8: *Any g-equivalence of a subspace of $\mathfrak{R}$ can be extended to a symplectic transformation.*

***12. Non-alternate skew-symmetric forms.** We suppose finally that $g(x, y)$ is a skew symmetric scalar product that is not alternate. Then Φ has characteristic two and $g(x, y)$ may also be regarded as symmetric. Moreover, we have seen that, if $(e_1, e_2, \cdots, e_n)$ is a basis for $\mathfrak{R}$, then $g(e_i, e_i) \neq 0$ for some i. We shall now show that it is possible to choose a basis $(u_1, u_2, \cdots, u_r, z_1, z_2, \cdots, z_{n-r})$ for $\mathfrak{R}$ such that the matrix determined by this basis is

$$\{\beta_1, \beta_2, \cdots, \beta_r, 0, 0, \cdots, 0\},$$

$\beta_i \neq 0$. Evidently u_1 can be chosen so that $(u_1, u_1) = \beta_1 \neq 0$. Now suppose that $u_1, u_2, \cdots, u_k$ have already been found such that $g(u_i, u_j) = \delta_{ij}\beta_i$, $\beta_i \neq 0$. As in the proof of Theorem 3 we can write $\mathfrak{R} = \mathfrak{S}_k \oplus \mathfrak{R}F_k$ where $\mathfrak{S}_k = [u_1, u_2, \cdots, u_k]$ and $\mathfrak{R}F_k \subseteq \mathfrak{S}_k^{\perp}$. If $g(x, y)$ is identically 0 in $\mathfrak{R}F_k$, we set $k = r$ and choose a basis $(z_1, z_2, \cdots, z_{n-r})$ in $\mathfrak{R}F_k$. If $g(x, y)$ is not alternate in $\mathfrak{R}F_k$, then we choose a vector u_{k+1} in this subspace so that $g(u_{k+1}, u_{k+1}) = \beta_{k+1} \neq 0$. We then repeat the argument with the $k + 1$ u's. It remains to consider now the case in which $g(x, y)$ is not identically 0 and alternate in $\mathfrak{R}F_k$. Here we can find two linearly independent vectors v, w such that

$$(v, v) = 0 = (w, w), \quad (v, w) = 1 = (w, v).$$

We now set $u = u_k$, $\beta = \beta_k$ and we consider the scalar product in the three-dimensional space $[u, v, w]$. Using these vectors as basis we obtain the matrix

$$\begin{bmatrix} \beta & 0 & 0 \\ 0 & 0 & 1 \\ 0 & 1 & 0 \end{bmatrix}.$$

Thus if $y = \xi u + \eta v + \zeta w$ and $y' = \xi' u + \eta' v + \zeta' w$, then $g(y, y') = \beta\xi\xi' + \eta\zeta' + \eta'\zeta$. Hence the following vectors

$$y_1 = u + v$$

$$y_2 = u + \beta w$$

$$y_3 = u + v + \beta w$$

are orthogonal in pairs and satisfy $g(y_i, y_i) = \beta$. It follows that, if we replace the original u_k by y_1 and call this vector u_k again, then $u_1, u_2, \cdots, u_k, u_{k+1} = y_2, u_{k+2} = y_3$ satisfy $g(u_i, u_j) = \delta_{ij}\beta_i$, $\beta_i \neq 0$. This proves

Theorem 10. *If Φ has characteristic two and $g(x, y)$ is a non-alternate symmetric scalar product in $\mathfrak{R}$ over Φ, then there exists a basis of $\mathfrak{R}$ relative to which the matrix of $g(x, y)$ is a diagonal matrix.**

The argument used in the proof of the preceding theorem shows that the matrices.

$$\begin{bmatrix} 1 & 0 & 0 \\ 0 & 0 & 1 \\ 0 & 1 & 0 \end{bmatrix}, \quad \begin{bmatrix} 1 & 0 & 0 \\ 0 & 1 & 0 \\ 0 & 0 & 1 \end{bmatrix}$$

are cogredient. On the other hand these submatrices

$$\begin{bmatrix} 0 & 1 \\ 1 & 0 \end{bmatrix}, \quad \begin{bmatrix} 1 & 0 \\ 0 & 1 \end{bmatrix}$$

are not cogredient. These observations show that Witt's theorem does not hold in the characteristic two case.

* The canonical form given in this theorem is useful in that it simplifies matrix calculations with symmetric matrices over a field of characteristic two. A more geometric discussion of symmetric scalar products in the characteristic two case has been indicated by Dieudonné, *Sur les Groupes Classiques*, p. 62.

Chapter VI

EUCLIDEAN AND UNITARY SPACES

As we have pointed out at the beginning of these Lectures, Euclidean geometry is concerned with the study of a real vector space relative to the scalar product $\sum_1^n \xi_i\eta_i$ determined by writing $x = \Sigma\xi_i u_i$, $y = \Sigma\eta_i u_i$ in terms of basic unit vectors that are mutually orthogonal. Since this scalar product is fixed, it is customary to denote it simply as (x, y) instead of $g(x, y)$ as in the preceding chapter. The geometric meaning of (x, y) is clear. It gives the product of the cosine of the angle between x and y by the lengths of the two vectors. The length of x can also be expressed in terms of the scalar product, namely, $|x| = (x, x)^{1/2}$.

From our point of view the characteristic properties of (x, y) can be stated by saying that this function is positive definite in the sense that $(x, x) > 0$ for all $x \neq 0$. In fact, it is customary nowadays to axiomatize Euclidean geometry in the following way. We suppose $\mathfrak{R}$ is a finite dimensional vector space over the field of real numbers, and we take in $\mathfrak{R}$ a positive definite symmetric scalar product (x, y). The space $\mathfrak{R}$ over Φ, together with the fundamental scalar product, constitutes a *Euclidean space.* In a similar manner we define the complex analogue of a Euclidean space as a vector space over the field of complex numbers together with a positive definite hermitian scalar product defined in this space. In this chapter we shall study properties of these spaces and of certain special types of linear transformations in these spaces. We shall also consider briefly the theory of analytic functions of matrices (or linear transformations).

1. Cartesian bases. Let $\mathfrak{R}$ be a Euclidean space in the sense that $\mathfrak{R}$ is an n-dimensional vector space over the field Φ of real numbers in which a positive definite symmetric scalar product (x, y) is defined. Thus the basic properties of this real valued function are

$$(1) \qquad (x_1 + x_2, y) = (x_1, y) + (x_2, y),$$

$$(x, y_1 + y_2) = (x, y_1) + (x, y_2)$$

$$(2) \qquad (\alpha x, y) = \alpha(x, y) = (x, \alpha y)$$

$$(3) \qquad (x, y) = (y, x)$$

$$(4) \qquad (x, x) > 0 \quad \text{if} \quad x \neq 0.$$

Now we know that, if (x, y) is any symmetric scalar product in a real vector space, then there exists a basis $(u_1, u_2, \cdots, u_n)$ such that $(u_i, u_j) = \delta_{ij}\beta_j$ where the β_i are either 1, -1, or 0. If (x, y) is positive definite, clearly every $\beta_i = 1$. Thus we see that there exists a basis $(u_1, u_2, \cdots, u_n)$ for $\mathfrak{R}$ such that

$$(5) \qquad (u_i, u_j) = \delta_{ij}.$$

Then if $x = \Sigma\xi_i u_i$, $y = \Sigma\eta_i u_i$,

$$(x, y) = \Sigma\xi_i\eta_i$$

as usual.

A basis which satisfies (5) will now be called a *Cartesian basis* for the Euclidean space. The method of Lagrange for determining such a basis is capable of some refinements which we now indicate. Let $(e_1, e_2, \cdots, e_n)$ be any basis for $\mathfrak{R}$. Then $(e_1, e_1) > 0$. Hence if $u_1 = (e_1, e_1)^{-\frac{1}{2}}e_1$, $(u_1, u_1) = 1$. We next apply Lagrange's reduction and set

$$f_2 = e_2 - (e_2, u_1)u_1.$$

Then $(f_2, u_1) = (e_2, u_1) - (e_2, u_1)(u_1, u_1) = 0$. Hence u_1 and $u_2 = (f_2, f_2)^{-\frac{1}{2}}f_2$ satisfy

$$(u_1, u_1) = 1 = (u_2, u_2), \quad (u_1, u_2) = 0 = (u_2, u_1).$$

Suppose now that $(u_1, u_2, \cdots, u_k)$ have already been determined so that $(u_i, u_j) = \delta_{ij}$ and $[u_1, u_2, \cdots, u_i] = [e_1, e_2, \cdots, e_i]$ for $i \leq k$. Let

$$f_{k+1} = e_{k+1} - (e_{k+1}, u_1)u_1 - (e_{k+1}, u_2)u_2 - \cdots - (e_{k+1}, u_k)u_k.$$

Then $(f_{k+1}, u_i) = 0$ for $i = 1, 2, \cdots, k$. Hence if we set $u_{k+1} = (f_{k+1}, f_{k+1})^{-\frac{1}{2}} f_{k+1}$, then $(u_1, u_2, \cdots, u_{k+1})$ satisfy the condition stated for $(u_1, u_2, \cdots, u_k)$. Repetition of this method leads at length to a Cartesian basis for $\mathfrak{R}$. The process we have given for "orthogonalizing" the basis $(e_1, e_2, \cdots, e_n)$ is often called *E. Schmidt's orthogonalization process.*

It is worth noting that the matrix of $(u_1, u_2, \cdots, u_n)$ relative to $(e_1, e_2, \cdots, e_n)$ is triangular; for $[u_1, u_2, \cdots, u_i] = [e_1, e_2, \cdots, e_i]$. Hence u_i is a linear combination of the e_j with $j \leq i$. Thus $u_i = \sum_{j=1}^{i} \tau_{ij} e_j$, and the matrix of the u's relative to the e's is

$$(6) \qquad (\tau) = \begin{bmatrix} \tau_{11} & & & 0 \\ \tau_{21} & \tau_{22} & & \\ \cdot & \cdot & \cdot & \\ \cdot & \cdot & \cdot & \\ \cdot & \cdot & & \cdot \\ \tau_{n1} & \tau_{n2} & \cdots & \tau_{nn} \end{bmatrix}.$$

If (β) is the matrix of (x, y) relative to the originally chosen basis $(e_1, e_2, \cdots, e_n)$, then we know that the matrix of $(u_1, u_2, \cdots, u_n)$ is $(\tau)(\beta)(\tau)'$. On the other hand, by (5) the latter matrix is the identity. Hence $(\tau)(\beta)(\tau)' = 1$. The inverse (ν) of (τ) is also triangular of the same form as (τ) and $(\beta) = (\nu)(\nu)'$. This proves the following

Theorem 1. *Let (β) be a matrix of a positive definite symmetric scalar product. Then there exists a triangular matrix (ν) such that $(\beta) = (\nu)(\nu)'$.*

We consider now the relation between Cartesian bases and matrices. Let $(v_1, v_2, \cdots, v_n)$ be a definite Cartesian basis. Then if $(u_1, u_2, \cdots, u_n)$ is a second such basis, the matrices of (x, y) relative to these bases is 1. Hence if (σ) is the matrix of $(u_1, u_2,$

$\cdots, u_n)$ relative to $(v_1, v_2, \cdots, v_n)$, then $(\sigma)(\sigma)' = 1$. A matrix satisfying this condition:

$$(\sigma)(\sigma)' = 1 \tag{7}$$

is called an *orthogonal* matrix. Conversely, if (σ) is any orthogonal matrix and $u_i = \Sigma\sigma_{ij}v_j$, then the matrix of (x, y) relative to $(u_1, u_2, \cdots, u_n)$ is $(\sigma)1(\sigma)' = 1$. Hence the u's form a Cartesian basis. Thus we see that, if one Cartesian basis is known, all others can be obtained from it by applying orthogonal matrices. We have a 1–1 correspondence between the Cartesian bases and the orthogonal matrices.

If (σ) is orthogonal, then $(\det(\sigma))^2 = 1$. Hence $\det(\sigma) = \pm 1$. If $\det(\sigma) = 1$, we shall say that (σ) is *proper*, otherwise (σ) is *improper*. It is readily verified that the orthogonal matrices constitute a subgroup $O(\Phi, n)$ of the group $L(\Phi, n)$. The proper orthogonal matrices form an invariant subgroup $O_1(\Phi, n)$ in $O(\Phi, n)$ and the index of O_1 in O is two.

If two Cartesian bases are related by a proper orthogonal matrix, then we say that the bases have the *same orientation* and, if the matrix relating the bases is improper, we say that the bases have *opposite orientation*. Thus the Cartesian bases fall into two mutually exclusive classes, namely, those that have the same orientation and those that have opposite orientation to a particular Cartesian basis $(v_1, v_2, \cdots, v_n)$. We note that a change of sign of one vector or an odd permutation of the vectors changes the orientation.

Suppose again that $(e_1, e_2, \cdots, e_n)$ is any basis and let (γ) be the matrix of this basis relative to the Cartesian basis $(v_1, v_2, \cdots, v_n)$. Since the e's form an arbitrary basis, the matrix (γ) is an arbitrary matrix in $L(\Phi, n)$. Now we have seen that there exists a Cartesian basis $(u_1, u_2, \cdots, u_n)$ whose matrix relative to $(e_1, e_2, \cdots, e_n)$ is a triangular matrix (τ). The matrix of $(u_1, u_2, \cdots, u_n)$ relative to $(v_1, v_2, \cdots, v_n)$ is the product $(\tau)(\gamma)$. Since $(u_1, u_2, \cdots, u_n)$ is Cartesian, $(\tau)(\gamma) = (\sigma)$ is orthogonal. Hence we have the following

Theorem 2. *If* $(\gamma) \in L(\Phi, n)$, Φ *the field of real numbers, then* (γ) *may be factored as* $(\nu)(\sigma)$ *where* (ν) *is triangular and* (σ) *is orthogonal.*

EXERCISES

1. Show that the matrix (ν) in the preceding theorem can be taken to have positive diagonal elements. Show that, if this normalization is made, then (ν) and (σ) are uniquely determined by (γ).

2. Decompose

$$\begin{bmatrix} -1 & 0 & 2 & -2 \\ 3 & 4 & -1 & 7 \\ 1 & 2 & 1 & -1 \\ 0 & 2 & 0 & 1 \end{bmatrix}$$

as a product of a triangular matrix and an orthogonal matrix.

3. Show that, if (β) is the matrix of a positive definite symmetric scalar product, then $\det(\beta) > 0$. Generalize this to prove that every diagonal minor of (β) is positive.

4. Prove that, if (x, y) is a positive definite symmetric scalar product, then $(u, v)^2 \leq (u, u)(v, v)$ for any u and v and that equality holds only if u and v are linearly dependent.

5. Prove the *triangle inequality* for lengths of vectors:

$$(u + v, u + v)^{1/2} \leq (u, u)^{1/2} + (v, v)^{1/2}.$$

2. Linear transformations and scalar products. If A is a linear transformation of $\mathfrak{R}$ over Φ into itself, then $g(x, y) = (xA, y)$ is a second scalar product in the space. Conversely let $g(x, y)$ be any scalar product in $\mathfrak{R}$. Then if we hold x fixed, the function $g(x, y)$ is linear in y. Hence, as we have shown in Chapter V, there is a uniquely determined vector xA in $\mathfrak{R}$ such that $g(x, y) = (xA, y)$ for all y. The mapping A is linear. This shows that any scalar product in $\mathfrak{R}$ can be obtained from the fundamental scalar product (x, y) by applying a linear transformation A in the above manner. The theory of scalar products in $\mathfrak{R}$ is therefore equivalent to the theory of linear transformations in $\mathfrak{R}$.

For the most part we shall adopt the linear transformation point of view. We recall first the definition of the transpose A' relative to (x, y) of the linear transformation A: A' is the linear transformation in $\mathfrak{R}$ which satisfies the condition

$$(x, yA') = (xA, y) \tag{8}$$

for all x, y in $\mathfrak{R}$. There is only one linear transformation that satisfies this condition, and the requirement (8) can be reduced to the n^2 equations

$$(u_i, u_jA') = (u_iA, u_j), \quad i, j = 1, 2, \cdots, n$$

for a basis $(u_1, u_2, \cdots, u_n)$. If the basis is Cartesian these conditions imply that A' is the linear transformation whose matrix relative to $(u_1, u_2, \cdots, u_n)$ is the transpose of that of A. If

$$u_iA = \Sigma\alpha_{ik}u_k \quad \text{and} \quad u_jA' = \Sigma\beta_{jl}u_l,$$

then

$$(u_i, u_jA') = \Sigma\beta_{jl}(u_i, u_l) = \beta_{ji}$$

$$(u_iA, u_j) = \Sigma\alpha_{ik}(u_k, u_j) = \alpha_{ij}.$$

Hence $(\beta) = (\alpha)'$. (Cf. p. 145.)

We recall the fundamental algebraic properties of the mapping $A \rightarrow A'$:

$$(A + B)' = A' + B', \quad (AB)' = B'A'.$$

Also it is clear from (8) and the symmetry of (x, y) that

$$A'' = A.$$

3. Orthogonal complete reducibility. If $\mathfrak{S}$ is any subspace $\neq 0$ of $\mathfrak{R}$, (x, y) is non-degenerate in $\mathfrak{S}$. In fact, $(x, x) \neq 0$ if $x \neq 0$. It follows from this that the orthogonal complement $\mathfrak{S}^\perp$ of $\mathfrak{S}$ is a true complement, that is, $\mathfrak{R} = \mathfrak{S} \oplus \mathfrak{S}^\perp$ (cf. p. 151). It is easy to determine a basis for $\mathfrak{S}^\perp$. For this purpose one needs to have a Cartesian basis $(u_1, u_2, \cdots, u_r)$ for $\mathfrak{S}$. This can be supplemented to a basis $(u_1, u_2, \cdots, u_r; e_{r+1}, \cdots, e_n)$ for $\mathfrak{R}$. Then Schmidt's orthogonalization process yields the Cartesian basis $(u_1, u_2, \cdots, u_n)$, and it is clear that $\mathfrak{S}^\perp = (u_{r+1}, u_{r+2}, \cdots, u_n)$.

Suppose now that Ω is an arbitrary set of linear transformations in $\mathfrak{R}$ over Φ. We shall call Ω *orthogonally completely reducible* if the orthogonal complement $\mathfrak{S}^\perp$ of any subspace $\mathfrak{S}$ invariant under Ω is also invariant under Ω. Since $\mathfrak{R} = \mathfrak{S} \oplus \mathfrak{S}^\perp$, it is clear that orthogonal complete reducibility implies ordinary complete reducibility. We shall see that this property is enjoyed by many important types of linear transformations in $\mathfrak{R}$. If Ω is orthogonally completely reducible, we can decompose $\mathfrak{R}$ as a direct sum $\mathfrak{R}_1 \oplus \mathfrak{R}_2 \oplus \cdots \oplus \mathfrak{R}_t$ of subspaces that are invariant and irreducible relative to Ω and that are, moreover, mutually orthogonal. Thus let $\mathfrak{R}_1$ be an irreducible invariant subspace relative to Ω.

Then $\mathfrak{R}_1^{\perp}$ is invariant and $\mathfrak{R} = \mathfrak{R}_1 \oplus \mathfrak{R}_1^{\perp}$. Let $\mathfrak{R}_2$ be an irreducible invariant subspace contained in $\mathfrak{R}_1^{\perp}$. Then $\mathfrak{R}_1 + \mathfrak{R}_2 = \mathfrak{R}_1 \oplus \mathfrak{R}_2$ and these two spaces are orthogonal. Next we may write $\mathfrak{R} = \mathfrak{R}_1 \oplus \mathfrak{R}_2 \oplus (\mathfrak{R}_1 + \mathfrak{R}_2)^{\perp}$. Continuing in this way we obtain the required decomposition.

There is a very useful test for orthogonal complete reducibility which is based on the following

Theorem 3. *If Ω is any set of linear transformations and $\mathfrak{S}$ is invariant under Ω, then the orthogonal complement $\mathfrak{S}^{\perp}$ is invariant under Ω' the set of transposes of the linear transformations in Ω.*

Proof. Let $x \in \mathfrak{S}$ and $y \in \mathfrak{S}^{\perp}$. Then $xA \in \mathfrak{S}$ for any A in Ω. Hence $(xA, y) = 0$ and also $(x, yA') = 0$. Since this holds for all x in $\mathfrak{S}$, $yA' \in \mathfrak{S}^{\perp}$. Since y is arbitrary in $\mathfrak{S}^{\perp}$, this shows that $\mathfrak{S}^{\perp}$ is invariant under Ω'.

We can now state the following important criterion.

Corollary. *A set Ω is orthogonally completely reducible if and only if Ω and Ω' have the same invariant subspaces.*

4. Symmetric, skew and orthogonal linear transformations. Of special interest in Euclidean geometry are the following types of linear transformations: *Symmetric*, defined by the condition $A' = A$, *skew* defined by $A' = -A$ and *orthogonal* defined by $A' = A^{-1}$. These conditions can also be given in terms of the associated scalar products. Thus A is symmetric if and only if (xA, y) is symmetric. This is readily verified. Similarly A is skew if and only if (xA, y) is a skew scalar product in $\mathfrak{R}$. If (α) is the matrix of A relative to a Cartesian basis $(u_1, u_2, \cdots, u_n)$, then A is symmetric (skew) if and only if (α) is symmetric (skew).

In order to see the geometric meaning of orthogonal linear transformations we must recall that (u, u) gives the square of the length of the vector u. The condition $AA' = 1 = A'A$ implies that

$$(uA, uA) = (u, uAA') = (u, u). \tag{9}$$

Hence A preserves the length of any vector. We now prove the converse, namely, if A is any linear transformation that leaves the length of every vector unaltered, then A is orthogonal. For if (9) holds for all u, then

$$((x+y)A, (x+y)A) = (x+y, x+y)$$

for all x and y. Expanding and cancelling off equal terms,

$$2(xA, yA) = 2(x, y).$$

Hence $(x, yAA') = (x, y)$ for all x, y. Since (x, y) is non-degenerate this implies that $AA' = 1$.

It is clear that A is orthogonal if and only if its matrix (α) relative to a Cartesian basis is orthogonal. Another way of stating this result is that A is orthogonal if and only if the transform $(u_1A, u_2A, \cdots, u_nA)$ of a Cartesian basis is a Cartesian basis for $\mathfrak{R}$. The matrix (α) has determinant 1 or determinant -1. In the former case $(u_1A, u_2A, \cdots, u_nA)$ has the same orientation as $(u_1, u_2, \cdots, u_n)$. Then A is called a *rotation* in $\mathfrak{R}$.

5. Canonical matrices for symmetric and skew linear transformations. If $\mathfrak{S}$ is a subspace invariant under a linear transformation A that is symmetric, then clearly $\mathfrak{S}$ is invariant under A'. Hence the set consisting of A alone is orthogonally completely reducible. Our discussion of orthogonally completely reducible sets suggests the following procedure for obtaining a canonical matrix for A.

Let x be a non-zero vector in $\mathfrak{R}$ and let $\mu_x(\lambda)$ be its order (see p. 67). If $\pi(\lambda)$ is an irreducible factor (leading coefficient 1) of $\mu_x(\lambda)$ and $\mu_x(\lambda) = \pi(\lambda)\nu(\lambda)$, then $y = x\nu(A)$ has the order $\pi(\lambda)$. Since Φ is the field of real numbers the irreducible polynomial $\pi(\lambda)$ is either linear or quadratic. We shall now show that the symmetry of A assures that $\pi(\lambda)$ is linear. Otherwise (y, yA) is a basis for the cyclic space $\{y\}$, and this basis yields the matrix

$$\begin{bmatrix} 0 & 1 \\ \beta & \alpha \end{bmatrix} \tag{10}$$

where $\lambda^2 - \alpha\lambda - \beta = \pi(\lambda)$. On the other hand, if we choose a Cartesian basis in $\{y\}$, then we obtain a symmetric matrix

$$\begin{bmatrix} \gamma & \delta \\ \delta & \epsilon \end{bmatrix} \tag{11}$$

for A in $\{y\}$. Since (10) and (11) are similar,

$$\lambda^2 - \alpha\lambda - \beta = \lambda^2 - (\gamma + \epsilon)\lambda + (\gamma\epsilon - \delta^2).$$

The discriminant of this quadratic is

$$(\gamma + \epsilon)^2 - 4(\gamma\epsilon - \delta^2) = (\gamma - \epsilon)^2 + 4\delta^2 \geq 0$$

and this contradicts the irreducibility of $\pi(\lambda)$. Thus we see that $\pi(\lambda)$ must be linear so that $\{y\} = [y]$ is one dimensional and $yA = \rho y$.

We now replace y by a multiple y_1 that has length 1. Then $y_1A = \rho_1 y_1$, $\rho_1 \equiv \rho$. Also $\mathfrak{R} = [y_1] \oplus [y_1]^\perp$ and A induces a symmetric transformation in $[y_1]^\perp$. Hence we can find a y_2 of length 1 in $[y_1]^\perp$ such that $y_2A = \rho_2 y_2$. Next we use the decomposition $\mathfrak{R} = [y_1, y_2] + [y_1, y_2]^\perp$ to obtain a y_3 of length 1 in $[y_1, y_2]^\perp$ such that $y_3A = \rho_3 y_3$. We remark that the y_i thus obtained are orthogonal in pairs. Hence when we have finished our process, we get a Cartesian basis $(y_1, y_2, \cdots, y_n)$. The matrix of A determined by this basis is

$$\text{diag}\,\{\rho_1, \rho_2, \cdots, \rho_n\}. \tag{12}$$

If we recall that the passage from one Cartesian basis to another is given by an orthogonal matrix, we see that the following theorem holds.

Theorem 4. *If (α) is a real symmetric matrix, then there exists a real orthogonal matrix (σ) such that $(\sigma)(\alpha)(\sigma)^{-1}$ is a diagonal matrix.*

We can also arrange to have (σ) proper. This can be done by changing the sign, if necessary, of one of the y_i. We note also that the ρ_i in the canonical matrix (12) are the roots of the characteristic polynomial of this matrix. Hence they are also the roots of the characteristic polynomial of (α). We have therefore proved incidentally the

Corollary. *The characteristic roots of a real symmetric matrix are real.*

We pass now to the theory of a skew linear transformation in Euclidean space. Let A be skew and, as above, let y be a vector $\neq 0$ whose order is an irreducible polynomial $\pi(\lambda)$. If $\pi(\lambda)$ is linear, then $yA = \rho y$. Hence $\rho(y, y) = (yA, y) = -(y, yA) = -\rho(y, y)$, and this implies that $\rho = 0$. If $\pi(\lambda)$ is quadratic, we choose a Cartesian basis in $\{y\}$. We then obtain a skew symmetric matrix

$$\begin{bmatrix} 0 & \delta \\ -\delta & 0 \end{bmatrix}.$$

The polynomial $\pi(\lambda)$ is the characteristic polynomial of this matrix. Hence $\pi(\lambda) = \lambda^2 + \delta^2$ where $\delta \neq 0$. If we apply the method used above for symmetric mappings, we can decompose the space as a direct sum of mutually orthogonal spaces $\mathfrak{S}_i$, $i = 1, 2, \cdots, h$, such that each $\mathfrak{S}_i = \{y_i\}$, and the minimum polynomial $\pi_i(\lambda)$ of y_i is either λ or it is of the form $\lambda^2 + \delta_i^2$, $\delta_i \neq 0$. We can arrange the $\mathfrak{S}_i$ so that $\pi_i(\lambda) = \lambda^2 + \delta_i^2$ for $i = 1, 2, \cdots, k$ and $\pi_i(\lambda) = \lambda$ for $i > k$. If we choose Cartesian bases in the $\mathfrak{S}_i$, we obtain a Cartesian basis for $\mathfrak{R}$ by stringing these bases together. The matrix of A relative to this basis is

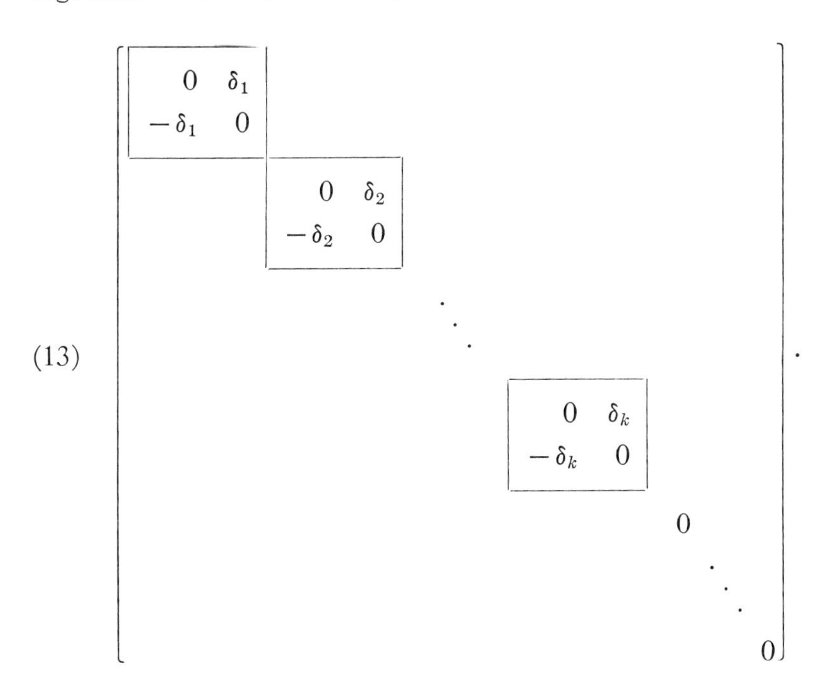

$$(13) \qquad \begin{bmatrix} \begin{matrix} 0 & \delta_1 \\ -\delta_1 & 0 \end{matrix} & & & & & & \\ & \begin{matrix} 0 & \delta_2 \\ -\delta_2 & 0 \end{matrix} & & & & & \\ & & \ddots & & & & \\ & & & \begin{matrix} 0 & \delta_k \\ -\delta_k & 0 \end{matrix} & & & \\ & & & & 0 & & \\ & & & & & \ddots & \\ & & & & & & 0 \end{bmatrix}.$$

This proves the following

Theorem 5. *If* (α) *is a real skew symmetrix matrix, then there exists a real orthogonal matrix* (σ) *such that* $(\sigma)(\alpha)(\sigma)^{-1}$ *has the form* (13).

As for symmetric matrices the canonical form is completely determined by the characteristic polynomial

$$\lambda^{n-2k} \prod_{1}^{k} (\lambda^2 + \delta_i^2), \quad \delta_i \neq 0$$

of (α). We remark also that the characteristic roots are pure imaginaries.

EXERCISES

1. If

$$(\alpha) = \begin{bmatrix} 1 & -2 & 0 \\ -2 & 2 & -2 \\ 0 & -2 & 3 \end{bmatrix},$$

find a proper orthogonal matrix (σ) such that $(\sigma)(\alpha)(\sigma)^{-1}$ is diagonal.

2. Prove that, if $g(x, y)$ is a symmetric scalar product in Euclidean space, then there exists a Cartesian basis for $\mathfrak{R}$ relative to which the matrix of g is diagonal.

3. Prove that, if (α) is a real symmetric or skew symmetric matrix of rank ρ, then there is a non-zero ρ-rowed diagonal minor in (α).

6. Commutative symmetric and skew linear transformations. If ρ is a root of the characteristic polynomial $f(\lambda)$ of the symmetric linear transformation, A, then the subspace $\mathfrak{R}_{\lambda-\rho}$ of vectors y such that $yA = \rho y$ is $\neq 0$. We call $\mathfrak{R}_{\lambda-\rho}$ the *characteristic space* corresponding to the root ρ or to the factor $\lambda - \rho$ of $f(\lambda)$. If A is skew and λ is a factor of $f(\lambda)$, we define in a similar manner the characteristic space $\mathfrak{R}_\lambda$. The other irreducible factors of $f(\lambda)$ have the form $\pi(\lambda) = \lambda^2 + \delta^2$, $\delta \neq 0$. Corresponding to such a factor we define the *characteristic space* $\mathfrak{R}_{\pi(\lambda)}$ to be the totality of vectors y such that $y\pi(A) = 0$. As we have seen in the preceding section, $\mathfrak{R}_{\pi(\lambda)} \neq 0$.

If B is any linear transformation which commutes with A, it is clear that B maps each characteristic space of A into itself. For, evidently, $y\pi(A) = 0$ implies that $y\pi(A)B = (yB)\pi(A) = 0$.*

* The ideas which occur here have been discussed in a more general form in § 9 of Chapter IV.

Now let Ω be a set of linear transformations such that 1) any two linear transformations of Ω commute, 2) every transformation of Ω is either symmetric or skew. In order to determine canonical matrices for the set Ω, we consider first the case in which Ω is an irreducible set. The result that we shall establish is the following

Theorem 6. *If a Euclidean space $\mathfrak{R}$ is irreducible relative to a commutative set of linear transformations which are either symmetric or skew, then dim $\mathfrak{R} \leq 2$.*

Proof. Let $\pi(\lambda)$ be an irreducible factor of the characteristic polynomial of any $A \varepsilon \Omega$. Then the characteristic space $\mathfrak{R}_{\pi(\lambda)}$ of A is invariant relative to Ω. Hence $\mathfrak{R}_{\pi(\lambda)} = \mathfrak{R}$ so that $\pi(A) = 0$. If A is symmetric, then we know that $\pi(\lambda) = \lambda - \rho$; hence, in this case, A is a scalar multiplication. If A is skew, either $\pi(\lambda) = \lambda$, in which case $A = 0$, or $\pi(\lambda) = \lambda^2 + \delta^2$, $\delta \neq 0$. In the latter case $A^2 = -\delta^2 1$; hence A^{-1} exists and is skew. Then if B is any other skew transformation in the set,

$$M = BA^{-1} = A^{-1}B = (-A^{-1})'(-B') = (A^{-1})'B' = M'$$

is symmetric. Since M commutes with every member of Ω, the argument used before shows that M is a scalar multiplication. Hence $B = \mu A$. Thus we see that either the set Ω consists of scalar multiplications only, or Ω consists of scalar multiplications and multiples of a single skew A in this set. In the former case every subspace of $\mathfrak{R}$ is invariant relative to Ω. Hence by the irreducibility of $\mathfrak{R}$, dim $\mathfrak{R} = 1$. In the second case we can find a two-dimensional subspace which is invariant and irreducible relative to the chosen skew transformation A. Clearly this space is also invariant relative to every member of Ω. Hence, here dim $\mathfrak{R} = 2$. This completes the proof.

Now suppose that Ω is an arbitrary commutative set of symmetric or skew linear transformations in a Euclidean space. It is clear by the Corollary to Theorem 3 that every set of symmetric and skew linear transformations is orthogonally completely reducible. This implies that we may write $\mathfrak{R} = \mathfrak{S}_1 \oplus \mathfrak{S}_2 \oplus \cdots \oplus \mathfrak{S}_h$ where the $\mathfrak{S}_i$ are mutually orthogonal and irreducible and invariant relative to Ω. We can now apply Theorem 6 to conclude that the spaces $\mathfrak{S}_i$ are either one-dimensional or two-dimensional. Moreover, the proof of Theorem 6 shows that the symmetric A are scalar multiplications in the $\mathfrak{S}_i$ and that the skew A are multiples of a particular one. If we choose Cartesian bases in the $\mathfrak{S}_i$, we obtain a Cartesian basis for $\mathfrak{R}$ relative to which the matrices of the $A \varepsilon \Omega$ all have the form

$$\begin{bmatrix} (\beta_1) & & & \\ & (\beta_2) & & \\ & & \ddots & \\ & & & (\beta_n) \end{bmatrix} \tag{14}$$

where (β_i) is either a one- or two-rowed scalar matrix or

$$(\beta_i) = \begin{bmatrix} 0 & \epsilon_i \\ -\epsilon_i & 0 \end{bmatrix}. \tag{15}$$

In matrix form our result is the following

Theorem 7. *Let ω be a set of commutative real matrices which are either symmetric or skew symmetric. Then there exists a real orthogonal matrix (σ) such that, for each $(\alpha) \varepsilon \omega$, $(\sigma)(\alpha)(\sigma)^{-1}$ has the form* (14).

EXERCISES

1. Prove that if A is symmetric, then any two distinct characteristic spaces of A are orthogonal. Also show that $\mathfrak{R}$ is a direct sum of the characteristic subspaces relative to A.
2. Prove that, if ω is a set of commutative real symmetric matrices, then there exists a real symmetric matrix (δ) such that each $(\alpha) \varepsilon \omega$ is a polynomial in (δ).

7. Normal and orthogonal linear transformations. A linear transformation A is called *normal* if it commutes with its transpose. Special cases of such mappings are the symmetric, skew and orthogonal linear transformations. If A is any linear transformation, we may write $A = \frac{1}{2}(A + A') + \frac{1}{2}(A - A') = B + C$ where $B = \frac{1}{2}(A + A')$ is symmetric and $C = \frac{1}{2}(A - A')$ is skew. This decomposition into a symmetric and a skew part is unique. For if $B + C = B_1 + C_1$ where $B_1' = B_1$ and $C_1' = -C_1$, then

$$(B - B_1) = (C_1 - C).$$

Since $B - B_1$ is symmetric and $C_1 - C$ is skew, $B - B_1 = 0 = C_1 - C$. We may therefore call B and C respectively *the symmetric* and *the skew* part of A. We note now that A is normal if and only if B and C commute. This is clear since $B = \frac{1}{2}(A + A')$, $C = \frac{1}{2}(A - A')$ and $A = B + C$, $A' = B - C$. This remark shows that the results on commutative symmetric and skew linear transformations can be applied to the study of normal linear transformations. Using this method we see that there exists a Cartesian basis relative to which B and C have matrices of the form (14). Since B is symmetric and C is skew we know that, if (β_i) is a one-rowed block, then it is 0 for C and if (β_i) is two-rowed, then it has the form (15) for C and is scalar for B. The matrix of $A = B + C$ is the sum of the matrices of B and of C. Hence it has the form (14) where each (β_i) is one-rowed or

$$(16) \qquad (\beta_i) = \begin{bmatrix} \rho_i & \epsilon_i \\ -\epsilon_i & \rho_i \end{bmatrix}$$

$\epsilon_i \neq 0$.

A matrix is called *normal* if it commutes with its transposed. A linear transformation A is normal if and only if its matrix relative to a Cartesian basis is normal. Our discussion therefore yields the following

Theorem 8. *If* (α) *is a real normal matrix, there exists an orthogonal matrix* (σ) *such that* $(\sigma)(\alpha)(\sigma)^{-1}$ *has the form* (14) *where each* (β_i) *is either one-rowed or is a two-rowed matrix of the form* (16).

We consider finally the special case of orthogonal transformations. Here the one-rowed blocks (β_i) are associated with vectors $y_i \neq 0$ such that $y_iA = \beta_i y_i$. Since A does not change the length of vectors, $\beta_i = \pm 1$. Also A is orthogonal in the two-dimensional subspaces $\mathfrak{S}_i$. Hence the blocks (16) are orthogonal. Thus

$$\begin{bmatrix} \rho_i & \epsilon_i \\ -\epsilon_i & \rho_i \end{bmatrix} \begin{bmatrix} \rho_i & -\epsilon_i \\ \epsilon_i & \rho_i \end{bmatrix} = \begin{bmatrix} 1 & 0 \\ 0 & 1 \end{bmatrix}.$$

This reduces to the single condition

$$(17) \qquad \rho_i^2 + \epsilon_i^2 = 1.$$

We can determine a number θ_i such that $\cos\theta_i = \rho_i$, $\sin\theta_i = \epsilon_i$. Then (16) becomes

$$\begin{bmatrix} \cos\theta_i & \sin\theta_i \\ -\sin\theta_i & \cos\theta_i \end{bmatrix}. \tag{18}$$

Theorem 9. *Every real orthogonal matrix is conjugate in the group of orthogonal matrices to a matrix* (14) *in which the* (β_i) *are either one-rowed matrices* (± 1) *or are of the form* (18).

EXERCISES

1. Prove that an orthogonal linear transformation is proper or improper according as the multiplicity of the root -1 of the characteristic polynomial is even or odd.
2. Prove that, if A is a rotation in an odd-dimensional Euclidean space, then there exists a vector $u \neq 0$ such that $uA = u$.
3. Prove that every set of orthogonal linear transformations is orthogonally completely reducible.
4. Show that, if O is orthogonal and does not have -1 as a characteristic root, then $S = (O - 1)(O + 1)^{-1}$ is skew and $O = (1 + S)(1 - S)^{-1}$. Show that, if O is orthogonal and 1 is not a characteristic root, then $S = (O + 1)(O - 1)^{-1}$ is skew and $O = (S + 1)(S - 1)^{-1}$.

8. Semi-definite transformations.

A linear transformation A is said to be *positive definite* if it is symmetric and if the associated symmetric bilinear form (xA, y) is positive definite. This, of course, means simply that $(xA, x) > 0$ for all $x \neq 0$. It is useful also to generalize this notion slightly by defining A to be (non-negative) *semi-definite* if $(xA, x) \geq 0$ for all x. It is evident from the definition that a positive definite transformation is semi-definite and 1–1. The converse holds also. This will follow from the following criterion for definiteness and semi-definiteness.

Theorem 10. *A symmetric transformation A is positive definite (semi-definite) if and only if its characteristic roots are all positive (non-negative).*

Proof. We know that there exists a Cartesian basis $(y_1, y_2, \cdots, y_n)$ for $\mathfrak{R}$ such that each y_i is a *characteristic vector* in the sense that $y_iA = \rho_i y_i$. The ρ_i are the characteristic roots. Now $(y_iA, y_i) = (\rho_i y_i, y_i) = \rho_i(y_i, y_i)$ has the same sign as ρ_i.

Hence if A is definite (semi-definite) each ρ_i is positive (non-negative). Conversely suppose that $\rho_i > 0$ (≥ 0) for all i and write $x = \Sigma \xi_i y_i$. Then

$$(xA, x) = \left(\sum_1^n \xi_i y_i A, \sum_1^n \xi_j y_j\right)$$
$$= \left(\sum_1^n \xi_i \rho_i y_i, \sum_1^n \xi_j y_j\right)$$
$$= \sum_1^n \rho_i \xi_i^2.$$

If all the ρ_i are > 0, this is > 0 unless each $\xi_i = 0$. Hence $\rho_i > 0$ insures definiteness. Also it is clear that, if the $\rho_i \geq 0$, then $(xA, x) \geq 0$.

If A is 1–1, then 0 is not a characteristic root of A. Hence if A is semi-definite and 1–1, all of its characteristic roots are positive. Theorem 10 then shows that A is definite.

If A is an arbitrary linear transformation in Euclidean space, then $B = AA'$ is semi-definite since $(xB, x) = (xAA', x) = (xA, xA) \geq 0$. We note also that B and A have the same null-space; for it is clear that, if $zA = 0$, then $zB = 0$. On the other hand, if $zB = 0$, then $0 = (zB, z) = (zA, zA)$, and this implies that $zA = 0$. As a consequence of this remark we see that B is positive definite if and only if A is 1–1.

We shall now prove the following useful result:

Theorem 11. *Any semi-definite transformation B has a semi-definite square root P (that is, $P^2 = B$), and P is unique.*

Proof. The determination of a semi-definite P such that $P^2 = B$ is easy. We choose a Cartesian basis $(y_1, y_2, \cdots, y_n)$ such that $y_i B = \rho_i y_i$ and we know that the ρ_i are ≥ 0. Hence we can define P to be the linear transformation such that $y_i P = \rho_i^{1/2} y_i$. Clearly $P^2 = B$. Also since the matrix of P relative to a Cartesian basis is symmetric, P is symmetric. Finally since the characteristic roots $\rho_i^{1/2}$ of P are non-negative, P is semi-definite. To prove the uniqueness of P we arrange the characteristic vectors of B in groups in such a way that the first n_1 are those that go with the root ρ_1, the next n_2 are those that go with ρ_2 $(\neq\rho_1)$, etc. Also we introduce the spaces

$$\mathfrak{R}_1 = [y_1, y_2, \cdots, y_{n_1}], \quad \mathfrak{R}_2 = [y_{n_1+1}, y_{n_1+2}, \cdots, y_{n_1+n_2}], \cdots$$

so that $\mathfrak{R} = \mathfrak{R}_1 \oplus \mathfrak{R}_2 \oplus \cdots \oplus \mathfrak{R}_h$ (h = number of distinct characteristic roots). If u_i is any vector in $\mathfrak{R}_i$, clearly $u_iB = \rho_iu_i$. On the other hand, let $u = u_1 + u_2 + \cdots + u_h$ where $u_i \in \mathfrak{R}_i$, be any characteristic vector belonging to the root ρ_i, that is, $uB = \rho_iu$. Then

$$\begin{aligned}\rho_iu &= uB = (u_1 + u_2 + \cdots + u_h)B = u_1B + u_2B + \cdots + u_hB \\ &= \rho_1u_1 + \rho_2u_2 + \cdots + \rho_hu_h.\end{aligned}$$

Hence $u_j = 0$ if $j \neq i$ and $u = u_i \in \mathfrak{R}_i$. Thus we see that the space $\mathfrak{R}_i$ is just the characteristic subspace of $\mathfrak{R}$ corresponding to the characteristic root ρ_i of B. As we have seen the characteristic subspaces of a linear transformation B are invariant with respect to any linear transformation C that commutes with B. If P is any linear transformation such that $P^2 = B$, then $BP = PB$ and $\mathfrak{R}_iP \subseteq \mathfrak{R}_i$. Thus if P is semi-definite, then P induces a semi-definite transformation in each $\mathfrak{R}_i$. Now we can find a Cartesian basis $(w_1^{(i)}, w_2^{(i)}, \cdots, w_{n_i}^{(i)})$ for $\mathfrak{R}_i$ such that $Pw_j^{(i)} = \gamma_jw_j^{(i)}$. Then $Bw_j^{(i)} = P^2w_j^{(i)} = \gamma_j^2w_j^{(i)}$. Hence $\gamma_j^2 = \rho_i$ and $\gamma_j = \rho_i^{1/2}$. This shows that P coincides with the scalar multiplication by $\rho_i^{1/2}$ in the space $\mathfrak{R}_i$. Hence P is the mapping we constructed before.

EXERCISES

1. Show that any symmetric A whose negative characteristic roots have even multiplicities has a square root.
2. Prove that any symmetric A has a unique symmetric cube root.

9. Polar factorization of an arbitrary linear transformation. Evidently any real number can be written as a product of a non-negative real number by one of the numbers 1, -1. This result can be generalized to linear transformations in Euclidean space as follows

Theorem 12. *Every linear transformation A in Euclidean space can be written as a product $A = PO$ where P is semi-definite and*

O is orthogonal. P is uniquely determined while O is unique if and only if A is 1–1.

Proof. The proof is somewhat simpler in the important case in which A is 1–1; hence we consider this case first. We form the positive definite transformation $B = AA'$ and its positive definite square root P. Set $O = P^{-1}A$. Then

$$OO' = P^{-1}AA'P^{-1} = P^{-1}BP^{-1} = P^{-1}P^2P^{-1} = 1$$

and this shows that O is orthogonal. Since $A = PO$, the existence of the representation is established for A 1–1.

The proof in the general case is basically the same as the foregoing. Again we define $B = AA'$, and we take P to be the semi-definite square root of B. Our task is to determine an orthogonal transformation O such that $A = PO$. We define first a mapping of the space $\mathfrak{S} = \mathfrak{R}P$ into the space $\mathfrak{S}_1 = \mathfrak{R}A$ by specifying that $xP \rightarrow xA$. If $xP = yP$, $(x - y)P = 0$ and $(x - y)B = 0$. Since $B = AA'$, this implies that $(x - y)A = 0$ so that $xA = yA$. This shows that our correspondence is single-valued. It is now clear that it is linear. Also this mapping preserves lengths of vectors, since

$$(xA, xA) = (xAA', x) = (xB, x)$$

and

$$(xP, xP) = (xP^2, x) = (xB, x).$$

Now A, B and P have the same null space; hence they have the same rank. It follows that the orthogonal complements $\mathfrak{S}^{\perp}$ and $\mathfrak{S}_1{}^{\perp}$ have the same dimensionality $= h$. Now let $(u_1, u_2, \cdots, u_h)$, $(v_1, v_2, \cdots, v_h)$ be Cartesian bases for these spaces. Then the linear transformation of $\mathfrak{S}^{\perp}$ into $\mathfrak{S}_1{}^{\perp}$ that sends u_i into v_i preserves lengths. It follows that the mapping

$$O\colon xP + \Sigma\alpha_i u_i \rightarrow xA + \Sigma\alpha_i v_i$$

is an orthogonal transformation. Clearly $xPO = xA$ for all x so that $A = PO$ as required.

If $A = PO$, $AA' = P^2$. Hence P is necessarily a square root of the semi-definite transformation $B = AA'$. Hence P is unique.

If A is 1–1, it follows that $O = P^{-1}A$ is also unique. On the other hand, if A is not 1–1, then $h > 0$ in the above notation. In this case there are many choices available for the transformation of $\mathfrak{S}^{\perp}$ into $\mathfrak{S}_1^{\perp}$ and these give different determinations of an orthogonal transformation O. This completes the proof.

In a similar manner we can establish a factorization $A = O_1P_1$ where O_1 is orthogonal and P_1 is semi-definite. This can also be deduced by applying Theorem 12 to A'.

EXERCISES

1. Decompose

$$\begin{bmatrix} 1 & -1 & 2 \\ 2 & 1 & 3 \\ 1 & 0 & -5 \end{bmatrix}$$

as a product of a positive definite matrix and an orthogonal matrix.

2. Show that $A = PO$ is normal if and only if $PO = OP$.

10. Unitary geometry. Unitary geometry is the study of a vector space over the field of complex numbers relative to a positive definite hermitian scalar product (x, y). It is evident at the outset that this geometry will have the essential features of Euclidean geometry. On the other hand, we can expect some simplifications here due to the fact that the underlying field is algebraically closed. Consequently the theory of canonical matrices will be simpler than in the real case. It is not necessary to duplicate our previous discussion in all detail. In the main we shall be content to state the principal results and will give proofs only when new methods yield simplifications over the corresponding proofs in the Euclidean case.

We suppose now that Φ is the field of complex numbers and that (x, y) is a hermitian scalar product relative to the usual mapping $\alpha \to \bar{\alpha}$. Then we know that (x, x) is real for any x. We shall assume, moreover, that (x, y) is positive definite in the sense that $(x, x) > 0$ if $x \neq 0$. This assumption implies that there exists a basis $(u_1, u_2, \cdots, u_n)$ that is *unitary* in the sense that $(u_i, u_j) = \delta_{ij}$, $i, j = 1, 2, \cdots, n$. The passage from one unitary basis to another is given by a matrix (σ) that is *unitary* in the sense that

$$(\sigma)(\bar{\sigma})' = 1 = (\bar{\sigma})'(\sigma).$$

These matrices constitute the *unitary subgroup* $U(\Phi, n)$ of $L(\Phi, n)$, Φ the complex field. The Schmidt orthogonalization process holds and this enables us to carry over all of the results of § 1. The analogue of Theorem 1 is that, *if* (β) *is the matrix of a positive definite hermitian scalar product, then* (β) *has the form* $(\nu)(\bar{\nu}')$ *for a suitable triangular matrix* (ν) *with complex elements.* The analogue of Theorem 2 is obtained by replacing the word "real" by "complex" and "orthogonal" by "unitary" in its statement.

The transpose of a linear transformation A is defined as usual. If $(u_1, u_2, \cdots, u_n)$ is a unitary basis, then the matrix of A' is the conjugate transpose $(\bar{\alpha})'$ of the matrix of A. Thus A is *hermitian*, $A' = A$, if and only if (α) is hermitian; A is *skew hermitian*, $A' = -A$, if and only if (α) is skew hermitian; A is *unitary*, $A'A = 1 = AA'$, if and only if (α) is unitary; and A is *normal*, $AA' = A'A$, if and only if $(\alpha)(\bar{\alpha})' = (\bar{\alpha})'(\alpha)$. Skew hermitian transformations need not be studied as a separate case since their theory can be reduced to that of hermitian transformations by observing that, if A is skew hermitian, then iA $(i^2 = -1)$ is hermitian. This follows from the fact that the transpose of the scalar multiplication $x \to \mu x$ is the scalar multiplication $x \to \bar{\mu} x$. Hence $x \to ix$ is skew hermitian and, since it commutes with A, iA is hermitian.

Suppose now that A is hermitian and let ρ be a root of the characteristic polynomial. Let y be a corresponding characteristic vector. Then

$$\rho(y, y) = (\rho y, y) = (yA, y) = (y, yA) = (y, \rho y) = (y, y)\bar{\rho}.$$

Since $(y, y) \neq 0$, this implies that $\bar{\rho} = \rho$ is real. We normalize y to obtain a multiple y_1 such that $(y_1, y_1) = 1$. Then also $y_1 A = \rho_1 y_1$, $\rho = \rho_1$. If $\mathfrak{S}$ is any subspace, we denote the orthogonal complement consisting of the vectors y' such that $(y, y') = 0$ for all $y \in \mathfrak{S}$ by $\mathfrak{S}^\perp$. This space is a complement of $\mathfrak{S}$ and, if $\mathfrak{S}$ is invariant under A, then so is $\mathfrak{S}^\perp$. If we apply this remark to $\mathfrak{S} = [y_1]$, then we see that $\mathfrak{R} = [y_1] \oplus [y_1]^\perp$ and that $[y_1]^\perp A \subseteq [y_1]^\perp$. Hence we can find a real number ρ_2 and a vector y_2 in $[y_1]^\perp$ such that $y_2 A = \rho_2 y_2$. Next we write $\mathfrak{R} = [y_1, y_2] \oplus [y_1, y_2]^\perp$, and we repeat the argument with $[y_1, y_2]^\perp$. This leads finally to a unitary basis relative to which the matrix of A is

$$\text{diag}\,\{\rho_1, \rho_2, \cdots, \rho_n\}$$

where the ρ_i are real.

Theorem 13. *If* (α) *is a complex hermitian matrix, there exists a unitary matrix* (σ) *such that* $(\sigma)(\alpha)(\sigma)^{-1}$ *is a real diagonal matrix.*

Evidently a similar result holds also for skew-hermitian matrices. As in the Euclidean case we may prove next that, if ω is a set of commutative hermitian and skew-hermitian matrices, then there exists a single unitary matrix (σ) such that every $(\sigma)(\alpha)(\sigma)^{-1}$ is diagonal for every α in ω. This leads as before to Theorem 13, with the word "real" omitted, for normal matrices and consequently also for unitary matrices. For the latter the characteristic roots are of absolute value 1. This follows from

$$|\rho|^2(y, y) = (yA, yA) = (y, y)$$

if $yA = \rho y$. The principal theorem on unitary matrices can also be derived directly by using the same argument that we used in the hermitian case. The important remark is that, if $\mathfrak{S}$ is a subspace invariant under a unitary transformation, then the orthogonal complement $\mathfrak{S}^\perp$ is also invariant.

If A is a hermitian linear transformation, the associated bilinear form (xA, y) is hermitian; for

$$(yA, x) = (y, xA) = \overline{(xA, y)}.$$

It follows that (xA, x) is real for any x. Now as in the Euclidean case we define A to be *positive definite* (*semi-definite*) if $(xA, x) > 0$ (≥ 0) for all $x \neq 0$. The discussion that we gave in the Euclidean case can be carried over without change. Thus we can prove that any semi-definite transformation has a unique square root. As before, this can be used to establish the polar factorization. *Every linear transformation in unitary space can be written as* $A = PU$ *where* P *is semi-definite and* U *is unitary.* P is unique and U is unique if and only if A is 1–1.

We shall prove now another theorem that is applicable to arbitrary linear transformations and matrices in a unitary space. This is the following

Theorem 14. *If* (α) *is a matrix with complex elements, then there exists a unitary matrix* (σ) *such that* $(\sigma)(\alpha)(\sigma)^{-1}$ *is triangular.*

Proof. In order to prove this we let A be the linear transformation whose matrix relative to some unitary basis $(v_1, v_2, \cdots, v_n)$ is the given matrix (α). If ρ_1 is a characteristic root, there exists a vector y_1 such that $(y_1, y_1) = 1$ and $y_1A = \rho_1 y_1$. We can find a unitary basis $(y_1, y_2, \cdots, y_n)$ that includes the vector y_1. Then, since $y_1A = \rho_1 y_1$, the matrix of A relative to this basis is

$$(\beta) = \begin{bmatrix} \rho_1 & 0 & \cdots & 0 \\ \rho_{21} & \rho_{22} & \cdots & \rho_{2n} \\ \vdots & \vdots & & \vdots \\ \rho_{n1} & \rho_{n2} & \cdots & \rho_{nn} \end{bmatrix}. \tag{19}$$

If (μ) is the matrix of $(y_1, y_2, \cdots, y_n)$ relative to $(v_1, v_2, \cdots, v_n)$, then (μ) is unitary and $(\mu)(\alpha)(\mu)^{-1}$ is the matrix (β) of (19). We may now assume that there exists a unitary matrix (ν) of $n - 1$ rows and columns such that

$$(\nu)\begin{bmatrix} \rho_{22} & \cdots & \rho_{2n} \\ \cdot & \cdots & \cdot \\ \rho_{n2} & \cdots & \rho_{nn} \end{bmatrix}(\nu)^{-1} = \begin{bmatrix} \rho_2 & & 0 \\ & \ddots & \\ * & & \rho_n \end{bmatrix}.$$

Then the matrix $(\tau) = \begin{bmatrix} 1 & 0 \\ 0 & (\nu) \end{bmatrix}$ is unitary and

$$(\tau)(\beta)(\tau)^{-1} = (\sigma)(\alpha)(\sigma)^{-1} = \begin{bmatrix} \rho_1 & & & 0 \\ & \rho_2 & & \\ & & \ddots & \\ * & & & \rho_n \end{bmatrix}$$

where $(\sigma) = (\tau)(\mu)$ is unitary.

EXERCISES

1. Show that Theorem 14 holds for any set ω of commutative complex matrices.
2. What is the analogue of Theorem 14 in the real case?
3. Show that a triangular matrix is normal if and only if it is diagonal. Use this to prove the theorem on canonical forms for normal matrices with complex elements.

11. Analytic functions of linear transformations. In treating analytic questions on matrices we shall take as our point of departure the notion of convergence of sequences of matrices. If $\{(\alpha^{(k)})\}$, $k = 1, 2, 3, \cdots$, is an infinite sequence of matrices in Φ_n, Φ the field of complex numbers, we say that $\{(\alpha^{(k)})\}$ *converges to* (α), $(\alpha^{(k)}) \to (\alpha)$, if the sequence of complex numbers $\alpha_{ij}{}^{(k)} \to \alpha_{ij}$ for every $i, j = 1, 2, \cdots, n$. The *limit* matrix (α) is unique since this is the case for sequences of complex numbers. Using our definition of convergence it is clear that addition and multiplication of matrices are continuous functions, that is, if $(\alpha^{(k)}) \to (\alpha)$ and $(\beta^{(k)}) \to (\beta)$, then

$$(\alpha^{(k)}) + (\beta^{(k)}) \to (\alpha) + (\beta) \tag{20}$$

$$(\alpha^{(k)})(\beta^{(k)}) \to (\alpha)(\beta). \tag{21}$$

To convince ourselves of the validity of these basic rules we have only to observe that the (i, j) element of the sum (product) of two matrices is a continuous function of the $2n^2$ coordinates α_{ij}, β_{ij}. An important special case of (21) is that, if $(\alpha^{(k)}) \to (\alpha)$ and (μ) is non-singular, then

$$(\mu)(\alpha^{(k)})(\mu)^{-1} \to (\mu)(\alpha)(\mu)^{-1}. \tag{22}$$

The last result can be used to define convergence for linear transformations. Let $\{A_k\}$, $k = 1, 2, \cdots$, be an infinite sequence of linear transformations in $\mathfrak{R}$ over Φ and let $(\alpha^{(k)})$ be the matrix of A_k relative to a basis $(e_1, e_2, \cdots, e_n)$ of $\mathfrak{R}$. We shall say that $\{A_k\}$ converges to the linear transformation A, $A_k \to A$ if $(\alpha^{(k)}) \to (\alpha)$ where (α) is the matrix of A. Because of (22) it is clear that the condition $A_k \to A$ is independent of the choice of basis in $\mathfrak{R}$.

We shall now consider power series in a linear transformation. As for ordinary series, we assign a meaning to

$$\gamma_0 1 + \gamma_1 A + \gamma_2 A^2 + \cdots, \quad \gamma_i \text{ in } \Phi, \tag{23}$$

if the sequence of partial sums $\{S_k\}$,

$$S_k = \gamma_0 1 + \gamma_1 A + \cdots + \gamma_k A^k$$

converges. The limit of $\{S_k\}$ is called *the sum* of (23), and we write

$$\gamma_0 1 + \gamma_1 A + \gamma_2 A^2 + \cdots = S.$$

The principal result we wish to prove is the following

Theorem 15. *Let $\gamma_0 + \gamma_1\lambda + \gamma_2\lambda^2 + \cdots$ be an ordinary power series with radius of convergence r. Then the power series* (23) *converges for every linear transformation A whose characteristic roots ρ_i satisfy $|\rho_i| < r$.*

Proof. We choose a coordinate system relative to which the matrix of A has the classical canonical form

$$\begin{bmatrix} (\alpha_1) & & & & \\ & (\alpha_2) & & & \\ & & \cdot & & \\ & & & \cdot & \\ & & & & \cdot \\ & & & & & (\alpha_h) \end{bmatrix}$$

where each diagonal block has the form

$$\begin{bmatrix} \rho & & & & & \\ 1 & \rho & & & & \\ & 1 & \rho & & & \\ & & \cdot & \cdot & & \\ & & & \cdot & \cdot & \\ & & & & \cdot & \cdot \\ & & & & 1 & \rho \end{bmatrix} \tag{24}$$

ρ a characteristic root. Evidently the matrix of A^m has the same block form as that of A. Moreover the block that is in the same position as (24) is

$$\begin{bmatrix} \rho^m & & & \\ \binom{m}{1}\rho^{m-1} & \rho^m & & \\ \binom{m}{2}\rho^{m-2} & \binom{m}{1}\rho^{m-1} & \rho^m & \\ & & & \ddots \end{bmatrix}.$$

This is immediate if we write (24) as $\rho 1 + z$ where

$$z = \begin{bmatrix} 0 & & & & \\ 1 & 0 & & & \\ & 1 & 0 & & \\ & & \ddots & \ddots & \\ & & & 1 & 0 \end{bmatrix}$$

and we note that

$$z^2 = \begin{bmatrix} 0 & & & & & \\ 0 & 0 & & & & \\ 1 & 0 & 0 & & & \\ & \ddots & \ddots & \ddots & & \\ & & & 1 & 0 & 0 \end{bmatrix}, \quad z^3 = \begin{bmatrix} 0 & & & & & & \\ 0 & 0 & & & & & \\ 0 & 0 & 0 & & & & \\ 1 & 0 & & \ddots & & & \\ 0 & 1 & & & & & \\ \cdot & \cdot & \cdot & \cdot & \cdot & 0 & \\ 0 & 0 & 0 & 1 & 0 & 0 & 0 \end{bmatrix},$$

etc. Hence if $S_k(\lambda)$ denotes the kth partial sum of $S(\lambda) = \gamma_0 + \gamma_1\lambda + \gamma_2\lambda^2 + \cdots$, then a typical block of the matrix of $S_k(A)$ is

$$\begin{bmatrix} S_k(\rho) & 0 & & \\ S_k'(\rho) & S_k(\rho) & 0 & \\ \dfrac{S_k''(\rho)}{2!} & S_k'(\rho) & S_k(\rho) & \\ \cdot & \cdot & \cdot & \ddots \end{bmatrix}.$$

If $|\rho| < r$, $S_k(\rho)$, $S_k'(\rho)$, $\cdots$ converge to $S(\rho)$, $S'(\rho)$, $\cdots$, respectively. Hence the matrix sequence determined by $\{S_k(A)\}$ converges to the matrix with typical diagonal block

$$\begin{bmatrix} S(\rho) & 0 & & & \\ S'(\rho) & S(\rho) & 0 & & \\ \dfrac{S''(\rho)}{2!} & S'(\rho) & S(\rho) & & \\ \cdot & \cdot & \cdot & \cdot & \\ & & & & \cdot \\ & & & & & \cdot \end{bmatrix}. \tag{25}$$

Consequently $\{S_k(A)\}$ converges.

An important special case is the exponential function

$$\exp A = 1 + A + A^2/2! + \cdots$$

which is defined for all A. A matrix of exp A has the diagonal block form with blocks

$$\begin{bmatrix} \exp \rho & & & & \\ \exp \rho & \exp \rho & & & \\ \dfrac{\exp \rho}{2!} & \exp \rho & \exp \rho & & \\ & & & \cdot & \\ & & & & \cdot \\ & & & & & \cdot \end{bmatrix}.$$

This formula can be used to calculate det (exp A). We find that

$$\det(\exp A) = \exp \rho_1 \exp \rho_2 \cdots \exp \rho_n = \exp(\Sigma\rho_i)$$

where $\rho_1, \rho_2, \cdots, \rho_n$ are the n characteristic roots. Hence

$$\det(\exp A) = \exp(\operatorname{tr} A) \tag{26}$$

where tr A as usual denotes the trace of A.

Power series in A are particularly easy to handle if A is a normal linear transformation; for in this case we can find a unitary basis for our vector space such that the matrix of A has the canonical form

$$\operatorname{diag}\{\rho_1, \rho_2, \cdots, \rho_n\}. \tag{27}$$

Then if $S(\lambda) = \gamma_0 + \gamma_1\lambda + \gamma_2\lambda^2 + \cdots$ is a power series with radius of convergence $r > |\rho_i|$, $i = 1, 2, \cdots, n$, $S(A)$ is defined and this linear transformation has the matrix

$$\text{diag }\{S(\rho_1), S(\rho_2), \cdots, S(\rho_n)\}$$

relative to the given unitary basis. Since $s(A)$ has a diagonal matrix relative to a unitary basis, this transformation is normal.

If U is unitary, the ρ_i in (27) are of absolute value 1. Hence $\rho_i = \exp \sqrt{-1}\theta_i$, θ_i real. Let H be the linear transformation with matrix

$$\text{diag }\{\theta_1, \theta_2, \cdots, \theta_n\}$$

relative to our unitary basis. Then H is hermitian, and our considerations show that

$$U = \exp (iH). \tag{28}$$

This result in conjunction with the polar factorization shows that every linear transformation has the form

$$A = P \exp (iH) \tag{29}$$

where P is a (positive) semi-definite hermitian transformation and H is hermitian. This factorization obviously generalizes the factorization $\alpha = |\alpha| \exp (i\eta)$, η real, of any complex number.

EXERCISES

1. Prove that a linear transformation A is positive definite hermitian if and only if $A = \exp H$ for some hermitian H.

2. Show that, if the characteristic roots of A are of absolute value <1, then $\log (1 + A) = A - A^2/2 + A^3/3 - \cdots$ is defined. Prove that $\exp (\log (1 + A)) = 1 + A$.

3. Prove that $A^k \to 0$ if and only if all the characteristic roots of A are of absolute value less than 1.

4. Prove that the sequence of powers $\{A^k\}$ of A possesses convergent subsequences if and only if 1) $|\rho| \leq 1$ for every characteristic root ρ and 2) the elementary divisors corresponding to the roots of absolute value 1 are simple.

Chapter VII

PRODUCTS OF VECTOR SPACES

In this chapter we consider a process for forming out of a pair consisting of a right vector space $\mathfrak{R}'$ and a left vector space $\mathfrak{S}$, a group $\mathfrak{R}' \times \mathfrak{S}$ called the direct product of the two spaces. The product $\mathfrak{R}' \times \mathfrak{S}$ is a commutative group, but in general there is no natural way of regarding this group as a vector space. Our process does lead to a vector space if one of the factors is a two-sided vector space. We define this concept here, and we remark that, if $\Delta = \Phi$ is a field, then any left or right space can be regarded, in a trivial fashion, as a two-sided space. This leads to the definition of the Kronecker product of two vector spaces over a field. We also discuss the elements of tensor algebra, and we consider the extension of a vector space over a field Φ to a vector space over a field P containing Φ. Finally we consider the concept of a (non-associative) algebra over a field, and we define the direct product of algebras.

1. Product groups of vector spaces. A bilinear form $g(x, y')$ connecting a left space and a right space may be regarded as a type of product of pairs of vectors, one chosen from each space, giving a result $g(x, y')$ in Δ. The basic properties of this product are the distributive laws and the homogeneity:

$$(1) \qquad g(\alpha x, y') = \alpha g(x, y'), \quad g(x, y'\alpha) = g(x, y')\alpha.$$

The fundamental concept of the present chapter is that of another kind of a product that we shall now define.

Let $\mathfrak{R}'$ be a right vector space and $\mathfrak{S}$ a left vector space over the same division ring Δ. Let $\mathfrak{P}$ be a commutative group (operation written as $+$) and suppose that, for each pair of vectors x', y, $x' \varepsilon \mathfrak{R}'$, $y \varepsilon \mathfrak{S}$, there is associated a unique element $x' \times y \varepsilon \mathfrak{P}$. Then we shall say that $\mathfrak{P}$ is a *product group* of $\mathfrak{R}'$ and $\mathfrak{S}$ *relative to the product* $\times$ if

1. $(x_1' + x_2') \times y = x_1' \times y + x_2' \times y$
   $x' \times (y_1 + y_2) = x' \times y_1 + x' \times y_2.$
2. $x'\alpha \times y = x' \times \alpha y.$
3. Every element of $\mathfrak{P}$ has the form $\Sigma x_i' \times y_i$.

We remark first that a non-zero bilinear form is a product of the present type if and only if $\Delta = \Phi$ is commutative. Thus if we adopt the present notation and write $x' \times y$ for $g(y, x')$, then by (1),

$$x' \times \alpha y = \alpha(x' \times y), \quad x'\alpha \times y = (x' \times y)\alpha.$$

On the other hand, $x' \times \alpha y = x'\alpha \times y$. Hence

$$\alpha(x' \times y) = (x' \times y)\alpha.$$

Now if the form is not the zero form, then it takes on all values in Δ. Hence the above equation shows that Δ is commutative. The converse is clear.

We shall give next a way of forming a product group for any pair of vector spaces $\mathfrak{R}'$ and $\mathfrak{S}$. We take first a left vector space $\mathfrak{R}$ dual to $\mathfrak{R}'$, and we suppose that this duality is given by the non-degenerate bilinear form $g(x, y')$, $x \varepsilon \mathfrak{R}$, $y' \varepsilon \mathfrak{R}'$. For value group $\mathfrak{P}$ we take the group $\mathfrak{L}(\mathfrak{R}, \mathfrak{S})$ of linear transformations of $\mathfrak{R}$ into $\mathfrak{S}$. Finally we define $x' \times y$ for x' in $\mathfrak{R}'$ and y in $\mathfrak{S}$ to be the linear transformation

$$x \to g(x, x')y. \tag{2}$$

Then **1.** and **2.** are immediate. Also we have seen (Chapter V, § 5) that any linear mapping of $\mathfrak{R}$ into $\mathfrak{S}$ has the form

$$x \to \Sigma g(x, x_i')y_i. \tag{3}$$

Hence according to our definition it is a sum $\Sigma x_i' \times y_i$. This proves **3.**

For any product we can prove as usual the rules:

$$0 \times y = 0 = x' \times 0$$

$$(-x') \times y = -(x' \times y) = x' \times (-y).$$

Suppose next that $(f_1, f_2, \cdots, f_m)$ is a basis for $\mathfrak{S}$. Then any $y = \Sigma\beta_i f_i$ and hence $x' \times y = \Sigma x' \times \beta_i f_i = \Sigma x'\beta_i \times f_i$. It follows that any element of $\mathfrak{P}$ can be written in the form $\Sigma x_i' \times f_i$. Similarly, if $(e_i', e_2', \cdots, e_n')$ is a basis for $\mathfrak{R}'$, then any element of $\mathfrak{P}$ can be written as $\Sigma e_i' \times y_i$, y_i in $\mathfrak{S}$. We shall call $\mathfrak{P}$ a direct product of $\mathfrak{R}'$ and $\mathfrak{S}$ if uniqueness holds for each of these ways of writing an element. A somewhat better way of putting this is the following. The group $\mathfrak{P}$ is a *direct product* of $\mathfrak{R}'$ and $\mathfrak{S}$ relative to $\times$ if

4. (a) $\Sigma x_i' \times y_i = 0$ implies that either the y_i are linearly dependent or that all the $x_i' = 0$;

(b) $\Sigma x_i' \times y_i = 0$ implies that either the x_i' are linearly dependent or that all the $y_i = 0$.

The group $\mathfrak{L}(\mathfrak{R}, \mathfrak{S})$ is a direct product in this sense. Suppose that $\Sigma x_i' \times y_i = 0$. Then by definition of the product this means that $\Sigma g(x, x_i')y_i = 0$ for all x. If the y_i are linearly independent, we have $g(x, x_i') = 0$ for all x. Hence by the non-degeneracy of the form, $x_i' = 0$ for all i. On the other hand, suppose that the x_i' are linearly independent. Then we know that we can find a set of x_i such that $g(x_j, x_i') = \delta_{ji}$. Then every

$$y_j = \Sigma g(x_j, x_i')y_i = 0.$$

A direct product of $\mathfrak{R}'$ and $\mathfrak{S}$ is the "most general" kind of product of these two spaces; for we have the following

Theorem 1. *Let $\mathfrak{P}$ be a direct product of $\mathfrak{R}'$ and $\mathfrak{S}$ relative to $\times$ and let $\mathfrak{P}_1$ be any product of these same spaces relative to the multiplication $\times_1$. Then the mapping $\Sigma x_i' \times y_i \to \Sigma x_i' \times_1 y_i$ is a homomorphism of $\mathfrak{P}$ onto $\mathfrak{P}_1$.*

Proof. Suppose we have two ways of writing an element $z \in \mathfrak{P}$ as a sum of products. We can suppose that these are $z = \sum_1^m x_i' \times y_i = \sum_{m+1}^q (-x_j') \times y_j$. Then $\sum_1^q x_k' \times y_k = 0$. Let us express the y_i in terms of a set of linearly independent elements $f_1, f_2, \cdots, f_r$ (a basis, for example). Then $y_k = \Sigma \alpha_{kl} f_l$ and

$$0 = \Sigma x_k' \times y_k = \sum_k x_k' \times \sum_l \alpha_{kl} f_l$$

$$= \sum_l \Big(\sum_k x_k' \alpha_{kl} \Big) \times f_l.$$

Hence $\Sigma x_k' \alpha_{kl} = 0$ for $l = 1, 2, \cdots, r$. This implies that in $\mathfrak{P}_1$ we have the relation $\sum_l \Big(\sum_k x_k' \alpha_{kl} \Big) \times {}_1 f_l = 0$. By retracing the steps we can conclude that $\Sigma x_k' \times_1 y_k = 0$ in $\mathfrak{P}_1$. Hence the two formally distinct images $\sum_1^m x_i' \times_1 y_i$ and $\sum_{m+1}^q (-x_j') \times_1 y_j$ of z are equal. This, of course, means that the correspondence

$$\Sigma x_i' \times y_i \rightarrow \Sigma x_i' \times_1 y_i$$

is single-valued. It can now be verified directly that this mapping is a homomorphism of $\mathfrak{P}$ onto $\mathfrak{P}_1$.

Suppose now that $\mathfrak{P}_1$, too, is a direct product. Then the argument that we have given shows that $\Sigma x_i' \times_1 y_i = 0$ implies that $\Sigma x_i' \times y_i = 0$. Thus the kernel of the homomorphism is 0, and hence the homomorphism is an isomorphism. This proves

Theorem 2. *If $\mathfrak{P}$ and $\mathfrak{P}_1$ are direct products of $\mathfrak{R}'$ and $\mathfrak{S}$, then the "natural" mapping $\Sigma x_i' \times y_i \rightarrow \Sigma x_i' \times_1 y_i$ is an isomorphism.*

This result shows that there is essentially only one direct product of two given vector spaces. We may therefore speak of *the* direct product and we denote this group as $\mathfrak{R}' \times \mathfrak{S}$.

EXERCISES

1. Write any vector in $\mathfrak{R}'$ as an $n_1 \times 1$ matrix and any vector in $\mathfrak{S}$ as a $1 \times n_2$ matrix. Let scalar multiplication in $\mathfrak{R}'$ and $\mathfrak{S}$ be respectively right and left multiplication of coordinates. Show that the group of $n_1 \times n_2$ matrices is a direct product of $\mathfrak{R}'$ and $\mathfrak{S}$ relative to $x' \times y = x'y$ the ordinary matrix product.
2. Show that either of the conditions in **4.** implies the other. [Hint: note that only one of these is used to prove Theorem 1.]

2. Direct products of linear transformations. We shall now consider an important generalization of Theorem 1. As in that theorem we let $\mathfrak{P} = \mathfrak{R}' \times \mathfrak{S}$, but for the second group we now take any product $\mathfrak{Q}_1$ of the right space $\mathfrak{U}'$ by the left space $\mathfrak{V}$ (both over Δ). Suppose A' is a linear mapping of $\mathfrak{R}'$ into $\mathfrak{U}'$ and B is a linear mapping of $\mathfrak{S}$ into $\mathfrak{V}$. Then we assert that the mapping

$$\Sigma x_i' \times y_i \rightarrow \Sigma x_i'A' \times_1 y_iB \tag{4}$$

is a homomorphism of $\mathfrak{P}$ into $\mathfrak{Q}_1$. The proof is obtained by repeating the argument used to prove Theorem 1. Thus, we have to show that a relation $\Sigma x_k' \times y_k = 0$ implies $\Sigma x_k'A' \times_1 y_kB = 0$. As before, we write $y_k = \Sigma \alpha_{kl} f_l$ where the f's are independent. Then we obtain $\Sigma x_k'\alpha_{kl} = 0$. Hence also $\Sigma (x_k'A')\alpha_{kl} = 0$ by the linearity of A'. Then

$$\begin{aligned} 0 &= \sum_k (x_k'A')\alpha_{kl} \times_1 f_lB = \sum_k x_k'A' \times_1 \sum_l \alpha_{kl}(f_lB) \\ &= \sum_k x_k'A' \times_1 \sum_k y_kB \end{aligned}$$

by the linearity of B. The rest of the proof is an exact repetition of the previous one.

Theorem 1 is obtained from the present result by specializing $\mathfrak{U}' = \mathfrak{R}'$, $\mathfrak{V} = \mathfrak{S}$, $A' = 1$, $B = 1$. Another important special case is that in which $\mathfrak{U}' = \mathfrak{R}'$, $\mathfrak{V} = \mathfrak{S}$ and $\mathfrak{Q}_1 = \mathfrak{P}$. Here we see that, if A' is any linear transformation in $\mathfrak{R}'$ and B is any linear transformation in $\mathfrak{S}$, then the mapping defined by (4) is an endomorphism in the direct product $\mathfrak{P}$. We shall call this the *direct product* of A' and B, and we denote it as $A' \times B$. It is immediate from the definition that the direct product $A' \times B$ is distributive:

$$\begin{aligned} (A_1' + A_2') \times B &= A_1' \times B + A_2' \times B \\ A' \times (B_1 + B_2) &= A' \times B_1 + A' \times B_2, \end{aligned} \tag{5}$$

and that

$$(A' \times B)(C' \times D) = A'C' \times BD. \tag{6}$$

If we now use $1'$ to denote the identity mapping in $\mathfrak{R}'$ and 1 that in $\mathfrak{S}$, then clearly $1' \times 1$ is the identity in $\mathfrak{P}$. We note also that by (6) any $A' \times B$ can be factored as

$$A' \times B = (A' \times 1)(1' \times B) = (1' \times B)(A' \times 1). \tag{7}$$

EXERCISES

1. Let $\mathfrak{L}(\mathfrak{R}, \mathfrak{S})$ be the group of linear mappings of $\mathfrak{R}$ into $\mathfrak{S}$ and regard $\mathfrak{L}$ as a product of $\mathfrak{R}'$ and $\mathfrak{S}$ in the manner defined in § 1. Show that, if A' is a linear transformation in $\mathfrak{R}'$ and B is a linear transformation in $\mathfrak{S}$, then the mapping $A' \times B$ in $\mathfrak{L} = \mathfrak{R}' \times \mathfrak{S}$ is identical with $L \to ALB$ where L is a general element of $\mathfrak{L}$ and A is the transpose in $\mathfrak{R}$ of A'.

2. Let $\mathfrak{P} = \mathfrak{R}' \times \mathfrak{S}$ and let $\mathfrak{M}$ be the totality of endomorphisms $\Sigma A_i' \times B_i$ where A_i' is a linear transformation in $\mathfrak{R}'$ and B_i is one in $\mathfrak{S}$. Show that $\mathfrak{M}$ is a subring of the ring of endomorphisms of $\mathfrak{P} = \mathfrak{R}' \times \mathfrak{S}$.

3. Show that the subset of elements of the form $1' \times B$ is a subring of $\mathfrak{M}$ isomorphic to $\mathfrak{L}(\mathfrak{S}, \mathfrak{S})$ and that the subset of elements of the form $A' \times 1$ is a subring isomorphic to $\mathfrak{L}(\mathfrak{R}', \mathfrak{R}')$.

3. Two-sided vector spaces. The process that we have given of "pasting together" two vector spaces to form their direct product has the serious defect that the final result is not a vector space but only a group. To get around this difficulty we are led to consider *two-sided vector spaces* instead of the one-sided ones that we have studied hitherto. We define such a system to consist of a commutative group $\mathfrak{R}$, a division ring Δ, and two functions αx and $x\alpha$ which satisfy the conditions for left and right scalar multiplication respectively, and that, in addition, satisfy the associative law

$$(\alpha x)\beta = \alpha(x\beta) \tag{8}$$

for all $\alpha, \beta \in \Delta$ and all $x \in \mathfrak{R}$. Thus our assumptions are that $\mathfrak{R}$ is at the same time a left and a right vector space and that any right (left) scalar multiplication is a linear transformation in $\mathfrak{R}$ regarded as a left (right) vector space.

If $\Delta = \Phi$ is a field, any right vector space can be regarded as a left vector space. One merely has to set $x\alpha = \alpha x$. Hence in dealing with vector spaces over fields we can suppose that all of these are left vector spaces. On the other hand, we can also consider these vector spaces as special types of two-sided spaces in which the left multiplication and the right multiplication determined by any $\alpha \in \Delta$ are identical. This "trivial" type of two-sided vector space is the one that will be our main concern in the remainder of this chapter. It is not difficult to construct other, non-trivial, types of two-sided vector spaces. We give one example here.

Example. Let $\mathfrak{R}$ be an n-dimensional left vector space over a field Φ and let $(e_1, e_2, \cdots, e_n)$ be a (left) basis for $\mathfrak{R}$. Let $S_1, S_2, \cdots, S_n$ be automorphisms in Φ and define the right multiplication by $\alpha \varepsilon \Phi$ to be the linear transformation in the left space $\mathfrak{R}$ defined by the matrix

$$\text{diag}\,\{\alpha^{S_1}, \alpha^{S_2}, \cdots, \alpha^{S_n}\}.$$

Then $e_i\alpha = \alpha^{S_i}e_i$ and $(\Sigma\xi_ie_i)\alpha = \Sigma\xi_i\alpha^{S_i}e_i$. Since

$$\begin{aligned} e_i(\alpha + \beta) &= (\alpha + \beta)^{S_i}e_i = \alpha^{S_i}e_i + \beta^{S_i}e_i = e_i\alpha + e_i\beta \\ e_i(\alpha\beta) &= (\alpha\beta)^{S_i}e_i = \alpha^{S_i}\beta^{S_i}e_i = \alpha^{S_i}(e_i\beta) \\ &= (\alpha^{S_i}e_i)\beta = (e_i\alpha)\beta, \\ e_i1 &= e_i, \end{aligned}$$

$\mathfrak{R}$ is a right vector space over Φ. Since the right multiplications are linear transformations in $\mathfrak{R}$ as left space, (8) holds. Hence $\mathfrak{R}$ is a two-sided vector space.

Suppose now $\mathfrak{R}$ is any two-sided vector space over a division ring Δ and $\mathfrak{S}$ is a left vector space over Δ. We can form the direct product $\mathfrak{P} = \mathfrak{R} \times \mathfrak{S}$ obtained by regarding $\mathfrak{R}$ as a right vector space. Since the mapping $x \to \alpha x$ is a linear transformation in $\mathfrak{R}$ regarded as a right vector space, we know that the mapping $\Sigma x_i \times y_i \to \Sigma\alpha x_i \times y_i$, $x_i \varepsilon \mathfrak{R}$, $y_i \varepsilon \mathfrak{S}$, is an endomorphism in $\mathfrak{P}$. We now set

$$\alpha\Sigma x_i \times y_i = \Sigma\alpha x_i \times y_i, \tag{9}$$

and we can verify that, relative to this scalar multiplication, $\mathfrak{P}$ is a left vector space over Δ.

We shall now show that the (left) dimensionality $\dim_l \mathfrak{P}$ is the product of the left dimensionality of $\mathfrak{R}$ and the left dimensionality of $\mathfrak{S}$. Thus, let $(e_1, e_2, \cdots, e_n)$ be a left basis for $\mathfrak{R}$ and let $(f_1, f_2, \cdots, f_m)$ be a left basis for $\mathfrak{S}$. Any $x \varepsilon \mathfrak{R}$ can be written as $\Sigma\xi_ie_i$ and any $y \varepsilon \mathfrak{S}$ can be written as $\Sigma\eta_jf_j$. Hence

$$\begin{aligned} x \times y &= \Sigma\xi_ie_i \times \eta_jf_j = \Sigma(\xi_ie_i)\eta_j \times f_j \\ &= \sum_j \sum_i (\xi_ie_i)\eta_j \times f_j. \end{aligned}$$

On the other hand, the vectors $\sum_i (\xi_ie_i)\eta_j$ belong to $\mathfrak{R}$. Hence we have $\sum_i (\xi_ie_i)\eta_j = \sum_k \mu_{jk}e_k$. Hence

$$x \times y = \sum_j \left(\sum_k \mu_{jk}e_k\right) \times f_j = \sum_{j,k} \mu_{jk}(e_k \times f_j).$$

This proves that any vector in $\mathfrak{P}$ is a left linear combination of the vectors $e_i \times f_j$. Suppose next that $\Sigma\gamma_{ij}(e_i \times f_j) = 0$. Then $0 = \Sigma(\gamma_{ij}e_i) \times f_j = \sum_j \left(\sum_i \gamma_{ij}e_i\right) \times f_j$. Since the f's are linearly independent, this gives $\sum_i \gamma_{ij}e_i = 0$; $j = 1, 2, \cdots, m$. Hence each $\gamma_{ij} = 0$.

A similar discussion applies if $\mathfrak{R}$ is a right space over Δ and $\mathfrak{S}$ is two-sided. Here $\mathfrak{P} = \mathfrak{R} \times \mathfrak{S}$ can be turned into a right space over Δ by defining

$$(\Sigma x_i \times y_i)\alpha = \Sigma x_i \times y_i\alpha. \tag{10}$$

Also the product relation for right dimensionalities holds.

If both $\mathfrak{R}$ and $\mathfrak{S}$ are two-sided vector spaces over Δ, then by using (9) and (10) we can convert $\mathfrak{P}$ into a left and a right vector space over Δ. As we have seen, any left multiplication defined in this way commutes with any right multiplication. Hence $\mathfrak{P}$ is a two-sided vector space.

It is natural to define a *linear transformation* of one two-sided vector space $\mathfrak{R}$ into a second one $\mathfrak{S}$ to be a mapping of $\mathfrak{R}$ into $\mathfrak{S}$ which is linear for the left vector spaces $\mathfrak{R}$, $\mathfrak{S}$ and is linear also for the right vector spaces $\mathfrak{R}$, $\mathfrak{S}$. Similarly, $\mathfrak{R}$ and $\mathfrak{S}$ are considered as *equivalent* if there exists a 1–1 linear transformation of $\mathfrak{R}$ onto $\mathfrak{S}$.

Suppose now that $\mathfrak{R}$, $\mathfrak{S}$ and $\mathfrak{T}$ are three two-sided Δ-spaces and consider the two-sided space $\mathfrak{R} \times (\mathfrak{S} \times \mathfrak{T})$ where the $\times$ here indicates the direct product regarded as a two-sided space in the manner indicated. It is clear that any element of this space has the form $\Sigma x_i \times (y_i \times z_i)$, $x_i \in \mathfrak{R}$, $y_i \in \mathfrak{S}$, $z_i \in \mathfrak{T}$. Similarly any element of $(\mathfrak{R} \times \mathfrak{S}) \times \mathfrak{T}$ has the form $\Sigma(x_i \times y_i) \times z_i$. We now wish to show that the rule $\Sigma x_i \times (y_i \times z_i) \to \Sigma(x_i \times y_i) \times z_i$ defines a 1–1 linear transformation of $\mathfrak{R} \times (\mathfrak{S} \times \mathfrak{T})$ onto $(\mathfrak{R} \times \mathfrak{S}) \times \mathfrak{T}$. For this purpose assume that we have a relation $\Sigma x_i \times (y_i \times z_i) = 0$ in $\mathfrak{R} \times (\mathfrak{S} \times \mathfrak{T})$. Then we can write $x_i = \Sigma e_j\alpha_{ji}$ and $z_i = \Sigma\beta_{ik}g_k$ where the e's are right linearly independent and the g's are left linearly independent. If we substitute in our relation we obtain

$$0 = \sum_j e_j \times \sum_i \alpha_{ji}(y_i \times z_i) = \sum_j e_j \times \sum_i (\alpha_{ji}y_i \times z_i).$$

Hence $\sum_i \alpha_{ji} y_i \times z_i = 0$ for $j = 1, 2, \cdots$. Then

$$0 = \sum_{i,k} (\alpha_{ji} y_i)\beta_{ik} \times g_k = \sum_k \sum_i (\alpha_{ji} y_i)\beta_{ik} \times g_k.$$

Hence $\sum_i (\alpha_{ji} y_i)\beta_{ik} = 0$ for all j, k. Thus $\sum_i \alpha_{ji}(y_i \beta_{ik}) = 0$ and this in turn implies that

$$0 = \sum_j e_j \times \sum_i \alpha_{ji}(y_i\beta_{ik}) = \sum_i \left(\sum_j e_j \alpha_{ji}\right) \times y_i\beta_{ik}$$
$$= \sum_i x_i \times y_i \beta_{ik}.$$

Similarly this relation yields the relation $\Sigma(x_i \times y_i) \times z_i = 0$. A direct verification now shows that the correspondence

$$\Sigma x_i \times (y_i \times z_i) \to \Sigma(x_i \times y_i) \times z_i$$

is a linear transformation of $\mathfrak{R} \times (\mathfrak{S} \times \mathfrak{T})$ onto $(\mathfrak{R} \times \mathfrak{S}) \times \mathfrak{T}$. Moreover, by symmetry, the kernel of this mapping is 0; hence it is an equivalence. Because of this equivalence we need not distinguish between the two direct products $\mathfrak{R} \times (\mathfrak{S} \times \mathfrak{T})$ and $(\mathfrak{R} \times \mathfrak{S}) \times \mathfrak{T}$. We therefore use the notation $\mathfrak{R} \times \mathfrak{S} \times \mathfrak{T}$ for either of these products, and we shall call this space the *direct product* of the three spaces $\mathfrak{R}$, $\mathfrak{S}$, $\mathfrak{T}$. Also we write $x \times y \times z$ for either of the products $x \times (y \times z)$ or $(x \times y) \times z$.

In a similar fashion we can define direct products of more than three two-sided spaces: Any two direct products obtained by taking $\mathfrak{R}, \mathfrak{S}, \cdots, \mathfrak{U}$ in this order and associating the factors are equivalent under a natural equivalence of the type given for three factors. We can therefore denote any of the resulting product spaces as $\mathfrak{R} \times \mathfrak{S} \times \cdots \times \mathfrak{U}$. The product of vectors in this space will be written as $x \times y \times \cdots \times u$.

EXERCISES

1. Let $\alpha \to \alpha^S$ be an isomorphism of the field Φ into itself and let $\mathfrak{R}$ be a left vector space over Φ. Define $x\alpha = \alpha^S x$ and verify that this turns $\mathfrak{R}$ into a two-sided vector space. Show that if the subfield Φ^S of image elements α^S is properly contained in Φ, then the left and the right dimensionalities of $\mathfrak{R}$ over Φ are different.

2. Let $\mathfrak{R}$ be a two-sided vector space over Δ such that the left and the right dimensionalities over Δ are finite and equal. Prove that there exists a set of

vectors $e_1, e_2, \cdots, e_n$ which is at the same time a left basis and a right basis for $\mathfrak{R}$ over Δ.

3. Let $\mathfrak{R}$ be defined as in Ex. 1 with S an automorphism in Φ and let $\mathfrak{S}$ be a second space of this type defined by the automorphism T. Show that, if $ST \neq TS$, then $\mathfrak{R} \times \mathfrak{S}$ is not equivalent to $\mathfrak{S} \times \mathfrak{R}$.

4. The Kronecker product. In the remainder of this chapter we shall suppose that $\Delta = \Phi$ is a field. Moreover, we shall be interested only in one-sided spaces and we shall write all of these as left spaces. The conjugate space of a space $\mathfrak{R}$ will be denoted as usual as $\mathfrak{R}^*$, but this, too, will be regarded as a left vector space.

Suppose now that $\mathfrak{R}$ and $\mathfrak{S}$ are two spaces over Φ. We consider $\mathfrak{R}$ and $\mathfrak{S}$ momentarily as two-sided vector spaces in the trivial way that $\alpha x = x\alpha$ for all α and x, and we form the direct product $\mathfrak{R} \times \mathfrak{S}$. This is a two-sided vector space, too, but of the trivial sort since

$$\begin{aligned}\alpha(\Sigma x_i \times y_i) &= \Sigma \alpha x_i \times y_i = \Sigma x_i\alpha \times y_i = \Sigma x_i \times \alpha y_i \\ &= \Sigma x_i \times y_i\alpha = \Sigma(x_i \times y_i)\alpha.\end{aligned}$$

Hence we may regard $\mathfrak{R} \times \mathfrak{S}$ simply as a left vector space. The (left) vector space thus obtained is called the *Kronecker product* of $\mathfrak{R}$ and $\mathfrak{S}$. Of the equations above, the significant ones from the point of view of left vector spaces are

$$\alpha(\Sigma x_i \times y_i) = \Sigma \alpha x_i \times y_i = \Sigma x_i \times \alpha y_i.$$

Thus, the space $\mathfrak{R} \times \mathfrak{S}$ is characterized by the following properties: There is defined a product $x \times y \in \mathfrak{R} \times \mathfrak{S}$ for every $x \in \mathfrak{R}$ and $y \in \mathfrak{S}$ such that

1′. $(x_1 + x_2) \times y = x_1 \times y + x_2 \times y$
$x \times (y_1 + y_2) = x \times y_1 + x \times y_2.$

2′. $\alpha(x \times y) = \alpha x \times y = x \times \alpha y.$

3′. Every element of $\mathfrak{R} \times \mathfrak{S}$ is of the form $\Sigma x_i \times y_i$, $x_i \in \mathfrak{R}$, $y_i \in \mathfrak{S}$.

4′. **(a)** $\Sigma x_i \times y_i = 0$ implies that either the y_i are linearly dependent or all the x_i are 0.
(b) $\Sigma x_i \times y_i = 0$ implies that either the x_i are linearly dependent or all the y_i are 0.

The foregoing discussion has established the existence of the Kronecker product of any two (finite dimensional) vector spaces.* Most of the results which we have obtained are applicable in the present situation. In particular, we recall that $\dim (\mathfrak{R} \times \mathfrak{S}) = \dim \mathfrak{R} \dim \mathfrak{S}$. We shall now show that this equation can be used as a substitute for the independence conditions **4'.**; for we have the following

Theorem 3. *Let $\mathfrak{R}$, $\mathfrak{S}$ and $\mathfrak{P}$ be vector spaces over a field Φ and suppose that there is defined a product $x \times y$, $x \varepsilon \mathfrak{R}$, $y \varepsilon \mathfrak{S}$, $x \times y \varepsilon \mathfrak{P}$, such that* **1'.**, **2'.** *and* **3'.** *hold. Then $\mathfrak{P}$ is a Kronecker product of $\mathfrak{R}$ and $\mathfrak{S}$ if and only if $\dim \mathfrak{P} = \dim \mathfrak{R} \dim \mathfrak{S}$.*

Proof. The necessity of this condition has already been proved. Conversely, let $\dim \mathfrak{P} = \dim \mathfrak{R} \dim \mathfrak{S}$. Let $(e_1, e_2, \cdots, e_n)$ be a basis for $\mathfrak{R}$ and let $(f_1, f_2, \cdots, f_m)$ be a basis for $\mathfrak{S}$. Then it follows easily that the vectors $e_i \times f_j$ are generators of $\mathfrak{P}$. Since their number is nm, they constitute a basis. Now suppose that $e_1, e_2, \cdots, e_r$ are any linearly independent vectors in $\mathfrak{R}$. Then we can suppose that these are part of a basis for $\mathfrak{R}$. If $\Sigma e_i \times y_i = 0$, we can write $y_i = \Sigma \beta_{ij} f_j$ and we obtain $\Sigma \beta_{ij} e_i \times f_j = 0$. Hence each $\beta_{ij} = 0$, and this means that each $y_i = 0$. In a similar manner we establish the second independence condition.

We have seen that direct multiplication of two-sided vector spaces is associative and, in particular, this holds for Kronecker products. Thus, if $\mathfrak{R}$, $\mathfrak{S}$ and $\mathfrak{T}$ are three vector spaces over a field, then we have the natural equivalence $\Sigma x_i \times (y_i \times z_i) \rightarrow \Sigma (x_i \times y_i) \times z_i$ of $\mathfrak{R} \times (\mathfrak{S} \times \mathfrak{T})$ onto $(\mathfrak{R} \times \mathfrak{S}) \times \mathfrak{T}$. We therefore identify these two spaces and we identify the elements $\Sigma x_i \times (y_i \times z_i)$ and $\Sigma (x_i \times y_i) \times z_i$. Also as before we may simplify our notation and write $\mathfrak{R} \times \mathfrak{S} \times \mathfrak{T}$ and $x \times y \times z$. More generally we can define the Kronecker product $\mathfrak{R} \times \mathfrak{S} \times \cdots \times \mathfrak{U}$ of any finite number of spaces and denote its elements as $\Sigma x \times y \times \cdots \times u$.

Direct multiplication of two-sided spaces is not in general commutative (cf. Ex. 3, p. 208). However, in the special case of Kronecker products the commutative law does hold. In fact, we

* As we shall see in Chapter IX, it is easy to carry this over to infinite dimensional spaces.

can show that the rule $\Sigma x_i \times y_i \rightarrow \Sigma y_i \times x_i$ defines an equivalence of $\mathfrak{R} \times \mathfrak{S}$ onto $\mathfrak{S} \times \mathfrak{R}$. The proof of this result follows the pattern of our other proofs: By expressing the x_i and the y_i in terms of linearly independent elements we can show that $\Sigma x_i \times y_i = 0$ holds if and only if $\Sigma y_i \times x_i = 0$. The remainder of the argument is a verification. More generally the Kronecker product $\mathfrak{R}^{(1)} \times \mathfrak{R}^{(2)} \times \cdots \times \mathfrak{R}^{(s)}$ is equivalent to $\mathfrak{R}^{(j_1)} \times \mathfrak{R}^{(j_2)} \times \cdots \times \mathfrak{R}^{(j_s)}$ for any permutation $(j_1, j_2, \cdots, j_s)$ of $(1, 2, \cdots, s)$. The mapping $\Sigma x_i^{(1)} \times x_i^{(2)} \times \cdots \times x_i^{(s)} \rightarrow \Sigma x_i^{(j_1)} \times x_i^{(j_2)} \times \cdots \times x_i^{(j_s)}$ is an equivalence.

The notion of the Kronecker product of vector spaces leads to the definition of the useful concept of a free (associative) algebra. To define this concept we begin with an arbitrary vector space $\mathfrak{R}$ over a field Φ and we introduce the Kronecker products $\mathfrak{R}_0^i \equiv \mathfrak{R} \times \mathfrak{R} \times \cdots \times \mathfrak{R}$ (i factors). Also we adopt the convention that $\mathfrak{R}_0^1 = \mathfrak{R}$ and $\mathfrak{R}_0^0 = \Phi$; hence $\mathfrak{R}_0^i$ is defined for all $i = 0, 1, 2, \cdots$. By the above remarks on associativity $\mathfrak{R}_0^i \times \mathfrak{R}_0^j = \mathfrak{R}_0^{i+j}$ if $i, j \geq 1$. Hence any $x^{(i)} \in \mathfrak{R}_0^i$, $x^{(j)} \in \mathfrak{R}_0^j$ determine a vector $x^{(i)} \times x^{(j)}$ of $\mathfrak{R}_0^{i+j}$. If either $i = 0$ or $j = 0$, then we define $x^{(i)} \times x^{(j)}$ to be the product of $x^{(j)}$ (or $x^{(i)}$) by the field element $x^{(i)}$ (or $x^{(j)}$). We now form the *direct sum* $\mathfrak{F}$ of the spaces $\mathfrak{R}_0^i$. This space can be defined to be the set of sequences $x = (x^{(0)}, x^{(1)}, x^{(2)}, \cdots)$ where $x^{(i)} \in \mathfrak{R}_0^i$ and $x^{(i)} = 0$ for i sufficiently large. We consider $x = y \equiv (y^{(0)}, y^{(1)}, y^{(2)}, \cdots)$ if and only if $x^{(i)} = y^{(i)}$ for all i. For arbitrary x and y we define

$$\begin{aligned} x + y &= (x^{(0)} + y^{(0)}, x^{(1)} + y^{(1)}, \cdots) \\ \alpha x &= (\alpha x^{(0)}, \alpha x^{(1)}, \alpha x^{(2)}, \cdots). \end{aligned} \tag{11}$$

Then it is clear that $\mathfrak{F}$ is a vector space. If we identify the vector $(0, \cdots, 0, x^{(i)}, 0, \cdots)$ with $x^{(i)}$, then any vector in $\mathfrak{F}$ can be written in one and only one way as $\sum_{i=0}^{\infty} x^{(i)}$ where $x^{(i)} = 0$ for i sufficiently large. If the i_j are distinct and $x^{(i_j)} \neq 0$, then these vectors are linearly independent. It follows that, if $\mathfrak{R} \neq 0$, then $\mathfrak{F}$ is infinite dimensional.

We now introduce a multiplication $\times$ in $\mathfrak{F}$ by defining $(\Sigma x^{(i)}) \times (\Sigma y^{(j)}) = \Sigma z^{(k)}$ where

$$z^{(k)} = x^{(0)} \times y^{(k)} + x^{(1)} \times y^{(k-1)} + \cdots + x^{(k)} \times y^{(0)}. \tag{12}$$

It is easy to verify that $\mathfrak{F}$ is an associative algebra. We shall call this algebra the *free (associative) algebra* based on the vector space $\mathfrak{R}$.

EXERCISES

1. Show that, if $\mathfrak{R}$ is a one-dimensional space, then $\mathfrak{F}$ is essentially the polynomial algebra in one indeterminate (transcendental element) over Φ.
2. Let $\mathfrak{F}$ be the free algebra based on an n dimensional vector space $\mathfrak{R}$ and let $\mathfrak{B}$ be the (two-sided) ideal in $\mathfrak{F}$ generated by the vectors $x \times y - y \times x$, x and y in $\mathfrak{R}$. Show that $\mathfrak{F}/\mathfrak{B}$ is essentially the polynomial algebra in n algebraically independent transcendental elements.
3. Let $\mathfrak{F}$ be as in Ex. 2. Assume the characteristic of $\Phi \neq 2$ and let $\mathfrak{C}$ be the ideal in $\mathfrak{F}$ generated by the vectors $x \times y + y \times x$, x, y in $\mathfrak{R}$. Then $\mathfrak{F}/\mathfrak{C}$ is called the *Grassmann* (*or exterior*) *algebra* based on $\mathfrak{R}$. Show that $\dim \mathfrak{F}/\mathfrak{C} = 2^n$.
4. If $\mathfrak{R}, \mathfrak{S}, \cdots, \mathfrak{U}$ are vector spaces over Φ, we define a *multilinear function* on $\mathfrak{R}, \mathfrak{S}, \cdots, \mathfrak{U}$ as a function $f(x, y, \cdots, u)$, $x \in \mathfrak{R}$, $y \in \mathfrak{S}, \cdots, u \in \mathfrak{U}$, with values in Φ, such that f is a linear function of any one of the arguments for fixed values of the remaining ones. Show that this concept is equivalent to that of linear function on $\mathfrak{R} \times \mathfrak{S} \times \cdots \times \mathfrak{U}$ by proving: 1) if f is multilinear, then $\Sigma x_i \times y_i \times \cdots \times u_i \rightarrow \Sigma f(x_i, y_i, \cdots, u_i)$ is an element of the conjugate space of $\mathfrak{R} \times \mathfrak{S} \times \cdots \times \mathfrak{U}$; and 2) if f is a linear function on $\mathfrak{R} \times \mathfrak{S} \times \cdots \times \mathfrak{U}$, then the contraction of f to the subset of vectors of the form $x \times y \times \cdots \times u$ is a multilinear function.
5. Let $g(x), h(y), \cdots, k(u) \in \mathfrak{R}^*, \mathfrak{S}^*, \cdots, \mathfrak{U}^*$ respectively. Show that $f(x, y, \cdots, u) \equiv g(x)h(y) \cdots k(u)$ is multilinear. Let f also denote the associated element of $(\mathfrak{R} \times \mathfrak{S} \times \cdots \times \mathfrak{U})^*$ and show that the rule $\Sigma g \times h \times \cdots \times k \rightarrow \Sigma f$ defines an equivalence of $\mathfrak{R}^* \times \mathfrak{S}^* \times \cdots \times \mathfrak{U}^*$ onto $(\mathfrak{R} \times \mathfrak{S} \times \cdots \times \mathfrak{U})^*$.

5. Kronecker products of linear transformations and of matrices. If A is a linear transformation in $\mathfrak{R}$ and B is one in $\mathfrak{S}$, then we know that $\Sigma x_i \times y_i \rightarrow \Sigma x_i A \times y_i B$ is a group homomorphism in $\mathfrak{R} \times \mathfrak{S}$. It is evident also that this mapping commutes with the scalar multiplications; hence it is linear. We shall call this mapping the *Kronecker product* $A \times B$ of A and B. In a similar manner we can define the Kronecker products $A \times B \times \cdots \times D$ of a number of linear transformations. We shall now show that, relative to Kronecker multiplication, the vector space $\mathfrak{L}(\mathfrak{R} \times \mathfrak{S}, \mathfrak{R} \times \mathfrak{S})$ of linear transformations in $\mathfrak{R} \times \mathfrak{S}$ is a

Kronecker product $\mathfrak{L}(\mathfrak{R}, \mathfrak{R}) \times \mathfrak{L}(\mathfrak{S}, \mathfrak{S})$. Equation (5) shows that the distributive laws hold. Also

$$(x \times y)(\alpha(A \times B)) = (xA \times yB)\alpha = (xA)\alpha \times yB$$
$$= x(A\alpha) \times yB.$$

Hence $\alpha(A \times B) = \alpha A \times B$ and similarly $\alpha(A \times B) = A \times \alpha B$. Next let $(e_1, e_2, \cdots, e_n)$ be a basis for $\mathfrak{R}$ and $(f_1, f_2, \cdots, f_m)$ a basis for $\mathfrak{S}$. Then the nm vectors $e_i \times f_j$ form a basis for $\mathfrak{R} \times \mathfrak{S}$. Let $E_{ii',jj'}$ be the linear transformation in $\mathfrak{R} \times \mathfrak{S}$ which sends $e_i \times f_j$ into $e_{i'} \times f_{j'}$ and sends the remaining $e_k \times f_l$ into 0. Then we can verify that

$$E_{ii',jj'} = E_{ii'} \times F_{jj'} \tag{13}$$

where $E_{ii'}$ is the linear transformation in $\mathfrak{R}$ which sends e_i into $e_{i'}$ and the remaining e's into 0 and $F_{jj'}$ is the linear transformation in $\mathfrak{S}$ which sends f_j into $f_{j'}$ and the remaining f's into 0. It follows that every element of $\mathfrak{L}(\mathfrak{R} \times \mathfrak{S}, \mathfrak{R} \times \mathfrak{S})$ has the form $\Sigma A_i \times B_i$, $A_i \in \mathfrak{L}(\mathfrak{R}, \mathfrak{R})$, $B_i \in \mathfrak{L}(\mathfrak{S}, \mathfrak{S})$. Since $\dim \mathfrak{L}(\mathfrak{R} \times \mathfrak{S}, \mathfrak{R} \times \mathfrak{S}) = (mn)^2 = \dim \mathfrak{L}(\mathfrak{R}, \mathfrak{R}) \dim \mathfrak{L}(\mathfrak{S}, \mathfrak{S})$, our assertion follows from Theorem 3. We have therefore proved the following

Theorem 4. *If $\mathfrak{R}$ and $\mathfrak{S}$ are (finite dimensional) vector spaces, the space $\mathfrak{L}(\mathfrak{R} \times \mathfrak{S}, \mathfrak{R} \times \mathfrak{S}) = \mathfrak{L}(\mathfrak{R}, \mathfrak{R}) \times \mathfrak{L}(\mathfrak{S}, \mathfrak{S})$ relative to Kronecker multiplication of linear transformations.*

Assume now that

$$e_iA = \Sigma \alpha_{ij} e_j, \quad f_kB = \Sigma \beta_{kl} f_l \tag{14}$$

so that (α) and (β) are the matrices of A and B, respectively, relative to the chosen bases. Then we have

$$(e_i \times f_k)(A \times B) = \Sigma \alpha_{ij}\beta_{kl} e_j \times f_l. \tag{15}$$

We shall arrange the vectors $e_i \times f_k$ lexicographically as

$$(e_1 \times f_1, \cdots, e_1 \times f_m; \; e_2 \times f_1, \cdots, e_2 \times f_m; \; \cdots; \; \cdots, e_n \times f_m)$$

and we shall call this ordered basis the one *associated* with $(e_1, e_2, \cdots, e_n)$ and $(f_1, f_2, \cdots, f_m)$. Equation (15) shows that the matrix of $A \times B$ relative to the associated basis is

$$\left[\begin{array}{ccc|ccc|cc}
\alpha_{11}\beta_{11} & \cdots & \alpha_{11}\beta_{1m} & \alpha_{12}\beta_{11} & \cdots & \alpha_{12}\beta_{1m} & \cdots & \alpha_{1n}\beta_{1m} \\
\vdots & & \vdots & \vdots & & \vdots & & \vdots \\
\alpha_{11}\beta_{m1} & \cdots & \alpha_{11}\beta_{mm} & \alpha_{12}\beta_{m1} & \cdots & \alpha_{12}\beta_{mm} & \cdots & \alpha_{1n}\beta_{mm} \\
\hline
\alpha_{21}\beta_{11} & \cdots & \alpha_{21}\beta_{1m} & \alpha_{22}\beta_{11} & \cdots & \alpha_{22}\beta_{1m} & \cdots & \alpha_{2n}\beta_{1m} \\
\vdots & & \vdots & \vdots & & \vdots & & \vdots \\
\alpha_{21}\beta_{m1} & \cdots & \alpha_{21}\beta_{mm} & \alpha_{22}\beta_{m1} & \cdots & \alpha_{22}\beta_{mm} & \cdots & \alpha_{2n}\beta_{mm} \\
\hline
\vdots & & \vdots & \vdots & & \vdots & & \vdots
\end{array}\right].$$

We denote this matrix as $(\alpha) \times (\beta)$ and we shall call it the *Kronecker product* of the two matrices (α) and (β).

The basic rules governing Kronecker multiplication of linear transformations have their counterparts for matrices. In particular the preceding theorem shows that the space of $nm \times nm$ matrices over Φ is the Kronecker product of the space of $n \times n$ matrices and the space of $m \times m$ matrices relative to Φ. Also, we know that $(A \times B)(C \times D) = AC \times BD$ holds for linear transformations and this leads directly to the matrix rule

$$((\alpha) \times (\beta))((\gamma) \times (\delta)) = (\alpha)(\gamma) \times (\beta)(\delta). \tag{16}$$

EXERCISES

1. Show that the matrix of $A \times B$ relative to the basis $(e_1 \times f_1, e_2 \times f_1, \cdots, e_n \times f_1; e_1 \times f_2, e_2 \times f_2, \cdots, e_n \times f_2; \cdots; \cdots, e_n \times f_m)$ is $(\beta) \times (\alpha)$. Hence prove that $(\beta) \times (\alpha)$ is similar to $(\alpha) \times (\beta)$.
2. Prove that $(\alpha) \times ((\beta) \times (\gamma)) = ((\alpha) \times (\beta)) \times (\gamma)$.
3. Prove that if ρ is a root of the characteristic polynomial of A and σ has this property for B then $\rho\sigma$ has the property for $A \times B$.
4. Prove that $\det (A \times B) = (\det A)^m(\det B)^n$.

6. Tensor spaces. If $\mathfrak{R}$ is a vector space and $\mathfrak{R}^*$ is the conjugate of $\mathfrak{R}$, then any Kronecker product space $\mathfrak{R} \times \cdots \times \mathfrak{R} \times \mathfrak{R}^* \times \cdots \times \mathfrak{R}^*$ is called a *tensor space based on the space* $\mathfrak{R}$. The elements of such a space are called *tensors*, and if there are r factors $\mathfrak{R}$ and s factors $\mathfrak{R}^*$, then we say that these tensors are *contra-*

variant of rank r and covariant of rank s. We shall use the abbreviated notation $\mathfrak{R}_s{}^r$ for the space. Also, as in § 4, it will be convenient to regard Φ as the tensor space $\mathfrak{R}_0{}^0$ of ranks 0.

If $(e_1, e_2, \cdots, e_n)$ is a basis in the base space $\mathfrak{R}$ and $(e^1, e^2, \cdots, e^n)$ * is the complementary basis in $\mathfrak{R}^*$ in the sense that

$$e^j(e_i) = \delta_i{}^j,$$

the Kronecker delta, then the vectors

$$e_{i_1} \times e_{i_2} \times \cdots \times e_{i_r} \times e^{j_1} \times \cdots \times e^{j_s} \tag{17}$$

constitute a basis in the tensor space. We can write any vector in this space as

$$\Sigma \xi^{i_1 \cdots i_r}_{j_1 \cdots j_s} e_{i_1} \times \cdots \times e_{i_r} \times e^{j_1} \times \cdots \times e^{j_s}. \tag{18}$$

It is natural to give preference to bases of this type that are determined by the basis $(e_1, e_2, \cdots, e_n)$ in $\mathfrak{R}$. The element (18) of the tensor space will be called *the tensor whose coordinates are* $\xi^{i_1 \cdots i_r}_{j_1 \cdots j_s}$ relative to the (e)-basis in $\mathfrak{R}$. If $(f_1, f_2, \cdots, f_n)$ is a second basis in $\mathfrak{R}$ and $f_i = \Sigma \mu_i^j e_j$, then

$$e_j = \Sigma \nu_j^i f_i, \quad f^i = \Sigma \nu_j^i e^j, \quad e^j = \Sigma \mu_i^j f^i$$

where (ν) is the inverse of the matrix (μ). Hence

$$\Sigma \xi^{i_1 \cdots i_r}_{j_1 \cdots j_s} e_{i_1} \times \cdots \times e_{i_r} \times e^{j_1} \times \cdots \times e^{j_s}$$

$$\Sigma \xi^{i_1 \cdots i_r}_{j_1 \cdots j_s} \nu^{k_1}_{i_1} \cdots \nu^{k_r}_{i_r} \mu^{j_1}_{l_1} \cdots \mu^{j_s}_{l_s} f_{k_1} \times \cdots \times f_{k_r} \times f^{l_1} \times \cdots \times f^{l_s}.$$

Thus our tensor has the coordinates

$$\eta^{k_1 \cdots k_r}_{l_1 \cdots l_s} = \Sigma \xi^{i_1 \cdots i_r}_{j_1 \cdots j_s} \nu^{k_1}_{i_1} \cdots \nu^{k_r}_{i_r} \mu^{j_1}_{l_1} \cdots \mu^{j_s}_{l_s} \tag{19}$$

relative to thc (f) basis.

We consider next a generalization of the concept of the transpose of a linear transformation. We recall that, if A is any linear transformation in $\mathfrak{R}$, then the transpose A^* of A is the mapping $x^* \to y^*$ in $\mathfrak{R}^*$ where $y^*(x) = x^*(xA)$. We shall now associate with A the linear transformation $A_s{}^r \equiv \overbrace{A \times \cdots \times A}^{r} \times \overbrace{A^* \times \cdots \times A^*}^{s}$. It is natural to consider this transformation to be the transformation *induced* by A in the tensor space $\mathfrak{R}_s{}^r$. We

* This notation is somewhat more convenient than our former one: $(e_1^*, e_2^*, \ldots, e_n^*)$.

recall that, if (α) is the matrix of A relative to the basis $(e_1, e_2, \cdots, e_n)$, then $(\alpha)'$ is the matrix of A^* relative to the complementary basis $(e^1, e^2, \cdots, e^n)$. It follows that the matrix of $A_s{}^r$ relative to the natural basis (17) ordered lexicographically is the Kronecker product $\overbrace{(\alpha) \times \cdots \times (\alpha)}^{r} \times \overbrace{(\alpha)' \times \cdots \times (\alpha)'}^{s}$.

There is an important process called *contraction* for associating with any tensor which is contravariant of rank $r > 0$ and covariant of rank $s > 0$ a tensor contravariant of rank $r - 1$ and covariant of rank $s - 1$. We recall first of all that, if $x \varepsilon \mathfrak{R}$ and $y^* \varepsilon \mathfrak{R}^*$, then the mapping $(x, y^*) \to y^*(x)$ is a bilinear form connecting $\mathfrak{R}$ and $\mathfrak{R}^*$. Since Φ is commutative, this form defines a product of the vector spaces $\mathfrak{R}$ and $\mathfrak{R}^*$ in the sense of § 1. Hence by Theorem 1 we know that the mapping $\Sigma x_i \times y_i^* \to \Sigma y_i^*(x_i)$ is a homomorphism of $\mathfrak{R}_1{}^1 = \mathfrak{R} \times \mathfrak{R}^*$ onto the additive group Φ. Also it is immediate that our correspondence is a linear transformation of $\mathfrak{R}_1{}^1$ onto the one-dimensional vector space Φ $(= \mathfrak{R}_0{}^0)$.

The considerations of § 1 show also that, if $\mathfrak{S}$ is a second vector space, then the linear mapping of $\mathfrak{R}_1{}^1$ onto $\mathfrak{R}_0{}^0$ can be combined with the identity mapping in $\mathfrak{S}$ to give a linear mapping of $\mathfrak{R}_1{}^1 \times \mathfrak{S}$ onto $\mathfrak{R}_0{}^0 \times \mathfrak{S}$. The resulting mapping is defined by the following rule

$$\Sigma x_i \times y_i^* \times z_i \to \Sigma y_i^*(x_i) \times z_i,$$

$x_i \varepsilon \mathfrak{R}$, $y_i^* \varepsilon \mathfrak{R}^*$, $z_i \varepsilon \mathfrak{S}$. It is easy to see that the space $\mathfrak{R}_0{}^0 \times \mathfrak{S}$ is equivalent to $\mathfrak{S}$ under the mapping $\Sigma \alpha_i \times z_i \to \Sigma \alpha_i z_i$. Hence we see that

$$\Sigma x_i \times y_i^* \times z_i \to \Sigma y_i^*(x_i) z_i \tag{20}$$

is a linear transformation of $\mathfrak{R}_1{}^1 \times \mathfrak{S}$ onto $\mathfrak{S}$.

The process of contraction is obtained by specializing $\mathfrak{S}$ to be the tensor space $\mathfrak{R}_{s-1}{}^{r-1}$. This specialization shows directly that the mapping

$$\Sigma x \times y^* \times x_1 \times \cdots \times x_{r-1} \times y_1^* \times y_2^* \times \cdots \times y_{s-1}^*$$
$$\to \Sigma y^*(x) x_1 \times \cdots \times x_{r-1} \times y_1^* \times \cdots \times y_{s-1}^* \dagger$$

† We have omitted the summation subscript here. It is understood that we have a sum of terms of the type indicated.

is a linear mapping of $\mathfrak{R}_1{}^1 \times \mathfrak{R}_{s-1}{}^{r-1}$ onto $\mathfrak{R}_s{}^r$. Now let k be chosen in $1, 2, \cdots, r$ and l in $1, \cdots, s$. Then we know that the mapping

$$\Sigma x_1 \times \cdots \times x_r \times y_1{}^* \times \cdots \times y_s{}^* \rightarrow$$

$$\Sigma x_k \times y_l{}^* \times x_1 \times \cdots \times x_{k-1} \times x_{k+1} \times \cdots \times x_r \times y_1{}^* \times \cdots \times y_{l-1}{}^* \times y_{l+1}{}^* \times \cdots \times y_s{}^*$$

is an equivalence of $\mathfrak{R}_s{}^r$ onto $\mathfrak{R}_1{}^1 \times \mathfrak{R}_{s-1}{}^{r-1}$. Hence the mapping

$$\Sigma x_1 \times \cdots \times x_r \times y_1{}^* \times \cdots \times y_s{}^* \rightarrow \tag{21}$$

$$\Sigma y_l{}^*(x_k) x_1 \times \cdots \times x_{k-1} \times x_{k+1} \times \cdots \times x_r \times y_1{}^* \times \cdots \times y_{l-1}{}^* \times y_{l+1}{}^* \times \cdots \times y_s{}^*$$

is a linear transformation of $\mathfrak{R}_s{}^r$ onto $\mathfrak{R}_{s-1}{}^{r-1}$. This mapping is called *contraction* of $\mathfrak{R}_s{}^r$ *with respect to the kth contravariant index and the lth covariant index.*

We now employ a basis $e_{i_1} \times \cdots \times e_{i_r} \times e^{j_1} \times \cdots \times e^{j_s}$ in $\mathfrak{R}_s{}^r$ and consider the tensor (18). Contraction with respect to the kth contravariant and the lth covariant index yields the tensor

$$\Sigma \xi^{i_1 \cdots i_r}_{j_1 \cdots j_s} \delta^{j_l}_{i_k} e_{i_1} \times \cdots \times e_{i_{k-1}} \times e_{i_{k+1}} \times \cdots \times e_{i_r} \times e^{j_1} \times \cdots \times e^{j_{l-1}} \times e^{j_{l+1}} \times \cdots \times e^{j_s}.$$

Hence the coordinates of the contracted tensor relative to the basis determined by the (e)-basis in $\mathfrak{R}$ are

$$\xi^{i_1 \cdots i_{k-1} i_{k+1} \cdots i_r}_{j_i \cdots j_{l-1} j_{l+1} \cdots j_s} = \sum_q \xi^{i_1 \cdots i_{k-1} q i_{k+1} \cdots i_r}_{j_1 \cdots j_{l-1} q j_{l+1} \cdots j_s}. \tag{22}$$

The notion of contraction can be used to give a definition of the trace of a linear transformation which is independent of bases. We note first that the space $\mathfrak{L}(\mathfrak{R}, \mathfrak{R})$ of linear transformations in $\mathfrak{R}$ can be regarded as the tensor space $\mathfrak{R}_1{}^1$. This is clear from our construction of direct products, (§ 1). Thus, if $x \varepsilon \mathfrak{R}$ and $y^* \varepsilon \mathfrak{R}^*$, then we have defined $x \times y^*$ to be the linear transformation of $\mathfrak{R}$ into itself that sends the vector u into $y^*(u)x$. We know that, relative to this definition of $\times$, $\mathfrak{L}(\mathfrak{R}, \mathfrak{R}) = \mathfrak{R} \times \mathfrak{R}^*$. Any element of $\mathfrak{L}$ can be written in the form $\Sigma x_i \times y_i{}^*$. Now the contraction which we have defined is a linear mapping of $\mathfrak{L}$ into

Φ which sends $\Sigma x_i \times y_i^*$ into $\Sigma y_i^*(x_i)$. We shall now show that this mapping coincides with the mapping $A \to \mathrm{tr}\, A$ defined by bases for $\mathfrak{R}$.

To see this, let $(e_1, e_2, \cdots, e_n)$ be a basis for $\mathfrak{R}$ and write $A = \Sigma u_i^* \times e_i$. Then

$$e_j A = \sum_i u_i^*(e_j) e_i.$$

Hence the matrix of A relative to $(e_1, e_2, \cdots, e_n)$ is $(\alpha) \equiv (u_i^*(e_j))$. Hence $\mathrm{tr}\, A = \Sigma u_i^*(e_i)$. This is the same result which is obtained by contraction. A change of basis yields a second matrix of A similar to (α). Its trace coincides with the contraction of the tensor. Thus, without verifying this fact by direct computation (as we have done previously) we see that similar matrices have the same trace.

7. Symmetry classes of tensors. We consider in this section the space $\mathfrak{R}_0^r$ of contravariant tensors of rank r. The elements of this space are the tensors $\Sigma x_1 \times x_2 \times \cdots \times x_r$. Let $i \to i'$ be any permutation of the numbers $1, 2, \cdots, r$. Then we know that the mapping

$$\Sigma x_1 \times x_2 \times \cdots \times x_r \to \Sigma x_{1'} \times x_{2'} \times \cdots \times x_{r'}$$

is a (1–1) linear transformation in $\mathfrak{R}_0^r$. We shall call this transformation a *symmetry* in $\mathfrak{R}_0^r$. There is one of these associated with every element of the symmetric group $\mathfrak{S}_r$ on the r letters $1, 2, \cdots, r$. If $\sigma \in \mathfrak{S}_r$, we denote the associated mapping in $\mathfrak{R}_0^r$ by $P(\sigma)$. Then it is clear from our definition that

$$P(\sigma\tau) = P(\sigma)P(\tau). \tag{23}$$

A linear transformation of the form $\Sigma \alpha_\sigma P(\sigma)$ where α_σ denotes the scalar multiplication by the element α_σ is called a *symmetry operator* in $\mathfrak{R}_0^r$. Because of (23) these operators form a subalgebra of the complete algebra of linear transformations in $\mathfrak{R}_0^r$. If Q is a symmetry operator, then the set of vectors annihilated by Q is a subspace; more generally if $\{Q\}$ is any set of symmetry operators then the vectors z such that $zQ = 0$ for all Q is a subspace. We shall call such a subspace a *symmetry class* of tensors. For example, let the set $\{Q\}$ be the totality of symmetry operators

$P(\sigma) - 1 \equiv P(\sigma) - P(1)$, $\sigma \in \mathfrak{S}_r$. A tensor in this symmetry class is *symmetric* in the sense that

$$\Sigma x_{1'} \times x_{2'} \times \cdots \times x_{r'} = \Sigma x_1 \times x_2 \times \cdots \times x_r$$

for any permutation $i \to i'$ of $1, 2, \cdots, r$. Similarly if $\{Q\}$ is the set of symmetry operators $P(\sigma) - \epsilon_\sigma$ where $\epsilon_\sigma = 1$ if σ is an even permutation and $\epsilon_\sigma = -1$ if σ is odd, then the set $\{Q\}$ determines the class of tensors which are *skew symmetric* in the sense that

$$\Sigma x_{1'} \times x_{2'} \times \cdots \times x_{r'} = \epsilon_\sigma \Sigma x_1 \times x_2 \times \cdots \times x_r$$

for $\sigma: i \to i'$.

The symmetry operators are of particular importance in studying the linear transformations in $\mathfrak{R}_0{}^r$ which are induced by the linear transformations in the base space $\mathfrak{R}$. If A is such a mapping in $\mathfrak{R}$, we have defined the "induced" linear transformation $A_0{}^r \equiv A \times A \times \cdots \times A$ by the formula

$$(\Sigma x_1 \times x_2 \times \cdots \times x_r)A_0{}^r = \Sigma x_1 A \times x_2 A \times \cdots \times x_r A.$$

Evidently we have the relation

$$(AB)_0{}^r = A_0{}^r B_0{}^r. \tag{24}$$

Since $1_0{}^r = 1$, if A has an inverse A^{-1}, then $A_0{}^r(A^{-1})_0{}^r = 1_0{}^r = 1$. Hence if A is non-singular, then $A_0{}^r$ is non-singular in the ring $\mathfrak{L}_0{}^r$ of linear transformations of $\mathfrak{R}_0{}^r$. The equation (24) shows that the mapping $A \to A_0{}^r$ is a homomorphism of the units of $\mathfrak{L}$ in the group of units of $\mathfrak{L}_0{}^r$.

It is clear from the definition that $A_0{}^r$ commutes with every symmetry and hence with every symmetry operator Q. If $zQ = 0$, then also $(zA_0{}^r)Q = 0$. This remark shows that any symmetry class of tensors is an invariant subspace relative to the totality of mappings of the form $A_0{}^r$. If $U(A)$ denotes the linear transformation induced by $A_0{}^r$ in a particular symmetry class, then by (24) we have the relation $U(AB) = U(A)U(B)$.

We shall consider now the special case of the symmetry class of skew symmetric tensors. We shall assume also that the characteristic of Φ is not two. We determine first a basis for our symmetry class. Let

$$\Sigma \xi^{i_1 i_2 \cdots i_r} e_{i_1} \times e_{i_2} \times \cdots \times e_{i_r}$$

be skew symmetric. Then

$$\Sigma\xi^{i_1 i_2 \cdots i_r} e_{i_{1'}} \times e_{i_{2'}} \times \cdots \times e_{i_{r'}} = \epsilon_\sigma \Sigma \xi^{i_1 i_2 \cdots i_r} e_{i_1} \times e_{i_2} \times \cdots \times e_{i_r} \tag{25}$$

and, since $\epsilon_\sigma^{-1} = \epsilon_\sigma$, this implies that

$$\xi^{i_{1'} i_{2'} \cdots i_{r'}} = \epsilon_\sigma \xi^{i_1 i_2 \cdots i_r}. \tag{26}$$

Hence an interchange of any two subscripts changes the sign of the component $\xi^{i_1 i_2 \cdots i_r}$. As a corollary we see that $\xi^{i_1 i_2 \cdots i_r} = 0$ if any two of the i_j are equal. In particular, if $r > n$ every $\xi^{i_1 i_2 \cdots i_r} = 0$, that is, the only skew symmetric tensor of rank $r > n$ is the 0 tensor. Also we see that, whatever the value of r, the expression $\Sigma\xi^{i_1 i_2 \cdots i_r} e_{i_1} \times \cdots \times e_{i_r}$ for a skew symmetric tensor has zero coefficients for each term $e_{i_1} \times e_{i_2} \times \cdots \times e_{i_r}$ in which two i_j are equal. Hence these terms can be dropped. We suppose next that $i_1 < i_2 < \cdots < i_r$. Then the coefficient $\xi^{i_{1'} i_{2'} \cdots i_{r'}} = \epsilon_\sigma \xi^{i_1 i_2 \cdots i_r}$. Hence the terms involving all the base tensors $e_{i_{1'}} \times e_{i_{2'}} \times \cdots \times e_{i_{r'}}$ add up to

$$\xi^{i_1 i_2 \cdots i_r} [e_{i_1} e_{i_2} \cdots e_{i_r}], \quad i_1 < i_2 < \cdots < i_r \tag{27}$$

where

$$[e_{i_1} e_{i_2} \cdots e_{i_r}] = \sum_\sigma \epsilon_\sigma e_{i_{1'}} \times e_{i_{2'}} \times \cdots \times e_{i_{r'}}. \tag{28}$$

It is clear that $[e_{i_1} e_{i_2} \cdots e_{i_r}]$ is skew symmetric. Moreover, if $j_1 < j_2 < \cdots < j_r$ and $(j_1, j_2, \cdots, j_r) \neq (i_1, i_2, \cdots, i_r)$, then $[e_{j_1} e_{j_2} \cdots e_{j_r}]$ involves a set of base tensors $e_{j_{1'}} \times e_{j_{2'}} \times \cdots \times e_{j_{r'}}$ that has vacuous intersection with the set $e_{i_{1'}} \times e_{i_{2'}} \times \cdots \times e_{i_{r'}}$. It follows that the vectors $[e_{i_1} e_{i_2} \cdots e_{i_r}]$ determined by all possible choices of the indices $i_1 < i_2 < \cdots < i_r$ from 1 to n are linearly independent. Hence these tensors form a basis for the space of skew symmetric tensors. The number of elements in this basis is the same as the number of combinations of r distinct objects that can be selected from n distinct objects. This number $\binom{n}{r}$ is the dimensionality of our space.

We arrange the vectors $[e_{i_1} e_{i_2} \cdots e_{i_r}]$ lexicographically with respect to the indices $i_1, i_2, \cdots, i_r$. For example, if $n = 3$ and $r = 2$, the order is $[e_1 e_2]$, $[e_1 e_3]$, $[e_2 e_3]$. If (α) is the matrix of a linear transformation A relative to the basis $(e_1, e_2, \cdots, e_n)$, then the matrix relative to the lexicographically ordered basis $[e_{i_1}e_{i_2} \cdots e_{i_r}]$, of the linear transformation induced by A in the space of skew symmetric tensors of rank r is called the rth *compound* of (α). We denote this matrix as $C_r(\alpha)$. Evidently we have the relation

$$C_r((\alpha)(\beta)) = C_r(\alpha)C_r(\beta). \tag{29}$$

We shall now obtain the explicit form of $C_r(\alpha)$. Since $e_iA = \Sigma\alpha_{ij}e_j$,

$$\begin{aligned}
[e_{i_1}e_{i_2} \cdots e_{i_r}]A_0{}^r & \\
&= \sum_\sigma \epsilon_\sigma e_{i_{1'}}A \times e_{i_{2'}}A \times \cdots \times e_{i_{r'}}A \\
&= \sum_\sigma \sum_j \epsilon_\sigma \alpha_{i_{1'}j_{1'}}\alpha_{i_{2'}j_{2'}} \cdots \alpha_{i_{r'}j_{r'}} e_{j_{1'}} \times e_{j_{2'}} \times \cdots \times e_{j_{r'}} \\
&= \sum_j \sum_\sigma \epsilon_\sigma \alpha_{i_{1'}j_{1'}}\alpha_{i_{2'}j_{2'}} \cdots \alpha_{i_{r'}j_{r'}} e_{j_{1'}} \times e_{j_{2'}} \times \cdots \times e_{j_{r'}} \\
&= \sum_j | \alpha_{i_1j_{1'}}\alpha_{i_2j_{2'}} \cdots \alpha_{i_rj_{r'}} | (e_{j_{1'}} \times e_{j_{2'}} \times \cdots \times e_{j_{r'}})
\end{aligned} \tag{30}$$

where $| \alpha_{i_1j_{1'}}\alpha_{i_2j_{2'}} \cdots \alpha_{i_rj_{r'}} |$ denotes the determinant

$$\begin{vmatrix}
\alpha_{i_1j_{1'}} & \alpha_{i_1j_{2'}} & \cdots & \alpha_{i_1j_{r'}} \\
\alpha_{i_2j_{1'}} & \alpha_{i_2j_{2'}} & \cdots & \alpha_{i_2j_{r'}} \\
\vdots & \vdots & \cdots & \vdots \\
\alpha_{i_rj_{1'}} & \alpha_{i_rj_{2'}} & \cdots & \alpha_{i_rj_{r'}}
\end{vmatrix}. \tag{31}$$

If any two j's are equal, this determinant is 0 and for unequal j's

$$| \alpha_{i_1j_{1'}}\alpha_{i_2j_{2'}} \cdots \alpha_{i_rj_{r'}} | = \epsilon_\sigma | \alpha_{i_1j_1}\alpha_{i_2j_2} \cdots \alpha_{i_rj_r} |$$

where $j_1 < j_2 < \cdots < j_r$. Hence (30) can be written as

$$[e_{i_1}e_{i_2} \cdots e_{i_r}]A_0{}^r = \sum_{j_1<j_2<\cdots<j_r} | \alpha_{i_1j_1}\alpha_{i_2j_2} \cdots \alpha_{i_rj_r} | \; [e_{j_1}e_{j_2} \cdots e_{j_r}].$$

This shows that $C_r(\alpha)$ is the matrix whose elements are the r-rowed minors $| \alpha_{i_1j_1}\alpha_{i_2j_2} \cdots \alpha_{i_rj_r} |$ of (α).

If $r = n$, the skew symmetric tensors have the basis $[e_1e_2 \cdots e_n]$. The matrix $C_n(\alpha) = \det(\alpha)$. Thus (29) specializes in this case to the multiplication rule for determinants.

These remarks will serve as an introduction to the rich but somewhat involved theory of symmetry classes of tensors. A thorough treatment of this subject is given in Weyl's *The Classical Groups*, Chapters 3 and 4. (Cf. also Wedderburn's *Lectures on Matrices*, Chapter 5.)

EXERCISES

1. Show that $[x_1x_2 \cdots x_n] \equiv \Sigma\epsilon_\sigma x_{1'}x_{2'} \cdots x_{n'} = |\xi_{11}\xi_{22} \cdots \xi_{nn}|[e_1e_2 \cdots e_n]$ where $|\xi_{11} \cdots \xi_{nn}|$ is the determinant of the coordinates of the x's relative to the e's.
2. Show that, for the scalar matrix $\alpha 1$, $C_r(\alpha 1) = \alpha^r 1$.
3. Prove that $\det C_r(\alpha) = (\det(\alpha))^{\binom{n-1}{r-1}}$.

8. Extension of the field of a vector space. If P is a field which contains Φ as a subfield, then P can be regarded as a vector space over Φ. The addition in the vector space is taken to be the field addition, and scalar multiplication $\alpha\xi$ for α in Φ and ξ in P is defined to be the field product. The axioms of a vector space are immediate consequences of the associative and distributive laws and the fact that Φ contains the identity of P. The vector space may be infinite dimensional or finite dimensional. For example, if $P = \Phi(\lambda)$ where λ is an indeterminate, then P is infinite over Φ. On the other hand, if P is the field of complex numbers and Φ is the subfield of real numbers, then P is two-dimensional over Φ.

Now let $\mathfrak{R}$ be a vector space over Φ and let $\mathfrak{P} = P \times \mathfrak{R}$ the Kronecker products of P regarded as a vector space over Φ with the vector space $\mathfrak{R}$.* The elements of $\mathfrak{P}$ have the form $\Sigma\rho_i \times x_i$ where $\rho_i \in P$ and $x_i \in \mathfrak{R}$. Let σ be any element of P; then, since $\xi \to \sigma\xi$ is a linear transformation in P over Φ, the mapping

$$\Sigma\rho_i \times x_i \to \Sigma\sigma\rho_i \times x_i$$

is an endomorphism. We use this to define a scalar multiplication in $\mathfrak{P}$ by setting

$$\sigma(\Sigma\rho_i \times x_i) = \Sigma\sigma\rho_i \times x_i. \tag{32}$$

* Strictly speaking the existence of the Kronecker product has thus far been proved only in the case in which P is finite over Φ. The considerations of Chapter IX will enable us to extend our construction to the infinite case.

Then we have shown that the function σz is single-valued and that $\sigma(z_1 + z_2) = \sigma z_1 + \sigma z_2$. The other axioms for scalar multiplication can be verified directly. Hence $\mathfrak{P}$ can be regarded as a vector space over P. We shall denote $\mathfrak{P}$ as $\mathfrak{R}_P$, and we call this space obtained from $\mathfrak{R}$ by *extending the base field* Φ *to the field* P.

It is easy to see that the dimensionality of $\mathfrak{R}_P$ over P is the same as that of $\mathfrak{R}$ over Φ. For let $(e_1, e_2, \cdots, e_n)$ be a basis for $\mathfrak{R}$ over Φ and let $\bar{e}_i = 1 \times e_i$. Then $\rho_i \times e_i = \rho_i(1 \times e_i) = \rho_i\bar{e}_i$ and, since any $z \in \mathfrak{R}_P$ has the form $\Sigma\rho_i \times e_i$, it also has the form $\Sigma\rho_i\bar{e}_i$. Similarly, if $\Sigma\rho_i\bar{e}_i = 0$, then $\Sigma\rho_i \times e_i = 0$ and, since the e_i are linearly independent over Φ, every $\rho_i = 0$.

For each $x \in \mathfrak{R}$ we set $\bar{x} = 1 \times x$. Then it is easy to check that the correspondence $x \to \bar{x}$ is an equivalence between the vector space $\mathfrak{R}$ over Φ and a subspace $\bar{\mathfrak{R}}$ of $\mathfrak{R}_P$ over Φ. If we like we can replace the space $\mathfrak{R}$ by the subset $\bar{\mathfrak{R}}$ of $\mathfrak{R}_P$. The relation between $\mathfrak{R}_P$ and $\bar{\mathfrak{R}}$ can be described by the following statements:

1. Any vector in $\mathfrak{R}_P$ has the form $\Sigma\rho_i\bar{x}_i$ where $\rho_i \in P$ and $\bar{x}_i \in \bar{\mathfrak{R}}$.
2. If the vectors $(\bar{e}_1, \bar{e}_2, \cdots, \bar{e}_n)$ are in $\bar{\mathfrak{R}}$ and are linearly independent over Φ, then they are also linearly independent over P.

Suppose now that A is a linear transformation in $\mathfrak{R}$ over Φ; then we know that the mapping $\Sigma\rho_i \times x_i \to \Sigma\rho_i \times x_iA$ is an endomorphism in $\mathfrak{R}_P$. It is also immediate that it is a linear transformation in $\mathfrak{R}_P$ over P. We shall call this mapping *the extension* of A in $\mathfrak{R}_P$. We use the same letter to denote a mapping and its extension. If (α) is a matrix of A relative to the basis $(e_1, e_2, \cdots, e_n)$ of $\mathfrak{R}$, then

$$\begin{aligned} \bar{e}_iA &= (1 \times e_i)A = 1 \times e_iA = \Sigma 1 \times \alpha_{ij}e_j \\ &= \Sigma\alpha_{ij}(1 \times e_j) = \Sigma\alpha_{ij}\bar{e}_j. \end{aligned}$$

Hence the matrix of the extension A relative to the basis $(\bar{e}_1, \bar{e}_2, \cdots, \bar{e}_n)$ is (α) also.

9. A theorem on similarity of sets of matrices. In this section we shall prove a theorem on sets of matrices which is of interest in itself and which illustrates a useful method in the theory of matrices and of algebras. The idea of this method is that it is sometimes easier to prove a result if the base field is "sufficiently large" (e.g., algebraically closed). We can, therefore, extend the base field to obtain one which has the necessary properties and then we are confronted with the problem of showing that the final result is valid in the original field.

The particular problem we shall consider is the following. Let ω_1 and ω_2 be two sets of matrices with elements in Φ. Suppose they are similar by means of a matrix whose elements belong to an extension field P. Then can we conclude that the sets are similar by a matrix with elements in Φ?

More precisely, we begin with a set Ω of linear transformations in $\mathfrak{R}$ over Φ, and we consider their extensions in $\mathfrak{R}$ over P. Let ω be the set of matrices determined by these extensions relative to a basis $(\bar{e}_1, \bar{e}_2, \cdots, \bar{e}_n)$, $\bar{e}_i = 1 \times e_i$. As we have seen, the matrices in ω all have elements in the field Φ. If $(u_1, u_2, \cdots, u_n)$ is any basis in $\mathfrak{R}_P$ over P and $u_i = \Sigma\mu_{ij}\bar{e}_j$, then the set of matrices of Ω relative to this basis is $(\mu)\omega(\mu)^{-1}$. In general some of these matrices will have elements which do not belong to Φ. However, it may happen that even though $(\mu) \not\in \Phi_n$ the set $(\mu)\omega(\mu)^{-1} \subseteq \Phi_n$. We shall now show that, if this is the case, then there exists a matrix (γ) in $L(\Phi, n)$ such that for every $(\alpha) \in \omega$

$$(\gamma)(\alpha)(\gamma)^{-1} = (\mu)(\alpha)(\mu)^{-1}.$$

We shall prove this result under the assumption that Φ has an infinite number of elements.*

Theorem 5. *Let Φ be infinite and let ω be a set of matrices with elements in Φ. Suppose that (μ) is a matrix in $L(\mathrm{P}, n)$, P an extension of Φ, such that $(\mu)\omega(\mu)^{-1} \subseteq \Phi_n$. Then there exists a matrix $(\gamma) \in L(\Phi, n)$ such that $(\gamma)(\alpha)(\gamma)^{-1} = (\mu)(\alpha)(\mu)^{-1}$ for every $(\alpha) \in \omega$.*

Proof. Denote $(\mu)(\alpha)(\mu)^{-1}$ as $(\alpha)_\mu$. Then

$$(\mu)(\alpha) = (\alpha)_\mu(\mu)$$

and these equations are equivalent to a system (possibly infinite) of homogeneous linear equations for the elements μ_{ij} of (μ).

* A proof for the finite case is given in Deuring, *Galoische Theorie und Darstellungstheorie*, Math. Annalen, v. 107, pp. 140–144.

Moreover, these equations have coefficients in Φ, that is, they have the form

$$\Sigma\beta_{ij}\mu_{ij} = 0, \quad \beta_{ij} \text{ in } \Phi.$$

We regard the set $\{(\beta)\}$ first as a subset of Φ_n, and we suppose that r is the rank of this set in the sense that this is the largest number of linearly independent (relative to Φ) matrices in $\{(\beta)\}$. Then the equations $\Sigma\beta_{ij}\xi_{ij} = 0$ have $n^2 - r$ linearly independent solutions for ξ_{ij} in Φ such that every solution is a linear combination of these. Let $(\gamma_{ij}^{(1)}), (\gamma_{ij}^{(2)}), \cdots, (\gamma_{ij}^{(h)}), h = n^2 - r$, be such a set of solutions.

We remark next that a set of matrices with coefficients in Φ that are linearly independent over Φ are also linearly independent over P. This is clear since the space P_n is the extension space $\Phi_{n\mathrm{P}}$. Hence the matrices $(\gamma_{ij}^{(1)}), \cdots, (\gamma_{ij}^{(h)})$ are P independent. We now note that r is also the rank of the set $\{(\beta)\}$ relative to P and that the maximum number of linearly independent solutions of the equations $\Sigma\beta_{ij}\xi_{ij}$ in P is also $n^2 - r$. Hence the particular solutions $(\gamma_{ij}^{(1)}), (\gamma_{ij}^{(2)}), \cdots, (\gamma_{ij}^{(h)})$ which we selected form a basic set for the solutions in P. In particular

$$\mu_{ij} = \nu_1\gamma_{ij}^{(1)} + \nu_2\gamma_{ij}^{(2)} + \cdots + \nu_h\gamma_{ij}^{(h)}$$

where the ν's are in P.

We now replace the ν_i by independent indeterminates λ_i, and we consider the polynomial

$$\det\left(\sum_k \lambda_k\gamma_{ij}^{(k)}\right).$$

This polynomial in $\Phi[\lambda_1, \lambda_2, \cdots, \lambda_k]$ is not the 0 polynomial since its value $\det(\mu)$ for $\lambda_k = \mu_k$ is not 0. Since Φ is infinite, we can select values $\lambda_k = \beta_k$ in Φ such that $\det(\Sigma\beta_k\gamma_{ij}^{(k)}) \neq 0$.* Let $\gamma_{ij} = \Sigma\beta_k\gamma_{ij}^{(k)}$ and let $(\gamma) = (\gamma_{ij})$. Then $(\gamma) \in L(\Phi, n)$ and $\Sigma\beta_{ij}\gamma_{ij} = 0$ for all β. Hence

$$(\gamma)(\alpha) = (\alpha)_\mu(\gamma)$$

for all (α), and (γ) satisfies the requirements of the theorem.

EXERCISE

1. Prove Theorem 5 for ω a set consisting of a single matrix without any restriction on Φ.

* See Theorem 10, p. 112, of Volume I of these Lectures.

10. Alternative definition of an algebra. Kronecker product of algebras. The concept of an associative algebra has been introduced in Chapter II. We shall now see that a simple alternative definition can be given in terms of the notion of the Kronecker product. We find it convenient to generalize the former discussion by dropping the associative law. Then our previous definition takes the following form: *A (not necessarily associative) algebra* $\mathfrak{A}$ is a vector space over a field Φ together with a binary multiplication xy in $\mathfrak{A}$ such that

$$(33) \qquad (x_1 + x_2)y = x_1y + x_2y, \quad x(y_1 + y_2) = xy_1 + xy_2$$

$$(34) \qquad \alpha(xy) = (\alpha x)y = x(\alpha y).$$

Conditions (33) and the last equality in (34) state that $\mathfrak{A}$ is a product space of $\mathfrak{A}$ and $\mathfrak{A}$ relative to the multiplication xy. Hence we know that the mapping $\Sigma x \times y \rightarrow \Sigma xy$ is a linear transformation of the Kronecker product $\mathfrak{A} \times \mathfrak{A}$ into $\mathfrak{A}$.

This remark serves as a basis for our second definition of an algebra. According to it we define an algebra to be a vector space $\mathfrak{A}$ over a field Φ together with a linear transformation P of $\mathfrak{A} \times \mathfrak{A}$ into $\mathfrak{A}$. In an algebra defined in this way we can introduce a binary product by means of the formula

$$(35) \qquad xy = (x \times y)P.$$

Then it is easy to see that (33) and (34) hold. Hence it is clear that this procedure leads to the same concept as the former definition.

Suppose next that $\mathfrak{A}_1$ and $\mathfrak{A}_2$ are arbitrary algebras over the same field Φ. Let P_i be the linear mapping of $\mathfrak{A}_i \times \mathfrak{A}_i$ into $\mathfrak{A}_i$. Thus $x_iy_i = (x_i \times y_i)P_i$. We now form the Kronecker product $\mathfrak{A} = \mathfrak{A}_1 \times \mathfrak{A}_2$ and we consider the Kronecker product $\mathfrak{A} \times \mathfrak{A}$ of this space with itself. We know that, if x_1, y_1 are in $\mathfrak{A}_1$ and x_2, y_2 are in $\mathfrak{A}_2$, then the mapping

$$(36) \qquad \Sigma x_1 \times x_2 \times y_1 \times y_2 \rightarrow \Sigma x_1 \times y_1 \times x_2 \times y_2$$

is an equivalence of $\mathfrak{A} \times \mathfrak{A} = \mathfrak{A}_1 \times \mathfrak{A}_2 \times \mathfrak{A}_1 \times \mathfrak{A}_2$ onto $\mathfrak{A}_1 \times \mathfrak{A}_1 \times \mathfrak{A}_2 \times \mathfrak{A}_2$. Also

$$(37) \qquad \Sigma x_1 \times y_1 \times x_2 \times y_2 \rightarrow \Sigma (x_1 \times y_1)P_1 \times (x_2 \times y_2)P_2$$

is a linear mapping of $\mathfrak{A}_1 \times \mathfrak{A}_1 \times \mathfrak{A}_2 \times \mathfrak{A}_2$ into $\mathfrak{A} = \mathfrak{A}_1 \times \mathfrak{A}_2$.

Combining (36) and (37) we obtain a linear mapping P of $\mathfrak{A} \times \mathfrak{A}$ into $\mathfrak{A}$. Thus $\mathfrak{A}$, together with the mapping P, constitutes an algebra over the field Φ. If $\Sigma x_1 \times x_2$ and $\Sigma y_1 \times y_2$ are two elements of $\mathfrak{A}$, then their product in $\mathfrak{A}$ is

$$(\Sigma x_1 \times x_2)(\Sigma y_1 \times y_2) = \Sigma x_1 y_1 \times x_2 y_2; \tag{38}$$

for, by definition

$$\begin{aligned}(\Sigma x_1 \times x_2 \times y_1 \times y_2)P &= \Sigma (x_1 \times y_1)P_1 \times (x_2 \times y_2)P_2 \\ &= \Sigma x_1 y_1 \times x_2 y_2.\end{aligned}$$

The algebra $\mathfrak{A}$ thus defined is called the *Kronecker* or *direct product* of $\mathfrak{A}_1$ and $\mathfrak{A}_2$. It is an immediate consequence of (38) that, if $\mathfrak{A}_1$ and $\mathfrak{A}_2$ are associative algebras, then the Kronecker product $\mathfrak{A} = \mathfrak{A}_1 \times \mathfrak{A}_2$ is associative.

As an illustration of the concept of the Kronecker product of algebras we note the following extension of Theorem 4.

Theorem 6. *The algebra* $\mathfrak{L}(\mathfrak{R} \times \mathfrak{S}, \mathfrak{R} \times \mathfrak{S}) = \mathfrak{L}(\mathfrak{R}, \mathfrak{R}) \times \mathfrak{L}(\mathfrak{S}, \mathfrak{S})$.

Proof. We have already seen that this relation holds in the vector space sense. Also we have the relation $(A \times B)(C \times D) = AC \times BD$, and this shows that the ordinary product in $\mathfrak{L}(\mathfrak{R} \times \mathfrak{S}, \mathfrak{R} \times \mathfrak{S})$ coincides with the composition defined in the Kronecker product of the two algebras.

A slightly different form of our result is given in the

Corollary. $\Phi_{mn} = \Phi_m \times \Phi_n$.

EXERCISES

1. Let a be an element of an algebra $\mathfrak{A}$ and let R_a denote the mapping $x \rightarrow xa$ determined by a. Show that R_a is a linear transformation in $\mathfrak{A}$. Also show that the mapping $a \rightarrow R_a$ is a linear transformation of $\mathfrak{A}$ into the space $\mathfrak{L}$ consisting of all the linear transformations in $\mathfrak{A}$.
2. Let L_a denote the mapping $x \rightarrow ax$. Show that L_a is a linear transformation in $\mathfrak{A}$ and that $a \rightarrow L_a$ is linear.
3. Prove that $\mathfrak{A}$ is associative if and only if $L_a R_b = R_b L_a$ for all a, b. Also prove that $\mathfrak{A}$ is associative if and only if $R_a R_b = R_{ab}$ $(L_a L_b = L_{ba})$.
4. Let $\mathfrak{A}$ be the algebra of real quaternions with basis $1, i, j, k$ where

$$i^2 = j^2 = k^2 = -1, \quad ij = -ji = k, \quad jk = -kj = i, \quad ki = -ik = j.$$

Show that $\mathfrak{A} \times \mathfrak{A} \cong \Phi_4$.

Chapter VIII

THE RING OF LINEAR TRANSFORMATIONS

In this chapter we obtain the basic properties of the ring $\mathfrak{L}$ of all the linear transformations in a finite dimensional vector space. We determine the ideals, two-sided, left and right, of this ring. Also we show that two rings $\mathfrak{L}_1$ and $\mathfrak{L}_2$ determined respectively by the vector spaces $\mathfrak{R}_1$ and $\mathfrak{R}_2$ are not isomorphic unless the spaces in which they act are isomorphic. In studying the automorphisms of $\mathfrak{L}$ we are led to consider an important type of transformation called semi-linear which generalizes the concept of a linear transformation.

1. Simplicity of $\mathfrak{L}$. As we shall see, there are a number of ways of studying the ring $\mathfrak{L}$ of all the linear transformations in a vector space $\mathfrak{R}$ over a division ring Δ. The first of these that we consider consists in treating $\mathfrak{L}$ as a matrix ring. We introduce a set of "matrix units" E_{ij} determined by a particular basis $(e_1, e_2, \cdots, e_n)$ for $\mathfrak{R}$. We define E_{ij} to be the linear mapping such that

$$e_r E_{ij} = \delta_{ir} e_j, \quad r = 1, 2, \cdots, n. \tag{1}$$

By checking for the base vectors we verify that

$$E_{ij}E_{kl} = \delta_{jk}E_{il}, \tag{2}$$
$$\Sigma E_{ii} = 1.$$

Next we introduce a set $\bar{\Delta}$ of linear transformations which correspond to the set of matrices diag $\{\alpha, \alpha, \cdots, \alpha\}$. These are obtained by associating with each $\alpha \in \Delta$ the linear mapping $\bar{\alpha}$ such that

$$e_i\bar{\alpha} = \alpha e_i. \tag{3}$$

It should be noted that, although $\bar{\alpha}$ has the same effect on the e_i as the scalar multiplication α_l, these transformations are not in general identical; for if $x = \Sigma\xi_i e_i$, then $\alpha x = \Sigma(\alpha\xi_i)e_i$, while $x\bar{\alpha} = \Sigma(\xi_i e_i)\bar{\alpha} = \Sigma\xi_i(e_i\bar{\alpha}) = \Sigma(\xi_i\alpha)e_i$. Hence $\bar{\alpha} = \alpha_l$ if and only if α is in the center of Δ. We remark also that $\bar{\alpha}$ depends on the choice of the basis as well as on α. For the most part we shall stick to a single basis; hence the dependence on the basis need not be indicated.

The fundamental correspondence between linear transformations and matrices associates with $\bar{\alpha}$ the matrix diag $\{\alpha, \alpha, \cdots, \alpha\}$. Clearly the totality of these matrices is a subring of Δ_n isomorphic to Δ. Hence $\bar{\Delta}$ is a subring of $\mathfrak{L}$ which is isomorphic to Δ, the correspondence $\alpha \rightarrow \bar{\alpha}$ being an isomorphism.

It is easy to verify that

$$\bar{\alpha}E_{ij} = E_{ij}\bar{\alpha} \tag{4}$$

and that, if $e_iA = \Sigma\alpha_{ij}e_j$, then

$$A = \Sigma\bar{\alpha}_{ij}E_{ij} = \Sigma E_{ij}\bar{\alpha}_{ij}. \tag{5}$$

Thus any $A \in \mathfrak{L}$ can be written as a linear combination with coefficients in $\bar{\Delta}$ of the E_{ij}. Also we have the important formula

$$\bar{\alpha}_{ij} = \sum_k E_{ki}AE_{jk} \tag{6}$$

for the coefficients in the expression $A = \Sigma\bar{\alpha}_{ij}E_{ij}$.

This formula yields a simple proof of the fact that $\mathfrak{L}$ is a *simple* ring. By this we mean that the only two-sided ideals in $\mathfrak{L}$ are 0 and $\mathfrak{L}$ itself. Thus let $\mathfrak{B}$ be a two-sided ideal $\neq 0$ in $\mathfrak{L}$ and let $B = \Sigma\bar{\beta}_{ij}E_{ij}$ be an element $\neq 0$ in $\mathfrak{B}$. Then each $\bar{\beta}_{ij} = \Sigma E_{ki}BE_{jk}$ is in $\mathfrak{B}$. If $\bar{\beta}_{pq} \neq 0$, then $1 = \bar{\beta}_{pq}\bar{\beta}_{pq}{}^{-1}$ is also in $\mathfrak{B}$, and this implies that every $A \in \mathfrak{L}$ is contained in $\mathfrak{B}$. Hence $\mathfrak{B} = \mathfrak{L}$.

Theorem 1. *The ring $\mathfrak{L}$ of linear transformations in a finite dimensional vector space is simple.*

EXERCISE

1. Prove that $\bar{\Delta}$ is the totality of linear transformations that commute with the E_{ij}.

2. Operator methods. We shall now give a second proof of the simplicity of $\mathfrak{L}$ using operator methods. Let $\mathfrak{B}$ be a two-sided ideal $\neq 0$ in $\mathfrak{L}$. We show first that, if $[x]$ is any one-dimensional subspace of $\mathfrak{R}$, then $\mathfrak{B}$ contains a projection E of $\mathfrak{R}$ on $[x]$. Let $B \neq 0$ belong to $\mathfrak{B}$ and let $\mathfrak{R}B = [y_1, y_2, \cdots, y_r]$ where the y_i are linearly independent. Let x_1 be a vector such that $x_1B = y_1$. We can find a linear transformation A_1 such that $y_1A_1 = x_1$ and $y_iA_1 = 0$ for $i > 1$. Then $x_1BA_1 = y_1A_1 = x_1$. Hence $E_1 \equiv BA_1$ is a projection of $\mathfrak{R}$ on $[x_1]$. Now let C and D be linear transformations such that $xC = x_1$ and $x_1D = x$. Then if $E = CE_1D$, $\mathfrak{R}E = \mathfrak{R}CE_1D = [x_1]D = [x]$ and $xE = xCE_1D = x_1E_1D = x_1D = x$. Hence E is a projection of $\mathfrak{R}$ on $[x]$. Since $E = CE_1D$, $E \in \mathfrak{B}$.

Now let E be a projection of maximum rank in $\mathfrak{B}$. We assert that $E = 1$. Otherwise $\mathfrak{S} = \mathfrak{R}E \subset \mathfrak{R}$, and there is a vector $z \neq 0$ such that $zE = 0$. As we have seen, $\mathfrak{B}$ contains a projection F of $\mathfrak{R}$ on $[z]$. Now set $G = E + F - EF$. If $x \in \mathfrak{S}$,

$$xG = xE + xF - xEF = x + xF - xF = x.$$

Moreover,

$$zG = zE + zF - zEF = zF = z.$$

Thus G acts as the identity in $\mathfrak{S} + [z]$. Then $\mathfrak{R}G \supseteq \mathfrak{R}E + \mathfrak{R}F = \mathfrak{S} + [z]$. It follows that G is a projection of greater rank than E. Since $G \in \mathfrak{B}$, this contradicts the choice of E. Hence we see that $E = 1 \in \mathfrak{B}$. Clearly this means that $\mathfrak{B} = \mathfrak{L}$.

We use a similar method to prove the following

Theorem 2. *If $\mathfrak{L}$ is the ring of linear transformations in $\mathfrak{R}$ over Δ, then reciprocally Δ is the complete set of endomorphisms in $\mathfrak{R}$ which commute with all the transformations contained in $\mathfrak{L}$.*

Proof. Let C be an endomorphism in $\mathfrak{R}$ which commutes with every element of $\mathfrak{L}$. If x is any vector in $\mathfrak{R}$, then x and xC are linearly dependent. Otherwise there exists an $A \in \mathfrak{L}$ such that $xA = 0$, but $(xC)A \neq 0$. This contradicts $0 = (xA)C = xCA$. Thus for each $x \neq 0$, $xC = \gamma_x x$. Now let x and y be any two non-zero vectors. Then

$$xC = \gamma_x x, \quad yC = \gamma_y y.$$

On the other hand, there exists a $B \varepsilon \mathfrak{L}$ such that $xB = y$. Hence

$$yC = xBC = xCB = \gamma_x(xB) = \gamma_x y.$$

Hence $\gamma_x = \gamma_y$. This proves that $xC = \gamma x$ for all $x \neq 0$. Since this holds also for $x = 0$, C is the scalar multiplication γ_l.

3. The left ideals of $\mathfrak{L}$. In order to determine the one-sided ideals of $\mathfrak{L}$ we shall make use of still another technique, namely, we consider $\mathfrak{L}$ as a product group. Let $\mathfrak{R}'$ be a right vector space dual to $\mathfrak{R}$ relative to the bilinear form $g(x, x')$. As in the preceding chapter we define $x' \times y$ for $x' \varepsilon \mathfrak{R}'$ and $y \varepsilon \mathfrak{R}$ to be the mapping $x \to g(x, x')y$. Then we know that $\mathfrak{L}$ is a direct product $\mathfrak{R}' \times \mathfrak{R}$ relative to this multiplication.

We consider now the representation of the product (resultant) of the two mappings $x' \times y$ and $u' \times v$. We have the following equations:

$$\begin{aligned} x[(x' \times y)(u' \times v)] &= g(x, x')y(u' \times v) = g(g(x, x')y, u')v \\ &= g(x, x')g(y, u')v \\ &= g(x, x'g(y, u'))v \end{aligned}$$

which show that

$$(x' \times y)(u' \times v) = x'g(y, u') \times v = x' \times g(y, u')v.$$

More generally, if we use the distributive laws, we see that

$$\begin{aligned} (\Sigma x_i' \times y_i)(\Sigma u_j' \times v_j) &= \Sigma x_i' g(y_i, u_j') \times v_j \\ &= \Sigma x_i' \times g(y_i, u_j')v_j. \end{aligned} \tag{7}$$

This fundamental multiplication rule shows us how to construct one-sided ideals in $\mathfrak{L} = \mathfrak{R}' \times \mathfrak{R}$. Let $\mathfrak{S}$ be any subspace in $\mathfrak{R}$ and set $\mathfrak{J}' = \mathfrak{R}' \times \mathfrak{S}$ the totality of linear transformations of the form $\Sigma u_j' \times v_j$ where $u_j' \varepsilon \mathfrak{R}'$ and $v_j \varepsilon \mathfrak{S}$. Clearly $\mathfrak{J}'$ is closed under addition and subtraction. Moreover, by (7), $\mathfrak{J}'$ is closed under left multiplication by arbitrary elements of $\mathfrak{L}$. Thus $\mathfrak{J}'$ is a left ideal. In a similar fashion we see that, if $\mathfrak{S}'$ is a subspace of $\mathfrak{R}'$, then $\mathfrak{J} = \mathfrak{S}' \times \mathfrak{R}$ the totality of mappings $\Sigma x_i' \times y_i$ where $x_i' \varepsilon \mathfrak{S}'$ and $y_i \varepsilon \mathfrak{R}$ is a right ideal in $\mathfrak{L}$.

The main result we shall establish is that the ideals $\mathfrak{R}' \times \mathfrak{S}$, $\mathfrak{S}' \times \mathfrak{R}$ are the only one-sided ideals in $\mathfrak{L}$. We consider left ideals first. Let $\mathfrak{J}'$ be any left ideal $\neq 0$ in $\mathfrak{L}$ and let $\mathfrak{S}$ be the join of all the rank subspaces $\mathfrak{R}B$ for B in $\mathfrak{J}'$. If $B \neq 0$, we can write $B = \sum_1^r u_j' \times v_j$ where r is the rank of B. Then it is easily seen that the set $(v_1, v_2, \cdots, v_r)$ is a basis for $\mathfrak{R}B$ and $(u_1', u_2', \cdots, u_r')$ is one for $\mathfrak{R}'B'$, B' the transpose of B (cf. Chapter V, § 5). Since the $v_j \varepsilon \mathfrak{R}B \subseteq \mathfrak{S}$, $B \varepsilon \mathfrak{R}' \times \mathfrak{S}$. Hence $\mathfrak{J}' \subseteq \mathfrak{R}' \times \mathfrak{S}$. Next let $y_1, y_2, \cdots, y_r$ be vectors such that $g(y_i, u_j') = \delta_{ij}$ and let x_1', $x_2', \cdots, x_r'$ be arbitrary in $\mathfrak{R}'$. Set $A = \Sigma x_i' \times y_i$. Then by (7)

$$AB = \Sigma x_i' g(y_i, u_j') \times v_j = \Sigma x_i' \times v_i \tag{8}$$

and this linear transformation is in $\mathfrak{J}'$. This shows that, if $v = \Sigma \beta_i v_i$ is any vector in $\mathfrak{R}B$ and x' is arbitrary in $\mathfrak{R}'$, then the mapping $x' \times v = \Sigma x' \beta_i \times v_i \varepsilon \mathfrak{J}'$. It follows also that, if B_1, B_2, $\cdots$ are any linear transformations in $\mathfrak{J}'$ and $v^{(k)}$ is any vector in $\mathfrak{R}B_k$, then $\Sigma x_k' \times v^{(k)}$ is in $\mathfrak{J}'$. Thus $\mathfrak{R}' \times \mathfrak{S} \subseteq \mathfrak{J}'$ and this proves the following

Theorem 3. *Every left ideal $\mathfrak{J}'$ in $\mathfrak{L}$ has the form $\mathfrak{R}' \times \mathfrak{S}$ where $\mathfrak{S}$ is a subspace of $\mathfrak{R}$. The subspace $\mathfrak{S}$ is in fact the join of all the rank spaces $\mathfrak{R}B$, B in $\mathfrak{J}'$.*

This result can also be formulated in another way. Let $\mathfrak{S}$ be any subspace of $\mathfrak{R}$. Define $\mathfrak{J}'(\mathfrak{S})$ to be the totality of $B \varepsilon \mathfrak{L}$ such that $\mathfrak{R}B \subseteq \mathfrak{S}$. Clearly $\mathfrak{J}'(\mathfrak{S})$ is a left ideal in $\mathfrak{L}$. On the other hand, let $\mathfrak{J}'$ be any left ideal in $\mathfrak{L}$ and, as before, let $\mathfrak{S}$ be the join of all the subspaces $\mathfrak{R}B$, B in $\mathfrak{J}'$. Obviously $\mathfrak{J}' \subseteq \mathfrak{J}'(\mathfrak{S})$, and Theorem 3 shows that $\mathfrak{J}' = \mathfrak{R}' \times \mathfrak{S}$. Now let $B \varepsilon \mathfrak{J}'(\mathfrak{S})$. Then $\mathfrak{R}B \subseteq \mathfrak{S}$ and, if $\mathfrak{R}B = [v_1, v_2, \cdots, v_r]$ where the v_i are linearly independent, $B = \Sigma u_j' \times v_j$. Hence the $v_i \varepsilon \mathfrak{S}$ and $B \varepsilon \mathfrak{R}' \times \mathfrak{S} = \mathfrak{J}'$. Thus $\mathfrak{J}' = \mathfrak{J}'(\mathfrak{S})$. We therefore have the following alternate form of Theorem 3.

Theorem 4. *Every left ideal $\mathfrak{J}'$ in $\mathfrak{L}$ has the form $\mathfrak{J}'(\mathfrak{S})$, the totality of linear transformations which map $\mathfrak{R}$ into a subspace $\mathfrak{S}$. The space $\mathfrak{S}$ is the join of the rank spaces of the mappings belonging to $\mathfrak{J}'$.*

Theorem 3 (or 4) establishes a 1–1 correspondence $\mathfrak{S} \to \mathfrak{R}' \times \mathfrak{S}$ $(\mathfrak{J}'(\mathfrak{S}))$ between the set of subspaces of $\mathfrak{R}$ and the set of left ideals of $\mathfrak{L}$. Clearly this correspondence preserves order: if $\mathfrak{S}_1 \supseteq \mathfrak{S}_2$, then $\mathfrak{R}' \times \mathfrak{S}_1 \supseteq \mathfrak{R}' \times \mathfrak{S}_2$. It follows that we have a lattice isomorphism of the lattice of subspaces of $\mathfrak{R}$ onto the lattice of left ideals of $\mathfrak{L}$.

EXERCISES

1. A left (right, two-sided) ideal $\mathfrak{J}'$ is called *minimal* if $\mathfrak{J}' \neq 0$ and there exists no left (right, two-sided) ideal $\mathfrak{J}''$ such that $\mathfrak{J}' \supset \mathfrak{J}'' \supset 0$. Show that $\mathfrak{L}$ possesses minimal left ideals and find their form.

2. A left (right, two-sided) ideal $\mathfrak{J}'$ is called *maximal* if $\mathfrak{J}' \neq \mathfrak{L}$ and there exists no left (right, two-sided) ideal $\mathfrak{J}''$ such that $\mathfrak{J}' \subset \mathfrak{J}'' \subset \mathfrak{L}$. Show that $\mathfrak{L}$ possesses maximal left ideals and find their form.

3. Prove that, if $\mathfrak{J}'$ is a left ideal in the matrix ring Δ_n, Δ a division ring, then there exists a matrix (μ) in $L(\Delta, n)$ such that $(\mu)\mathfrak{J}'(\mu)^{-1}$ is the set of matrices of the form

$$\begin{bmatrix} \alpha_{11} & \alpha_{12} & \cdots & \alpha_{1r} & 0 & \cdots & 0 \\ \alpha_{21} & \alpha_{22} & \cdots & \alpha_{2r} & 0 & \cdots & 0 \\ \cdot & \cdot & \cdots & \cdot & \cdot & \cdots & \cdot \\ \alpha_{n1} & \alpha_{n2} & \cdots & \alpha_{nr} & 0 & \cdots & 0 \end{bmatrix}$$

α_{ij} arbitrary.

4. Prove that every left ideal of $\mathfrak{L}$ is a principal ideal $\mathfrak{L}E$ generated by an idempotent element E.

4. Right ideals. If $\mathfrak{J}$ is a right ideal in $\mathfrak{L}$, the set $\mathfrak{J}'$ of transposes B', $B \in \mathfrak{J}$, is a left ideal in $\mathfrak{L}'$, the ring of linear transformations in the dual space $\mathfrak{R}'$. Moreover, $\mathfrak{J}$ is the set of transposes of the mappings in $\mathfrak{J}'$. We can therefore deduce the form of $\mathfrak{J}$ from that of $\mathfrak{J}'$. If we apply Theorem 3 to $\mathfrak{J}'$, we see that $\mathfrak{J}' = \mathfrak{S}' \times' \mathfrak{R}$, the totality of mappings $\Sigma x_i' \times' y_i$ where $x_i' \in \mathfrak{S}'$ and $y_i \in \mathfrak{R}$ and where $\Sigma x_i' \times' y_i$ denotes the mapping

$$x' \to \Sigma x_i' g(y_i, x'). \tag{9}$$

Hence $\mathfrak{J}$ is the totality of mappings

$$x \to \Sigma g(x, x_i')y_i \tag{10}$$

in $\mathfrak{L}$. Thus $\mathfrak{J} = \mathfrak{S}' \times \mathfrak{R}$. We therefore have the following

Theorem 5. *Every right ideal $\mathfrak{J}$ in $\mathfrak{L}$ has the form $\mathfrak{S}' \times \mathfrak{R}$ where $\mathfrak{S}'$ is a subspace of $\mathfrak{R}'$.*

We can obtain a correspondence also between subspaces of $\mathfrak{R}$ and right ideals of $\mathfrak{L}$. For this purpose we consider the subspace $\mathfrak{S} = j(\mathfrak{S}')$ of vectors y in $\mathfrak{R}$ such that $g(y, y') = 0$ for all y' in $\mathfrak{S}'$. If $y \in \mathfrak{S}$ and $B \in \mathfrak{J}$, then, by (10), $yB = 0$. Also if z is any vector such that $zB = 0$ for all B in $\mathfrak{J}$, then $\Sigma g(z, x_i')y_i = 0$ for all $x_i' \varepsilon \mathfrak{S}$ and all $y_i \varepsilon \mathfrak{R}$. If the y_i are chosen to be linearly independent, then this implies that $g(z, x_i') = 0$. Thus $z \varepsilon \mathfrak{S}$. Hence $\mathfrak{S}$ can be characterized as the totality of vectors annihilated by every B in $\mathfrak{J}$. Suppose next that C is any linear transformation such that $yC = 0$ for all $y \varepsilon \mathfrak{S}$. We write $C = \Sigma x_i' \times y_i$ where the y_i are linearly independent. Then $0 = yC = \Sigma g(y, x_i')y_i$, and so $g(y, x_i') = 0$ for all $y \varepsilon \mathfrak{S}$. Hence the $x_i' \varepsilon j(\mathfrak{S}) = j(j(\mathfrak{S}'))$ $= \mathfrak{S}'$ and $C \varepsilon \mathfrak{J}$. We therefore see also that $\mathfrak{J}$ is the totality, denoted as $\mathfrak{Z}(\mathfrak{S})$, of linear transformations which annihilate every vector in $\mathfrak{S}$. This proves

Theorem 6. *Every right ideal, $\mathfrak{J}$ in $\mathfrak{L}$ has the form $\mathfrak{Z}(\mathfrak{S})$, the set of linear transformations which annihilate a subspace $\mathfrak{S}$. The subspace $\mathfrak{S}$ is the totality of vectors annihilated by every $B \varepsilon \mathfrak{J}$.*

Since $\mathfrak{S}'$ is arbitrary in $\mathfrak{R}'$, $\mathfrak{S}$ is arbitrary in $\mathfrak{R}$. If $\mathfrak{S}$ is any subspace of $\mathfrak{R}$, it is clear at the start that the totality $\mathfrak{J} = \mathfrak{Z}(\mathfrak{S})$ of linear transformations which annihilate $\mathfrak{S}$ is a right ideal in $\mathfrak{L}$. The above argument shows that the only vectors annihilated by every mapping in $\mathfrak{Z}(\mathfrak{S})$ are those in $\mathfrak{S}$. Moreover, we have proved that every right ideal has the form $\mathfrak{Z}(\mathfrak{S})$. Thus the correspondence $\mathfrak{S} \rightarrow \mathfrak{Z}(\mathfrak{S})$ is 1–1 between the set of subspaces of $\mathfrak{R}$ and the set of right ideals in $\mathfrak{L}$. Clearly this correspondence reverses order: if $\mathfrak{S}_1 \supseteq \mathfrak{S}_2$, then $\mathfrak{Z}(\mathfrak{S}_1) \subseteq \mathfrak{Z}(\mathfrak{S}_2)$.

EXERCISES

1. State and prove the analogues of Exs. 1, 2, 3, 4 of p. 232 for right ideals.
2. Prove that, if $\mathfrak{J}'$ is a left ideal $\neq 0$, then the only vector which is annihilated by all the linear transformations in $\mathfrak{J}'$ is $z = 0$.
3. Use Theorem 6 and Ex. 2 to prove that $\mathfrak{L}$ is simple.

5. Isomorphisms of rings of linear transformations. Let $\mathfrak{J}$ be a right ideal $\neq 0$ in $\mathfrak{L}$. Then there exists a vector x such that $x\mathfrak{J}$, the set of images xB, B in $\mathfrak{J}$, is not 0. If B_1 and B_2 are in $\mathfrak{J}$, then

$$x(B_1 + B_2) = xB_1 + xB_2. \tag{11}$$

Hence the mapping $B \rightarrow xB$ is a homomorphism of the additive group of $\mathfrak{J}$ into $\mathfrak{R}$. It follows that the image $x\mathfrak{J}$ under this homomorphism is a subgroup of the group $\mathfrak{R}$. We shall now show that $x\mathfrak{J}$ is in fact the whole of $\mathfrak{R}$. We note first that $x\mathfrak{J}$ is mapped into itself by every $A \varepsilon \mathfrak{L}$; for if $y \varepsilon x\mathfrak{J}$, $y = xB$ for some $B \varepsilon \mathfrak{J}$, and $yA = xBA = xB'$ where $B' = BA$ is in $\mathfrak{J}$. Hence $yA \varepsilon x\mathfrak{J}$.

We note next that $x\mathfrak{J}$ is closed under scalar multiplication. If $\alpha \varepsilon \Delta$ there exists an A in $\mathfrak{L}$ such that $yA = \alpha y$. Then $\alpha y = yA$ is in $x\mathfrak{J}$. Thus we see that $x\mathfrak{J}$ is a non-zero subspace invariant under $\mathfrak{L}$. We have seen (Chapter IV, p. 116) that the only such subspace is $\mathfrak{R}$ itself. Hence we have the following

Lemma 1. *If $\mathfrak{J}$ is a right ideal $\neq 0$ and x is a vector such that $x\mathfrak{J} \neq 0$, then $x\mathfrak{J} = \mathfrak{R}$.*

We observe next that the mapping $B \rightarrow xB$ is more than a group homomorphism, namely, it is an operator or module homomorphism. This means simply that

$$BA \rightarrow (xB)A, \tag{12}$$

which is clear since $(xB)A = x(BA)$. Thus if we denote our mapping as χ, then (12) reads

$$(BA)\chi = (B\chi)A, \tag{13}$$

an equation which holds for all B in $\mathfrak{J}$ and all A in $\mathfrak{L}$.

Equation (13) shows that the kernel of the homomorphism χ is not merely a subgroup of the additive group of $\mathfrak{J}$ but a right ideal of the ring $\mathfrak{L}$; for if $B\chi = 0$, then $(BA)\chi = (B\chi)A = 0A = 0$. It follows that, if $\mathfrak{J}$ is a minimal right ideal, then the kernel of χ is 0. This, of course, means that χ is an isomorphism.

Lemma 2. *If $\mathfrak{J}$ is a minimal right ideal and $x\mathfrak{J} \neq 0$, then the homomorphism $\chi : B \rightarrow xB$ is an isomorphism of $\mathfrak{J}$ onto $x\mathfrak{J} = \mathfrak{R}$.*

If χ^{-1} denotes the inverse mapping of $\mathfrak{R}$ onto $\mathfrak{J}$ and A_r denotes the right multiplication $B \rightarrow BA$ in $\mathfrak{J}$, then, by (13), $A_r\chi = \chi A$ so that

$$A_r = \chi A \chi^{-1}. \tag{13'}$$

This relation between the vector space and any minimal right ideal of $\mathfrak{L}$ gives the underlying reason for some remarkable results on rings of linear transformations which we shall now derive. The first of these is the following

Theorem 7. *Let* $\mathfrak{R}_i$, $i = 1, 2$, *be a vector space over a division ring* Δ_i, *and let* $\mathfrak{L}_i$ *be the ring of linear transformations in* $\mathfrak{R}_i$ *over* Δ_i. *Then if* $\mathfrak{L}_1$ *is isomorphic to* $\mathfrak{L}_2$, Δ_1 *is isomorphic to* Δ_2 *and the two spaces have the same dimensionality.*

Proof. Let ϕ be an isomorphism of $\mathfrak{L}_1$ onto $\mathfrak{L}_2$ and let $\mathfrak{J}_1$ be a minimal right ideal in $\mathfrak{L}_1$. Then the image $\mathfrak{J}_2 = \mathfrak{J}_1\phi$ is a minimal right ideal in $\mathfrak{L}_2$. Let χ_i be an isomorphism of $\mathfrak{J}_i$ onto $\mathfrak{R}_i$ defined as above. Thus

$$(14) \qquad A_{1r} = \chi_1 A_1 \chi_1^{-1}, \quad A_{2r} = \chi_2 A_2 \chi_2^{-1}$$

if $A_i \varepsilon \mathfrak{L}_i$. Also $(B_1 A_1)\phi = (B_1\phi)(A_1\phi)$ if $B_1 \varepsilon \mathfrak{J}_1$. Hence $A_{1r}\phi = \phi(A_1\phi)_r$ where, as throughout this proof, $A_1\phi$ denotes the image of A_1 under ϕ. Thus

$$(15) \qquad (A_1\phi)_r = \phi^{-1} A_{1r} \phi.$$

Combining (14) and (15) we obtain

$$\begin{aligned} A_1\phi &= \chi_2^{-1}(A_1\phi)_r\chi_2 = \chi_2^{-1}\phi^{-1}A_{1r}\phi\chi_2 \\ &= (\chi_2^{-1}\phi^{-1}\chi_1)A_1(\chi_1^{-1}\phi\chi_2). \end{aligned}$$

Since χ_1^{-1} is an isomorphism of $\mathfrak{R}$, onto $\mathfrak{J}_1$, ϕ an isomorphism of $\mathfrak{J}_1$ onto $\mathfrak{J}_2$ and χ_2 an isomorphism of $\mathfrak{J}_2$ onto $\mathfrak{R}_2$, the mapping $U = \chi_1^{-1}\phi\chi_2$ is an isomorphism of $\mathfrak{R}_1$ onto $\mathfrak{R}_2$. Moreover, we have proved that

$$(16) \qquad A_1\phi = U^{-1}A_1U.$$

Now let α_{1l} denote a scalar multiplication on $\mathfrak{R}_1$. Since α_{1l} commutes with every $A_1 \varepsilon \mathfrak{L}_1$, $U^{-1}\alpha_{1l}U$ commutes with every $(A_1\phi)$; hence with every $A_2 \varepsilon \mathfrak{L}_2$. By Theorem 2 this implies that $U^{-1}\alpha_{1l}U$ is a scalar multiplication in $\mathfrak{R}_2$. Similarly if α_{2l} is any scalar multiplication in $\mathfrak{R}_2$, then $\alpha_{1l} = U\alpha_{2l}U^{-1}$ is a scalar multiplication in $\mathfrak{R}_1$. Clearly $U^{-1}\alpha_{1l}U = \alpha_{2l}$. Hence the mapping $\alpha_{1l} \rightarrow U^{-1}\alpha_{1l}U$ is an isomorphism of the ring Δ_{1l} of scalar multiplications in $\mathfrak{R}_1$ onto the ring Δ_{2l} of scalar multiplications in $\mathfrak{R}_2$. Now let α_1^u denote the element in Δ_2 such that

$$(\alpha_1^u)_l = U^{-1}\alpha_{1l}U.$$

Then the mapping u thus determined is an isomorphism of Δ_1 onto Δ_2. For any x_1 in $\mathfrak{R}_1$ and any α_1 in Δ_1, we have

$$(\alpha_1 x_1)U = (x_1\alpha_{1l})U = (x_1U)(U^{-1}\alpha_{1l}U)$$
$$= (x_1U)(\alpha_1{}^u)_l,$$

or

$$(\alpha_1x_1)U = \alpha_1{}^u(x_1U). \tag{17}$$

This equation shows that $\mathfrak{R}_1$ and $\mathfrak{R}_2$ have the same dimensionality. Let $(e_1, e_2, \cdots, e_n)$ be a basis for $\mathfrak{R}_1$ over Δ. Then any $x = \Sigma\xi_ie_i$, ξ_i in Δ_1. Hence by (17) $xU = \Sigma(\xi_ie_i)U = \Sigma\xi_i{}^u(e_iU)$ so that the vectors e_iU are generators of $\mathfrak{R}_2$. Moreover, if we have a relation between the e_iU, then we can write it in the form $\Sigma\xi_i{}^u(e_iU) = 0$. Then $(\Sigma\xi_ie_i)U = 0$ and the ξ_i and $\xi_i{}^u$ are all 0. This completes the proof.

In reality we have proved a good deal more than we set out to prove. In order to state the precise result we shall make use of an important generalization of the concept of a linear transformation which we define as follows: A mapping U of a vector space $\mathfrak{R}_1$ over Δ_1 into a vector space $\mathfrak{R}_2$ over Δ_2 is called a *semi-linear transformation* if 1) U is a homomorphism of the additive groups and 2) there exists an isomorphism u of Δ_1 onto Δ_2 such that

$$(\alpha x)U = \alpha^u(xU)$$

for all $x \,\varepsilon\, \mathfrak{R}_1$ and all $\alpha \,\varepsilon\, \Delta_1$. If u' is a second isomorphism such that the above equation holds, then $(\alpha^u - \alpha^{u'})(xU) = 0$ for all x. Hence, if $U \neq 0$, $\alpha^u = \alpha^{u'}$ and $u = u'$. Thus if $U \neq 0$, u is uniquely determined by U and in this case u will be called *the associated isomorphism* of U.

Let U_1 be a semi-linear transformation of $\mathfrak{R}_1$ into $\mathfrak{R}_2$ and U_2 a semi-linear transformation of $\mathfrak{R}_2$ into a third vector space $\mathfrak{R}_3$. Let u_1 and u_2 be the isomorphisms of U_1 and U_2 respectively. Then

$$(\alpha x)U_1U_2 = (\alpha^{u_1}(xU_1))U_2 = \alpha^{u_1u_2}(xU_1U_2).$$

Hence U_1U_2 is a semi-linear transformation of $\mathfrak{R}_1$ into $\mathfrak{R}_3$ with associated isomorphism u_1u_2. Suppose next that U_1 is 1–1 of $\mathfrak{R}_1$ onto $\mathfrak{R}_2$ and let $U_1{}^{-1}$ denote the inverse. Then from $(\alpha x)U_1 = \alpha^{u_1}(xU_1)$ we obtain $\alpha x = (\alpha^{u_1}(xU_1))U_1{}^{-1}$. Replacing xU_1 by

x, α^{u_1} by α, this becomes $(\alpha x)U_1^{-1} = \alpha^{u_1^{-1}}(xU_1^{-1})$. Hence U_1^{-1} is semi-linear with isomorphism u_1^{-1}.

These remarks show that, if $A_1 \in \mathfrak{L}_1$ and U is a 1–1 semi-linear transformation of $\mathfrak{R}_1$ onto $\mathfrak{R}_2$, then $U^{-1}A_1U \in \mathfrak{L}_2$. It is clear that $A_1 \rightarrow U^{-1}A_1U$ is an isomorphism of $\mathfrak{L}_1$ onto $\mathfrak{L}_2$. This is, of course, rather trivial, but it is striking that the converse holds and this is what equations (16) and (17) show.

Theorem 8. *Any isomorphism $A_1 \rightarrow A_1\phi$ of $\mathfrak{L}_1$ onto $\mathfrak{L}_2$ is given by a* 1–1 *semi-linear transformation U of $\mathfrak{R}_1$ onto $\mathfrak{R}_2$ in the sense that $A_1\phi = U^{-1}A_1U$.*

This, of course, gives the following

Corollary. *Any automorphism of $\mathfrak{L}$ is given by a* 1–1 *semi-linear transformation of $\mathfrak{R}$ onto itself.*

EXERCISES

1. Let $(e_1, e_2, \cdots, e_n)$ be a basis for $\mathfrak{R}$ and let U be a semi-linear transformation of $\mathfrak{R}$ into itself. If $e_iU = \Sigma\alpha_{ij}e_j$, then (α) is the *matrix* of U (relative to the given basis). Show that U is completely determined by (α) and the associated automorphism. Find the transformation formula giving the matrix of U relative to a second basis. Show that to any pair $((\alpha), u)$, (α) a matrix and u an automorphism, there corresponds a semi-linear transformation whose matrix relative to $(e_1, e_2, \cdots, e_n)$ is (α) and whose automorphism is u.

2. If U is semi-linear with matrix (α) and automorphism u, we write $U \rightarrow ((\alpha), u)$. Show that, if $U \rightarrow ((\alpha), u)$ and $V \rightarrow ((\beta), v)$, then $UV \rightarrow ((\alpha)^v(\beta), uv)$, $(\alpha)^v \equiv (\alpha_{ij}{}^v)$.

3. Prove that the image $\mathfrak{S}U$ of a subspace under a semi-linear transformation U is a subspace. Prove that the set $\mathfrak{Z}$ of vectors mapped into 0 by U is a subspace. Prove the dimensionality relation $\dim \mathfrak{R}U + \dim \mathfrak{Z} = n$.

4. Show that the scalar multiplications are semi-linear transformations.

5. Let $\mathfrak{L}$ be the ring of linear transformations in a vector space over a field Φ. Prove that every automorphism of $\mathfrak{L}$ which leaves the elements of the center fixed is inner, that is, has the form $A \rightarrow U^{-1}AU$ where $U \in \mathfrak{L}$.

Chapter IX

INFINITE DIMENSIONAL VECTOR SPACES

Up to this point we have directed our attention to the consideration of finite dimensional vector spaces exclusively. While it is true that the basic concepts of the finite case are applicable foı arbitrary spaces, it is not obvious that all of these are significant for spaces without finite bases.

In this chapter we give an introduction to the study of arbitrary vector spaces. The study of such spaces constitutes a comparatively new field of research whose development has been influenced to a considerable degree by demands of analysis. The most important applications make use of topological notions as well as of algebraic notions. The point of departure for these applications is the concept of a topological vector space. On the other hand, a number of interesting results have been found which deal with arbitrary (discrete) vector spaces and it is these which we shall discuss here. Notions from topology do not appear at the beginning of our discussion, but we shall encounter these in the consideration of certain natural topologies for sets of linear transformations. These will serve to give simple descriptions of purely algebraic results.

Many of the results which we shall give are direct generalizations of results in the finite case. However, there are some essential differences. The most important of these is the fact that the conjugate space of the conjugate space of a vector space cannot be identified with the original space. In fact, if $\mathfrak{R}$ is infinite dimensional, then $\dim \mathfrak{R}^* > \dim \mathfrak{R}$, and this precludes the equation $\mathfrak{R}^{**} = \mathfrak{R}$.

In the last part of this chapter we shall take up again the study of sets of linear transformations. As we shall see, a fundamental lemma of Schur permits the reduction of the theory of irreducible sets of endomorphisms to that of irreducible sets of linear transformations. In addition to this lemma we shall prove a density theorem for irreducible sets of linear transformations which generalizes a classical theorem of Burnside.

Naturally, the present considerations will require more powerful logical tools than those which served in the finite case. Among these we mention particularly Zorn's lemma and the theory of cardinal numbers. We shall assume a familiarity with these notions as well as with the elements of topology.

1. Existence of a basis. Let $\mathfrak{R}$ be an arbitrary vector space over a division ring Δ. We recall that a subset S of $\mathfrak{R}$ is a set of generators of $\mathfrak{R}$ if every vector in $\mathfrak{R}$ is a (finite) linear combination of the vectors belonging to S. The set S is *linearly independent* if every finite subset F of S is linearly independent. A subset B which is linearly independent and a set of generators is a *basis* in the sense that every vector can be written in one and only one way as a linear combination of elements in B.

We shall now prove the following two basic results: 1) any set of generators contains a basis for $\mathfrak{R}$ and 2) any linearly independent set of elements can be supplemented by elements from any basis to give a basis. For both of these results we shall use Zorn's lemma in the following form:

*Let P be a partially ordered set which has the property that every linearly ordered subset has an upper bound. Then P possesses a maximal element.**

The concept of a partially ordered set has been defined in Chapter I. We recall that this is a set in which a binary relation $a \leq b$ is defined such that: (i) $a \leq a$, (ii) if $a \leq b$ and $b \leq a$, then $a = b$; and (iii) if $a \leq b$ and $b \leq c$, then $a \leq c$. A *linearly ordered set* (or a *chain*) is a partially ordered set with the property that any two elements are comparable in the sense that either

* This maximum principle seems to have been discovered first by Hausdorff. Its importance in algebra was first recognized by Zorn. An adequate discussion of the principle can be found in Birkhoff's *Lattice Theory*, 2nd edition, p. 42.

$a \leq b$ or $b \leq a$. An *upper bound* of a subset S of a partially ordered set is an element $u \in P$ such that $s \leq u$ for every s in S. An element m of a subset S is *maximal* if no $s \neq m$ in S has the property $m \leq s$.

We give now the proof of statement 1) above. Let S be a set of generators for $\mathfrak{R}$. Let P be the collection of linearly independent subsets of S. Then P is a partially ordered set relative to the relation of inclusion. Let L be a linearly ordered subcollection of P, and let U be the logical sum of the sets belonging to L. We assert that U is linearly independent. Otherwise U contains a dependent set $u_1, u_2, \cdots, u_m$. Now $u_i \in A_i \in L$. Also for any i, j either $A_i \subseteq A_j$ or $A_j \subseteq A_i$. Hence one of the A's, say A_m, contains all the others. Thus every $u_i \in A_m$ and A_m contains a finite linearly dependent subset. This contradicts the assumption that $A_m \in P$; hence $U \in P$. It is clear that this element serves as an upper bound for all the $A \in L$. We can now apply Zorn's lemma to conclude that P contains a maximal element, and this means that the set of generators S contains a maximal linearly independent subset B. It is now easy to see that B is a basis for $\mathfrak{R}$. If y is any member of S not contained in B, then the set $B \cup y$ is a dependent set. This implies that y is a linear combination of elements of B (Lemma 1 on p. 11). Hence every $s \in S$ is a linear combination of elements of B. It follows that B is a set of generators and, since B is a linearly independent set, B is, in fact, a basis for $\mathfrak{R}$.

The proof of 2) is similar to the foregoing. Here let S be a linearly independent set and let B be a basis. Let P be the collection of linearly independent sets containing S and contained in $S \cup B$. Then a repetition of the above argument shows that P contains a maximal element C. It follows easily that C is a basis for $\mathfrak{R}$.

2. Invariance of dimensionality. We wish to prove next that the cardinal numbers of any two bases for $\mathfrak{R}$ are equal.

Let B and C be two bases for $\mathfrak{R}$. We label the elements of B by subscripts i belonging to a set I and the elements of C by subscripts k in a set K. If, say, B (or I) is finite, then $\mathfrak{R}$ has a finite basis. Then we know that C is also finite, and its cardinal number is the same as that of B (Theorems 2 and 3, Chapter I).

It therefore suffices to consider the case in which both B and C are infinite. Here the following argument due to Löwig can be applied. Express each e_i of B in terms of the f_k of C as, say,

$$e_i = \beta_1 f_{k_1} + \beta_2 f_{k_2} + \cdots + \beta_m f_{k_m}$$

where the $\beta_j \neq 0$. Now every f_k occurs in some such expression; for if a particular f_k does not occur, then, since this f_k is a linear combination of the e's and each e is a linear combination of f's $\neq f_k$, f_k is a linear combination of f's $\neq f_k$. This contradicts the linear independence of the f's.

We can now define a single-valued mapping ϕ of the set C into the set B. Let $f_k \varepsilon C$ and let $e_i \equiv \phi(f_k)$ be one of the e's in B whose expression involves f_k. We thus obtain a single-valued mapping of the whole of C into B. Let $B' = f(C)$ be the image set. If $e_{i'} \varepsilon B'$, the inverse image $\phi^{-1}(e_{i'})$ consists of f_k which occur in the expression for $e_{i'}$. Thus $\phi^{-1}(e_{i'})$ is a finite set. Now we have a 1–1 correspondence between the set B' and the collection Γ of inverse images $\phi^{-1}(e_{i'})$. The collection Γ gives a decomposition of the set C into non-overlapping finite sets. Moreover, since C is infinite, Γ is infinite. It follows easily from standard theorems in the theory of cardinal numbers that C and Γ have the same cardinal number.* Hence the cardinal number of B' is the same as that of C and so the cardinal number of B is greater than or equal to that of C. If we reverse the roles of B and C, we see that the cardinal number of B does not exceed that of C. Hence by Bernstein's theorem these two sets have the same cardinal number.

As in the finite case we shall call the cardinal number of any basis *the dimensionality* of $\mathfrak{R}$ over Δ. Also as in the finite case we can construct a vector space over any given Δ with dimensionality any given cardinal number. For this purpose let I be a set having the given cardinal number. Let $\mathfrak{R}$ be the set of functions defined on I with values $x(i)$ in Δ such that $x(i) = 0$ for all but a finite number of $i \varepsilon I$. We define the sum of two such functions by addition of coordinates and scalar product by left multiplication of the coordinates by the given element of Δ.

* Consult for example, Sierpinski, *Leçons sur les Nombres Transfinis*, Paris, 1928.

These compositions yield results in $\mathfrak{R}$, and it is clear that the postulates for a vector space are fulfilled. We now determine a special basis for $\mathfrak{R}$. For each $i \varepsilon I$, we define e_i to be the vector such that

$$e_i(j) = \delta_{ij}. \tag{1}$$

It is easy to see that these vectors are linearly independent. If x is any vector, let $i_1, i_2, \cdots, i_m$ be the i for which $x(i) \neq 0$. Then if $x(i_j) = \xi_j$, it is clear that $x = \xi_1 e_{i_1} + \xi_2 e_{i_2} + \cdots + \xi_m e_{i_m}$. Hence the e_i form a basis. Since $i \rightarrow e_i$ is a 1–1 correspondence, this basis has the same cardinal number as I.

We remark also that, as in the finite case, any two vector spaces which have the same dimensionality are equivalent. In particular, any vector space is equivalent to a space of functions of the type just constructed.

3. Subspaces. Almost all the properties, noted in Chapter I, of the lattice of subspaces of a finite dimensional vector space hold also in the infinite case. Exception must be made for the chain conditions. In fact, it is clear that neither of these conditions holds for vector spaces with infinite bases.

Our former considerations made use of bases only in the proof of the existence of a complement of a subspace. Property 2) can now be used to carry over the argument used to prove this result. Let $\mathfrak{S}$ be a subspace of $\mathfrak{R}$. Then we know that $\mathfrak{S}$ has a basis S. Since S is a linearly independent set, it can be imbedded in a basis B of $\mathfrak{R}$. We write $B = S \cup T$ where $S \cap T$ is vacuous, and we let $\mathfrak{S}'$ be the space spanned by the vectors in T. It is immediate that $\mathfrak{R} = \mathfrak{S} + \mathfrak{S}'$ and that $\mathfrak{S} \cap \mathfrak{S}' = 0$. Hence $\mathfrak{S}'$ is a complement of $\mathfrak{S}$.

The argument which we have used here can also be used to prove the existence of a special type of basis for any subspace of $\mathfrak{R}$. The result is the following

Theorem 1. *Let $\mathfrak{R}$ have a basis $B = (e_i)$. Then if $\mathfrak{S}$ is any subspace of $\mathfrak{R}$, we can divide B into two non-overlapping subsets $C = (e_j)$, $D = (e_k)$ such that $\mathfrak{S}$ has a basis of the form $f_j = e_j + u_j$ where the u_j are in the space spanned by D and $f_j \rightarrow e_j$ is a 1–1 mapping onto C.*

Proof. There exists a complement $\mathfrak{S}'$ of $\mathfrak{S}$ spanned by a subset $D = (e_k)$ of B. Let $C = (e_j)$ be the complement of D in B. Then each $e_j = f_j - u_j$ where $f_j \in \mathfrak{S}$ and $u_j \in \mathfrak{S}'$ and, as we proceed to show, (f_j) is a basis for $\mathfrak{S}$. The set (f_j) is linearly independent, for let $\Sigma\beta_j f_j = 0$. Then $\Sigma\beta_j e_j + \Sigma\beta_j u_j = 0$; hence every $\beta_j = 0$ by the linear independence of the e_j. The set (f_j) generates $\mathfrak{S}$; for if $y \in \mathfrak{S}$, then $y = \Sigma\beta_j e_j + \Sigma\gamma_k e_k = \Sigma\beta_j f_j - \Sigma\beta_j u_j + \Sigma\gamma_k e_k$. Then $y - \Sigma\beta_j f_j \in \mathfrak{S}'$, and hence $y - \Sigma\beta_j f_j = 0$. We note finally that, if e_j and $e_{j'}$ are distinct elements in C, then $f_j \neq f_{j'}$. Otherwise $e_j - e_{j'} \in \mathfrak{S}'$, contrary to the fact that B is a basis. We therefore see that the mapping $e_j \rightarrow f_j$ is 1–1 and this concludes the proof.

Theorem 1 is due to Emmy Noether. It is therefore appropriate to call a basis of the type described a *Noether basis* for the subspace $\mathfrak{S}$ relative to the basis B of $\mathfrak{R}$.

4. Linear transformations and matrices. The connection between linear transformations of finite dimensional vector spaces and finite matrices can be carried over directly to the infinite case.

Let $\mathfrak{R}$ and $\mathfrak{S}$ be vector spaces over Δ and let $B = (e_i)$ be a basis for $\mathfrak{R}$. Then we note first that any mapping $e_i \rightarrow y_i$ of B into $\mathfrak{S}$ can be extended in one and only one way to a linear transformation of $\mathfrak{R}$ into $\mathfrak{S}$. As is easily verified, this extension maps $\sum_1^m \xi_j e_{i_j}$ into $\sum_1^m \xi_j y_{i_j}$.* Next let $C = (f_k)$ be a basis for $\mathfrak{S}$ and write the image of e_i relative to the linear transformation A as

$$e_i A = y_i = \Sigma\alpha_{ik} f_k \tag{2}$$

where the sum is finite. Thus the matrix (α_{ik}) is *row finite* in the sense that for a fixed i $\alpha_{ik} \neq 0$ for only a finite number of k. In general, if I and J are any two sets, then a function on the product set $I \times J$ into Δ is called an *$I \times J$ matrix over* Δ. Thus we have established a correspondence between linear transformations of $\mathfrak{R}$ into $\mathfrak{S}$ and row finite $I \times J$ matrices over Δ. Of course, this correspondence depends on the choice of basis. In the special case $\mathfrak{R} = \mathfrak{S}$ it is natural to take C to be the same as the basis B.

* This establishes the existence of non-trivial linear transformations of $\mathfrak{R}$ into $\mathfrak{S}$. It seems to be impossible to do this without using bases.

We then obtain an $I \times I$ "square" matrix as the *matrix of A relative to the basis B.*

We can verify easily as in the finite case that the correspondence $A \to (\alpha)$ determined by (2) is 1–1 of the set $\mathfrak{L}(\mathfrak{R}, \mathfrak{S})$ of linear transformations onto the set of row finite $I \times J$ matrices. The matrix corresponding to the sum of two transformations is obtained from the two matrices by adding components $\alpha_{ii'}$, $\beta_{ii'}$. It follows that the row finite matrices form a group relative to this operation.

If $\mathfrak{L} = \mathfrak{L}(\mathfrak{R}, \mathfrak{R})$ and the matrix of A is determined by a single basis, then it can be seen that the element $\gamma_{i\lambda}$ of the matrix AB is

$$\gamma_{i\lambda} = \sum_k \alpha_{ik}\beta_{k\lambda} \tag{3}$$

where $A \to (\alpha)$ and $B \to (\beta)$ in our correspondence. The sum in (3) is finite, and the *product* matrix $(\gamma) = (\alpha)(\beta)$ is row finite. It follows now that the set Δ_I of $I \times I$ row finite matrices together with the indicated addition and multiplication is a ring isomorphic to the ring $\mathfrak{L}$.

Changes of bases can also be discussed as in the finite case. If we have a second basis for $\mathfrak{R}$, we can suppose that it has been put into 1–1 correspondence with the same set of indices I. If we denote the vectors in this basis by f_i, then we can write

$$f_k = \Sigma\mu_{ki}e_i, \quad e_i = \Sigma\nu_{ik}f_k \tag{4}$$

where (μ) and (ν) are row finite $I \times I$ matrices. It follows easily that $(\mu)(\nu) = 1 = (\nu)(\mu)$ where 1 is the matrix with elements δ_{ik}. Thus (μ) is a unit in Δ_I in the sense that is has a two-sided inverse $(\nu) = (\mu)^{-1}$. Conversely if (μ) is any unit, then the f_k defined by the first set of equations in (4) is a basis for $\mathfrak{R}$.

If A is a linear transformation with matrix (α) relative to the basis (e_i), then one verifies directly that the matrix of A relative to (f_i) is $(\mu)(\alpha)(\mu)^{-1}$.

5. Dimensionality of the conjugate space. If $\mathfrak{R}$ is a finite dimensional vector space, then we know that the right vector space $\mathfrak{R}^*$ of linear functions on $\mathfrak{R}$ has the same dimensionality as $\mathfrak{R}$. A fundamental difference between the finite and the infinite theories is that this does not hold for $\mathfrak{R}$ infinite dimensional.

In this case we shall show that dim $\mathfrak{R}^* >$ dim $\mathfrak{R}$. More precisely we shall show that if the cardinal number of Δ is d and the dimensionality of $\mathfrak{R}$ is b, then the dimensionality of $\mathfrak{R}^*$ is d^b, the cardinal number of the set of mappings of a set of cardinal number b into one of cardinal number d. We prove first the following

Lemma 1. *If dim $\mathfrak{R} = b$ is infinite and the cardinal number of Δ is d, then the cardinal number of $\mathfrak{R}$ is bd.*

Proof. If (e_i) is a basis for $\mathfrak{R}$, every non-zero vector x in $\mathfrak{R}$ has a unique representation as

$$x = \sum_{j=1}^{N} \xi_{i_j} e_{i_j}, \quad \xi_{i_j} \neq 0.$$

Thus with each $x \neq 0$ we can associate a uniquely determined subset $e_{i_1}, e_{i_2}, \cdots, e_{i_N}$ and a unique N-tuple $(\xi_{i_1}, \xi_{i_2}, \cdots, \xi_{i_N})$ with non-zero components in Δ. Since (e_i) is infinite, the cardinal number of the set of subsets containing N elements of (e_i) equals b.* On the other hand, the cardinal number of the set of N-tuples $(\xi_{i_1}, \xi_{i_2}, \cdots, \xi_{i_N})$ is either $d(= d^N)$ or it is finite. In either case the cardinal number of the set of pairs consisting of the set $e_{i_1}, e_{i_2}, \cdots, e_{i_N}$ and the N-tuple $(\xi_{i_1}, \xi_{i_2}, \cdots, \xi_{i_N})$ is db. It follows that the cardinal number of $\mathfrak{R}$ is $db + db + \cdots$; hence it is simply db.

Any linear function on $\mathfrak{R}$ is determined by its values on the basis (e_i) of $\mathfrak{R}$ and there exists a linear function such that $f(e_i)$ is any element of Δ. Thus we have a 1–1 correspondence between $\mathfrak{R}^*$ and the set of mappings of (e_i) into Δ. It follows that the cardinal number of $\mathfrak{R}^*$ is d^b. On the other hand, if $b^* =$ dim $\mathfrak{R}^*$, then Lemma 1 shows that the cardinal number of $\mathfrak{R}^*$ is db^*. Hence $db^* = d^b$. Since db^* is the maximum of d and b^*, the relation $b^* = d^b$ will follow if we can show that $b^* \geq d$. Since it is clear that b^* is infinite, the relation $b^* \geq d$ holds if $d \leq$ aleph null. Hence we shall assume from now on that $d >$ aleph null. In order to establish the required inequality in this case, we consider collections of denumerable sequences $(\gamma_1, \gamma_2, \cdots)$ with γ_i in Δ. A collection F of such sequences will be called

* This follows from the well-known result that $b^N = b$. See Sierpinski, *Leçons sur les Nombres Transfinis*, p. 217.

strongly independent if every finite square matrix chosen from the collection is non-singular. We choose a finite square matrix from F by selecting first a finite set of sequences

$$(\gamma_1^{(1)}, \gamma_2^{(1)}, \cdots), \quad (\gamma_1^{(2)}, \gamma_2^{(2)}, \cdots), \quad \cdots, \quad (\gamma_1^{(n)}, \gamma_2^{(n)}, \cdots)$$

and then a set $i_1, i_2, \cdots, i_n$ of indices, obtaining the matrix

$$\begin{bmatrix} \gamma_{i_1}^{(1)} & \gamma_{i_2}^{(1)} & \cdots & \gamma_{i_n}^{(1)} \\ \cdot & \cdot & \cdots & \cdot \\ \cdot & \cdot & \cdots & \cdot \\ \gamma_{i_1}^{(n)} & \gamma_{i_2}^{(n)} & \cdots & \gamma_{i_n}^{(n)} \end{bmatrix}.$$

We now prove the following

Lemma 2. *There exists a strongly independent collection of sequences with cardinal number* $\geq d$.*

Proof. We partially order the strongly independent collections F by inclusion. If a set $\{F\}$ of these collections is linearly ordered, clearly $\cup F$ is strongly independent. By Zorn's lemma, there exists a maximal strongly independent collection M. We shall prove, by induction, that if the cardinal number of M is less than d, then we can construct a sequence $X \notin M$ such that $M \cup \{X\}$ is strongly independent, thereby contradicting the maximality of M. Suppose that we have found the first p elements $\xi_1, \xi_2, \cdots, \xi_p$ of X so that every $q \times q$ matrix, $q \leq p$, chosen from $M \cup (\xi_1, \xi_2, \cdots, \xi_p)$ is non-singular. We shall determine ξ_{p+1} so that every $r \times r$ matrix, $r \leq p + 1$, chosen from $M \cup (\xi_1, \cdots, \xi_{p+1})$ is non-singular. The conditions that this imposes on ξ_{p+1} are that every matrix of the form

$$(5) \qquad \begin{bmatrix} & & & & * \\ & & & & \cdot \\ & & A & & \cdot \\ & & & & \cdot \\ & & & & * \\ \xi_{i_1} & \xi_{i_2} & \cdots & \xi_{i_{r-1}} & \xi_{p+1} \end{bmatrix},$$

* This lemma was communicated to the author by I. Kaplansky. It was proved by Kaplansky in collaboration with P. Erdös.

where the A is an $r-1 \times r-1$ matrix determined by M, be non-singular. Now for each matrix of the form (5), in which ξ_{p+1} is regarded as an indeterminate, there is just one choice of ξ_{p+1} in Δ making (5) singular. Thus, A is a non-singular matrix; hence its row vectors are left linearly independent. It follows that the row vector $(\xi_{i_1}, \xi_{i_2}, \cdots, \xi_{i_{r-1}})$ can be written in one and only one way as a linear combination of the rows of A. Then if μ represents the same linear combination of the elements of the last column of (5), (5) is non-singular, provided that $\xi_{p+1} \neq \mu$. We note next that the cardinal number of the collection of matrices (5) in which ξ_{p+1} is an indeterminate is less than d. By our assumption the cardinal number of M is less than d. The cardinal number of the set of matrices under consideration is the product of the cardinal number of sets of $r-1$ sequences chosen from M times the cardinal number of $r-1$ elements chosen in the integers $1, 2, \cdots, p$. The result is either finite or the cardinal number of M. In either case it is less than d. Since the cardinal number of Δ is d, we can choose a ξ_{p+1} so that all the matrices (5) are non-singular. This completes the proof by induction of the existence of X such that $M \cup \{X\}$ is strongly independent. We have therefore contradicted the maximality of M and established the lemma.

We can now prove the main result.

Theorem 2. *If $\mathfrak{R}$ is a vector space of infinite dimensionality b and the cardinal number of the division ring Δ is d, then the dimensionality b^* of $\mathfrak{R}^*$ is d^b.*

Proof. We have seen that it suffices to prove that $b^* \geq d$, and we may assume that d exceeds aleph null. Let (e_i) be a basis for $\mathfrak{R}$ and select a denumerable set (e_j) in (e_i). Let M be a strongly independent collection of denumerable sequences. For each $(\gamma_1, \gamma_2, \cdots) \in M$, we can define a linear function f such that $f(e_j) = \gamma_j$. The collection of linear functions thus obtained from the elements of M is a linearly independent one. Its cardinal number is the same as that of M. Hence by Lemma 2 we may suppose that it is $\geq d$. This proves that there exist at least d linearly independent elements in $\mathfrak{R}^*$; hence $b^* \geq d$ as required and $b^* = d^b$.

The relation $b^* = d^b$ implies, of course, that $b^* > b$. Moreover, if $b^{**} \equiv \dim \mathfrak{R}^{**}$, then $b^{**} > b^* > b$. This result shows that $\mathfrak{R}$ cannot be identified with the space of linear functions on $\mathfrak{R}^*$.

EXERCISE

1. (Mackey) Let $\mathfrak{R}$ be a vector space over a field Φ. Prove that the collection of sequences $\xi_\gamma = (\gamma, \gamma^2, \gamma^3, \cdots)$, $\gamma \neq 0$ in Φ, is linearly independent. (Note: This leads to an elementary proof of Theorem 2 for the commutative case.)

6. Finite topology for linear transformations. We shall now introduce a certain topology in the set $\mathfrak{L}(\mathfrak{R}, \mathfrak{S})$ of linear transformations of the vector space $\mathfrak{R}$ into the vector space $\mathfrak{S}$. Our topology will be a trivial (i.e., discrete) one if and only if $\mathfrak{R}$ is finite dimensional. Hence this gives another important point of difference between the finite and the infinite theories.

We recall that a *topological space* consists of a set E and a collection of subsets of E, called *open* sets, such that

1. The logical sum of any collection of open sets is open.
2. The intersection of any finite number of open sets is open.
3. The set E and the vacuous set are open.*

A subcollection $\mathfrak{B}$ of the set of open sets is called a *basis* if every open set is a logical sum of members of $\mathfrak{B}$. If $\mathfrak{B}$ is a basis, then the intersection of any two elements of $\mathfrak{B}$ is a logical sum of elements of $\mathfrak{B}$. Conversely, if E is any set and $\mathfrak{B}$ is a collection of its subsets such that their logical sum is E and the intersection of any two elements of $\mathfrak{B}$ is a logical sum of elements of $\mathfrak{B}$, then we can specify as open sets the logical sums of elements of $\mathfrak{B}$. The resulting collection satisfies **1.–3.**; hence it and the set E define a topological space. A set E is *topologized* if a collection of its subsets satisfying the above conditions is given. The collection of open sets is called *a topology* for E.

Now consider the set $\mathfrak{L}(\mathfrak{R}, \mathfrak{S})$. If $x_1, x_2, \cdots, x_m$ and $y_1, y_2, \cdots, y_m$ are finite subsets of $\mathfrak{R}$ and $\mathfrak{S}$ respectively, then we define $O(x_i; y_i)$ to be the set of linear mappings A of $\mathfrak{R}$ into $\mathfrak{S}$ such that $x_iA = y_i$, $i = 1, 2, \cdots, m$. It is clear that the intersection of any two $O(x_i; y_i)$ is another one. Hence this collection serves as

* Throughout our discussion we shall follow the terminology of Lefschetz' *Algebraic Topology*, Chapter 1.

a basis for a topology in $\mathfrak{L}(\mathfrak{R}, \mathfrak{S})$. We shall call this topology the *finite topology* of $\mathfrak{L}(\mathfrak{R}, \mathfrak{S})$. We note now that any open set $O(x_i; y_i)$ is either vacuous or it coincides with an open set $O(x_j; y_j)$ where the x_j are linearly independent. Thus, suppose $x_1, x_2, \cdots, x_r$ is a maximal linearly independent subset of $x_1, x_2, \cdots, x_m$ and let $x_k = \sum_{j=1}^{r} \beta_{kj}x_j$ hold for $k = r + 1, \cdots, m$. Then, unless $y_k = \Sigma\beta_{kj}y_j$, $O(x_i; y_i)$ is vacuous. On the other hand, if the conditions do hold, then $O(x_i; y_i) = O(x_j; y_j), j = 1, \cdots, r$. This remark shows that the sets $O(x_j; y_j)$, x_j linearly independent, constitute a basis for our topology.

We can now see that the topological space $\mathfrak{L}(\mathfrak{R}, \mathfrak{S})$ (endowed with the finite topology) is discrete in the sense that every subset is open, if and only if $\mathfrak{R}$ is finite dimensional. First, let $\mathfrak{R}$ have the finite basis $e_1, e_2, \cdots, e_n$ and let $A \, \varepsilon \, \mathfrak{L}(\mathfrak{R}, \mathfrak{S})$. Then $O(e_i; e_iA) = (A)$ so that A is an open set. It follows from **1.** that every subset of $\mathfrak{L}(\mathfrak{R}, \mathfrak{S})$ is open. Next let dim $\mathfrak{R}$ be infinite. Then if $x_1, x_2, \cdots, x_r$ is a linearly independent set, $O(x_j; y_j)$ is a non-denumerable set; for we can supplement the x_j to a basis and there exists a linear transformation mapping the elements of a basis into arbitrary elements of $\mathfrak{S}$. It follows now that every open set of $\mathfrak{L}(\mathfrak{R}, \mathfrak{S})$ is non-denumerable; hence the topology is not discrete.

We shall show next that $\mathfrak{L}(\mathfrak{R}, \mathfrak{S})$ is a topological group, that is, that $A - B$ is a continuous function of the two variables A, B in $\mathfrak{L}(\mathfrak{R}, \mathfrak{S})$. Finally, if $\mathfrak{R} = \mathfrak{S}$ so that $\mathfrak{L} = \mathfrak{L}(\mathfrak{R}, \mathfrak{R})$ is a ring, then $\mathfrak{L}$ is a topological ring in the sense that in addition to the continuity of $A - B$ we have the continuity of the multiplication composition. Let A and B be fixed elements of $\mathfrak{L}(\mathfrak{R}, \mathfrak{S})$ and let $O(x_i; y_i)$ be a member of the basis containing $A - B$. Such a set will be called a *neighborhood* of the point. Since $A - B \, \varepsilon \, O(x_i; y_i)$, we can write $y_i = x_i(A - B)$ so that $O(x_i; y_i) = O(x_i; x_i(A - B))$. Now $O(x_i; x_iA)$ and $O(x_i; x_iB)$ are neighborhoods of A and B respectively and it is clear that if $X \, \varepsilon \, O(x_i; x_iA)$ and $Y \, \varepsilon \, O(x_i; x_iB)$, then $X - Y \, \varepsilon \, O(x_i; x_i(A - B))$. This proves that the difference composition is continuous.

Next let $\mathfrak{R} = \mathfrak{S}$ and let $O(x_i; x_iAB)$ be any neighborhood of AB. Then $O(x_i; x_iA)$ and $O(x_iA; x_iAB)$ are neighborhoods of A

and B respectively. Moreover, if $X \varepsilon O(x_i; x_iA)$ and $Y \varepsilon O(x_iA; x_iAB)$, then $XY \varepsilon O(x_i; x_iAB)$. Hence multiplication is a continuous composition.

We recall that a subset of a topological space E is called *closed* if its complement is open. The closed sets satisfy conditions dual to **1.–3.** In particular, the intersection of any number of closed sets is closed. Hence any subset S of E has a *closure* Cl S defined to be the intersection of all the closed sets containing S. (Since E itself is closed, there exist closed sets containing S.) The closure of S can also be defined as the totality of points p (elements of E) with the property that every neighborhood of p has non-vacuous intersection with S. We shall now illustrate these concepts in the finite topology for linear transformations by considering the following

Example. Let $\mathfrak{R}$ be a vector space with a denumerable basis (e_i) over an infinite field Φ. Let (α_i) be a denumerable set of distinct elements in Φ and let A be the linear transformation such that $e_iA = \alpha_ie_i$. We wish to determine the closure Cl $\Phi[A]$ of the set $\Phi[A]$ of polynomials in A. We observe first that the sets $O^{(r)}(B) = O(e_j; e_jB), j = 1, 2, \cdots, r, r = 1, 2, \cdots$, form a complete set of neighborhoods of the point B, that is, any open set containing B contains one of these sets. Let $O(x_k; x_kB)$, x_k linearly independent, be a neighborhood of B. Then there exists an r such that the given $x_k \varepsilon [e_1, e_2, \cdots, e_r]$. It follows that $O^{(r)}(B) \subseteq O(x_k; x_kB)$ and this proves our assertion. It is now clear that an element B is in the closure of a set M if and only if every set $O(e_j; e_jB)$ meets M. Now let $B \varepsilon$ Cl $\Phi[A]$. If $\phi(\lambda)$ is a polynomial, then $e_i\phi(A) = \phi(\alpha_i)e_i$. Thus if $\phi(A) \varepsilon O^{(r)}(B)$, then $e_jB = e_j\phi(A) = \phi(\alpha_j)e_j$ for $j = 1, \cdots, r$. This shows that B is a diagonal linear transformation in the sense that $e_iB = \beta_ie_i$, $i = 1, 2, \cdots$. We wish to show that Cl $\Phi[A]$ is identical with the set of diagonal linear transformations. Let B be any diagonal linear transformation and let r be any integer. Consider the transformations induced by B and by the polynomials $\phi(A)$ in $[e_1, e_2, \cdots, e_r]$. Since the α_i are distinct, it is easy to see that the $\phi(A)$ induce every diagonal transformation in our subspace. Hence there exists a $\phi(A)$ such that $e_jB = e_j\phi(A)$ holds for $j = 1, 2, \cdots, r$. This shows that $B \varepsilon$ Cl $\Phi[A]$. Thus Cl $\Phi[A]$ is the set of diagonal transformations.

EXERCISES

1. Let $\mathfrak{R}$ be a vector space with a denumerable basis (e_i) over a field. Determine Cl $\Phi[A]$ where (a) $e_iA = e_{i+1}$, $i = 1, 2, \cdots$, and (b) $e_1A = 0$, $e_{i+1}A = e_i$, $i = 1, 2, \cdots$.

2. Show that the basic open sets $O(x_i; y_i)$ are also closed. (This implies that $\mathfrak{L}(\mathfrak{R}, \mathfrak{S})$ is a totally disconnected space in the sense that the only connected subsets of $\mathfrak{L}(\mathfrak{R}, \mathfrak{S})$ are points.)

3. Prove that every finite dimensional subspace of linear transformations in a vector space over a field is closed in the finite topology.

7. Total subspaces of $\mathfrak{R}^*$. Consider the subset $\mathfrak{R}_0^*$ of $\mathfrak{R}^*$ of linear functions f whose matrices relative to the basis (e_i) are column finite in the sense that $f(e_i) \neq 0$ for only a finite number of i. Evidently this collection of linear functions constitutes a subspace of the conjugate space. For each i we define a linear function e_i^* by the conditions

$$e_i^*(e_k) = \delta_{ik}. \tag{6}$$

Thus e_i^* is the linear function whose matrix relative to the bases (e_i) has a 1 in the i place and 0's elsewhere. Hence it is clear that the e_i^* form a basis for $\mathfrak{R}_0^*$. Consequently $\dim \mathfrak{R}_0^* = \dim \mathfrak{R}$. We now observe that $\mathfrak{R}_0^*$ is a *total subspace* of $\mathfrak{R}^*$ in the sense that, if u is any vector $\neq 0$ in $\mathfrak{R}$, then there exists an element $f \in \mathfrak{R}_0^*$ such that $f(u) \neq 0$. Write $u = \beta_1 e_{i_1} + \beta_2 e_{i_2} + \cdots + \beta_m e_{i_m}$ where $\beta_1 \neq 0$. Then $e_{i_1}^*(u) = \beta_1 \neq 0$ as required.

We recall that a subset S of a topological space is *dense* (in the space) if its closure Cl S is the whole space. We shall now show that a subspace $\mathfrak{R}'$ of $\mathfrak{R}^*$ is dense in the finite topology of $\mathfrak{R}^* = \mathfrak{L}(\mathfrak{R}, \Delta)$ if and only if $\mathfrak{R}'$ is total. First, it is clear that, if $\mathfrak{R}'$ is dense in $\mathfrak{R}^*$, then $\mathfrak{R}'$ is total. Otherwise there exists a vector $u \neq 0$ in $\mathfrak{R}$ such that $g(u) = 0$ for all $g \in \mathfrak{R}'$. Then $f(u) = 0$ for all f in Cl $\mathfrak{R}' = \mathfrak{R}^*$. On the other hand, we can use u as a vector in a basis, and then it is clear that there exists a linear function f such that $f(u) \neq 0$.

To prove the converse we make use of the following criterion for density of a subset of $\mathfrak{L}(\mathfrak{R}, \mathfrak{S})$: A subset $\mathfrak{A}$ of $\mathfrak{L}(\mathfrak{R}, \mathfrak{S})$ is dense in the finite topology if and only if for every ordered set $(x_1, x_2, \cdots, x_m)$ of linearly independent vectors in $\mathfrak{R}$ and every ordered set $(y_1, y_2, \cdots, y_m)$ in $\mathfrak{S}$, there exists an $A \in \mathfrak{A}$ such that $x_i A = y_i$ for $i = 1, 2, \cdots, m$. The sufficiency of this criterion is clear from the fact that the sets $O(x_i; x_i B)$ form a basis for the neighborhoods of B. The necessity follows from the fact that the complete set of linear transformations has the property stated in the criterion. If $\mathfrak{R}'$ is a subspace of $\mathfrak{R}^*$, we can simplify the criterion further and show that $\mathfrak{R}'$ is dense if, for any linearly independent finite set $x_1, x_2, \cdots, x_m$ of vectors in $\mathfrak{R}$, there exists a *complementary set* of vectors $x_1^*, x_2^*, \cdots, x_m^*$ in $\mathfrak{R}'$ such that

$$x_i^*(x_j) = \delta_{ij}, \quad i, j = 1, \cdots, m.$$

If this is the case and the β_i, $i = 1, \cdots, m$, are arbitrary elements of Δ, then the function $f = \Sigma x_i^* \beta_i \,\varepsilon\, \mathfrak{R}'$ and $f(x_i) = \beta_i$. Hence $\mathfrak{R}'$ is dense by the foregoing criterion. We now prove the following

Lemma. *If $\mathfrak{R}'$ is a total subspace of $\mathfrak{R}^*$ and $x_1, x_2, \cdots, x_m$ are any linearly independent vectors in $\mathfrak{R}$, then $\mathfrak{R}'$ contains a set of vectors $x_1^*, \cdots, x_m^*$ complementary to the x_i.*

Proof. If $m = 1$, this is clear; for, in this case, we can find an f in $\mathfrak{R}'$ such that $f(x_1) = \beta_1 \neq 0$. Then $x_1^* = f\beta_1^{-1}$ satisfies $x_1^*(x_1) = 1$. Assume the result for $m - 1$ vectors. Then, for $x_1, x_2, \cdots, x_{m-1}$ we can find $f_1, f_2, \cdots, f_{m-1}$ in $\mathfrak{R}'$ such that $f_k(x_l) = \delta_{kl}$ holds for $k, l = 1, 2, \cdots, m - 1$. For any $f \,\varepsilon\, \mathfrak{R}'$ we define

$$g(x) = f(x) - \sum_1^{m-1} f_k(x) f(x_k).$$

Then $g \,\varepsilon\, \mathfrak{R}'$ and $g(x_l) = 0$ for $l = 1, 2, \cdots, m - 1$. Also there exists an f in $\mathfrak{R}'$ such that the corresponding g has the property $g(x_m) \neq 0$. Otherwise we have

$$0 = g(x_m) = f(x_m) - \sum_1^{m-1} f_k(x_m) f(x_k)$$

$$= f(x_m - \sum_1^{m-1} f_k(x_m) x_k)$$

for all f in $\mathfrak{R}'$. This implies that $x_m = \sum_1^{m-1} f_k(x_m) x_k$ and contradicts the linear independence of the x's. Now choose f so that $g(x_m) = \gamma \neq 0$ and define

$$x_m^*(x) = g(x)\gamma^{-1}$$

$$x_k^*(x) = f_k(x) - x_m^*(x) f_k(x_m), \quad k = 1, 2, \cdots, m - 1.$$

Then we can verify that the x_j^* are complementary to the x_j.

This lemma completes the proof of the following

Theorem 3. *A necessary and sufficient condition that a subspace $\mathfrak{R}'$ of $\mathfrak{R}^*$ is total is that $\mathfrak{R}'$ is dense in the finite topology.*

If $\mathfrak{R}$ is finite dimensional, the finite topology in $\mathfrak{R}^*$ is discrete. Hence the closure $\text{Cl}\,\mathfrak{R}' = \mathfrak{R}'$ for every subspace $\mathfrak{R}'$ of $\mathfrak{R}^*$. It follows that the only total subspace of $\mathfrak{R}^*$ is $\mathfrak{R}^*$ itself. On the other hand, the example given at the beginning of this section shows that, for infinite dimensional $\mathfrak{R}$, $\mathfrak{R}^*$ contains proper total subspaces.

EXERCISES

1. Let $\mathfrak{R}$ have a denumerable basis and let $\mathfrak{R}'$ be a total subspace of $\mathfrak{R}^*$. Show that there exists a basis (u_i) for $\mathfrak{R}$ which has a complementary set of vectors (u_i^*) in $\mathfrak{R}'$ in the sense that $u_i^*(u_j) = \delta_{ij}$ holds for $i, j = 1, 2, \cdots$.
2. (Mackey) Let $\mathfrak{R}$ and $\mathfrak{R}'$ be as in 1. Assume, moreover, that $\mathfrak{R}'$ has a denumerable basis. Prove that $\mathfrak{R}$ and $\mathfrak{R}'$ have complementary bases.
3. Let $\mathfrak{R}$ and $\mathfrak{R}'$ be as in 2., and let (e_i), (e_i^*) be complementary bases. Show that the basis (f_i) where $f_1 = e_1$ and $f_i = e_i + e_{i-1}$ for $i > 1$ has no complementary set in $\mathfrak{R}'$.

8. Dual spaces. Kronecker products. Let $\mathfrak{R}'$ be a total subspace of $\mathfrak{R}^*$ and let $x \in \mathfrak{R}$. Then the mapping $f \to x(f) \equiv f(x)$ is a linear function on $\mathfrak{R}'$, that is, it is an element of the conjugate space $\mathfrak{R}'^*$ of $\mathfrak{R}'$. The mapping $x \to x(f)$ is a linear transformation of $\mathfrak{R}$ into $\mathfrak{R}'^*$ and, since $\mathfrak{R}'$ is total, this linear transformation is 1–1. Moreover, the image space is a total subspace of $\mathfrak{R}'^*$. Hence in a certain sense $\mathfrak{R}$ and $\mathfrak{R}'$ are interchangeable in our discussion. As in the finite case the symmetry which is implicit in this situation can be made explicit by introducing the notions of a bilinear form and of duality between a left vector space and a right vector space. Thus, we shall say that the left vector space $\mathfrak{R}$ and the right vector space $\mathfrak{R}'$ are *dual relative to the bilinear form* $g(x, y')$ if this form is *non-degenerate* in the sense that $g(z, y') = 0$ for all $y' \in \mathfrak{R}'$ implies that $z = 0$ and $g(x, z') = 0$ for all $x \in \mathfrak{R}$ implies that $z' = 0$. If $\mathfrak{R}'$ is a total subspace of $\mathfrak{R}^*$, then $\mathfrak{R}$ and $\mathfrak{R}'$ are dual relative to the bilinear form $s(x, f) = f(x) = x(f)$. On the other hand, if $\mathfrak{R}$ and $\mathfrak{R}'$ are dual relative to $g(x, y')$, then each $y' \in \mathfrak{R}'$ defines a linear function $g_{y'}(x)$ on $\mathfrak{R}$ and the mapping $y' \to g_{y'}$ is 1–1 linear transformation of $\mathfrak{R}'$ onto a total subspace of $\mathfrak{R}^*$. Similarly, $g(x, y')$ can be used to define a 1–1 linear mapping of $\mathfrak{R}$ onto a total subspace of $\mathfrak{R}'^*$.

The results on total subspaces can be carried over directly to dual spaces. In particular, each member of a pair of dual spaces defines a topology in the other which corresponds to the finite topology in the conjugate space. Thus let $\mathfrak{R}$ and $\mathfrak{R}'$ be dual and let $\phi(\mathfrak{R}')$ be the total subspace of $\mathfrak{R}^*$ corresponding to $\mathfrak{R}'$. Then $\phi(\mathfrak{R}')$ has a topology which is inherited from $\mathfrak{R}^*$: the open sets of $\phi(\mathfrak{R}')$ are the intersections of the open sets of $\mathfrak{R}^*$ with the set $\phi(\mathfrak{R}')$. (This is the standard subspace topology.) We can now use the 1–1 correspondence between $\phi(\mathfrak{R}')$ and $\mathfrak{R}'$ to transfer the topology of $\phi(\mathfrak{R}')$ to a topology in $\mathfrak{R}'$. The result that we obtain is that the open sets of $\mathfrak{R}'$ are the logical sums of the basic open sets $O(x_i; \beta_i)$, $i = 1, \cdots, r$, where $O(x_i; \beta_i)$ is the totality of $y' \varepsilon \mathfrak{R}'$ such that $g(x_i, y') = \beta_i$. We shall refer to this topology of $\mathfrak{R}'$ as the $\mathfrak{R}$-*topology*. In a similar fashion we can define an $\mathfrak{R}'$-*topology* for $\mathfrak{R}$.

Let $\mathfrak{R}$ and $\mathfrak{R}'$ be dual relative to $g(x, y')$. As in the finite dimensional case, if $\mathfrak{S}$ is a subspace of $\mathfrak{R}$, we denote by $j(\mathfrak{S})$ the subspace of $\mathfrak{R}'$ of vectors y' which are *incident* to every $x \varepsilon \mathfrak{S}$ in the sense that $g(x, y') = 0$. In a similar manner we can associate with every subspace $\mathfrak{S}'$ of $\mathfrak{R}'$ a subspace of $j(\mathfrak{S}')$ of $\mathfrak{R}$. It is easy to establish the following rules:

(i) If $\mathfrak{S}_1 \supseteq \mathfrak{S}_2$, then $j(\mathfrak{S}_1) \subseteq j(\mathfrak{S}_2)$,
(ii) $j(j(\mathfrak{S}_1)) \supseteq \mathfrak{S}_1$,
(iii) $j(j(j(\mathfrak{S}_1))) = j(\mathfrak{S}_1)$.

The first two of these are evident. To prove **(iii)** we use **(ii)** for $\mathfrak{S}_1$ replaced by $j(\mathfrak{S}_1)$. This gives $j(j(j(\mathfrak{S}_1))) \supseteq j(\mathfrak{S}_1)$. On the other hand, since $j(j(\mathfrak{S}_1)) \supseteq \mathfrak{S}_1$, **(i)** implies that $j(j(j(\mathfrak{S}_1))) \subseteq j(\mathfrak{S}_1)$. If $\mathfrak{R}$ (and hence $\mathfrak{R}'$) is finite dimensional, then the mapping $\mathfrak{S} \rightarrow j(\mathfrak{S})$ is 1–1 of the lattice of subspaces of $\mathfrak{R}$ onto the lattice of subspaces of $\mathfrak{R}'$. This need not hold for infinite dimensional spaces. For example, let $\mathfrak{R}' = \mathfrak{R}^*$, the complete conjugate space. Then if $\mathfrak{R}'$ is a total subspace of $\mathfrak{R}^*$ distinct from $\mathfrak{R}^*$, then $j(\mathfrak{R}') = 0 = j(\mathfrak{R}^*)$ in spite of $\mathfrak{R}' \neq \mathfrak{R}^*$. We shall show that the j-mapping induces a 1–1 mapping of the set of *closed* subspaces of $\mathfrak{R}$ onto the set of closed subspaces of $\mathfrak{R}'$.

We observe first that, if $\mathfrak{S}$ is any subspace of $\mathfrak{R}$, then the bilinear form g defines a bilinear form $\bar{g}$ connecting the space $\mathfrak{S}$ and the factor space $\mathfrak{R}' - j(\mathfrak{S})$. We define $\bar{g}(x, \bar{y}')$, where $\bar{y}' = y' + j(\mathfrak{S})$, by $\bar{g}(x, \bar{y}') = g(x, y')$. Since any two choices of y' are congruent modulo $j(\mathfrak{S})$, it is clear that we obtain in this way a single-valued mapping of the set of pairs $(x, \bar{y}')$ into Δ. We can verify directly that $\bar{g}$ is a bilinear form. If $x \neq 0$ in $\mathfrak{S}$, then we know that there exists a y' such that $g(x, y') \neq 0$. Then $\bar{g}(x, \bar{y}') \neq 0$. On the other hand, assume that $\bar{y}'$ is a vector in $\mathfrak{R}' - j(\mathfrak{S})$ such that $\bar{g}(x, \bar{y}') = 0$ holds for all x in $\mathfrak{S}$. Then $g(x, y') = 0$ for all x, and $y' \in j(\mathfrak{S})$. Thus $\bar{y}' = 0$. These remarks show that the pair of vector spaces $\mathfrak{S}$, $\mathfrak{R}' - j(\mathfrak{S})$ are dual relative to the bilinear form $\bar{g}$. We can now prove the following

Theorem 4. *If $\mathfrak{S}'$ is a subspace of $\mathfrak{R}'$, then $j(j(\mathfrak{S}'))$ is the closure of $\mathfrak{S}'$ in the $\mathfrak{R}$-topology of $\mathfrak{R}'$; in particular, $\mathfrak{S}'$ is closed if and only if $j(j(\mathfrak{S}')) = \mathfrak{S}'$.*

Proof. It follows directly from the definition of the topologies that subspaces of the form $j(\mathfrak{S})$, $j(\mathfrak{S}')$ are closed. Since $j(j(\mathfrak{S}')) \supseteq \mathfrak{S}'$, $j(j(\mathfrak{S}')) \supseteq \mathrm{Cl}\, \mathfrak{S}'$. Conversely, let $u' \in j(j(\mathfrak{S}'))$ and consider any neighborhood $O(x_i; g(x_i, u'))$ of u'. Let $\mathfrak{X}$ be the finite dimensional space spanned by the x_i and let $(y_1, y_2, \cdots, y_r)$ be a basis for this space such that $(y_1, y_2, \cdots, y_s)$ is a basis for $j(\mathfrak{S}') \cap \mathfrak{X}$. The cosets $\bar{y}_k = y_k + j(\mathfrak{S}')$, $k = s + 1, \cdots, r$, are linearly independent in $\mathfrak{R} - j(\mathfrak{S}')$. Since $\mathfrak{S}'$ and $\mathfrak{R} - j(\mathfrak{S}')$ are dual relative to $\bar{g}(x + j(\mathfrak{S}'), y') = g(x, y')$, we can conclude from the lemma of the preceding section that there exists a $v' \in \mathfrak{S}'$ such that $\bar{g}(y_k + j(\mathfrak{S}'), v') = g(y_k, u')$ holds for $k = s + 1, \cdots, r$. Thus $g(y_k, v') = g(y_k, u')$. On the other hand, $g(y_j, v') = 0 = g(y_j, u')$ for $j = 1, \cdots, s$. Hence $v' \in O(y_i; g(y_i, u'))$. Now $O(y_i; g(y_i, u')) = O(x_i; g(x_i, u'))$, $i = 1, \cdots, r$. Hence v' is in the given neighborhood $O(x_i; g(x_i, u'))$ of u'. This shows that $u' \in \mathrm{Cl}\, \mathfrak{S}'$; hence $\mathrm{Cl}\, \mathfrak{S}' = j(j(\mathfrak{S}'))$ as we wished to prove.

Now consider the mapping $\mathfrak{S} \to j(\mathfrak{S})$ where $\mathfrak{S}$ ranges over the set of closed subspaces of $\mathfrak{R}$. Since $j(j(\mathfrak{S})) = \mathfrak{S}$ for closed $\mathfrak{S}$, it is clear that our mapping is 1–1. Since every closed subspace of $\mathfrak{R}'$ has the form $\mathfrak{S}' = j(j(\mathfrak{S}'))$, the mapping is a mapping onto the set of closed subspaces of $\mathfrak{R}'$. This proves the assertion that we made before.

The existence of a dual for any given vector space enables us to carry over the proof given in the finite case for the existence of a direct product group for any right vector space $\mathfrak{R}'$ and any left vector space $\mathfrak{S}$ (§ 1, Chapter VII). As before, let $\mathfrak{R}$ be dual to $\mathfrak{R}'$ relative to the bilinear form g. Consider the set $\mathfrak{P}$ of linear transformations of $\mathfrak{R}$ into $\mathfrak{S}$ which have the form

$$x \to \sum_{1}^{m} g(x, x_i')y_i,$$

where the $x_i' \varepsilon \mathfrak{R}'$ and the $y_i \varepsilon \mathfrak{S}$. It is immediate that $\mathfrak{P}$ is a subgroup of $\mathfrak{L}(\mathfrak{R}, \mathfrak{S})$. If $x' \varepsilon \mathfrak{R}'$ and $y \varepsilon \mathfrak{S}$, then we define $x' \times y$ to be the linear transformation $x \to g(x, x')y$ belonging to the group $\mathfrak{P}$. Then we can verify as in the finite case that $\mathfrak{P}$ is a direct product group, $\mathfrak{P} = \mathfrak{R}' \times \mathfrak{S}$, of $\mathfrak{R}'$ and $\mathfrak{S}$.

This basic definition permits us to carry over much of our previous discussion of direct products. In particular we can use it to define the Kronecker product of arbitrary vector spaces over a field and the Kronecker product of arbitrary (non-associative) algebras. Also we can define the extension space $\mathfrak{R}_\Sigma$ for $\mathfrak{R}$ arbitrary and Σ any extension field of the base field Φ of $\mathfrak{R}$. The proofs of the elementary properties such as commutativity and associativity of Kronecker multiplication made no use of finiteness; hence these hold in the general case.

EXERCISES

1. Prove that, if $\mathfrak{S}$ is a closed subspace of $\mathfrak{R}$ (in the $\mathfrak{R}'$-topology) and $\mathfrak{F}$ is finite dimensional, then $\mathfrak{S} + \mathfrak{F}$ is closed.

2. Let $\mathfrak{F}$ be a finite dimensional vector space over a field Φ and let $\mathfrak{R}$ be an arbitrary vector space over Φ. Show that $\mathfrak{L}(\mathfrak{R} \times \mathfrak{F}, \mathfrak{R} \times \mathfrak{F})$ is the Kronecker product of $\mathfrak{L}(\mathfrak{R}, \mathfrak{R})$ and $\mathfrak{L}(\mathfrak{F}, \mathfrak{F})$ relative to $A \times B$ defined as on p. 211.

3. Show that the result of Ex. 2 does not hold for infinite dimensional $\mathfrak{R}$, $\mathfrak{F}$.

4. Let $\{\mathfrak{R}_\alpha\}$ be a collection of subspaces of a vector space $\mathfrak{R}$ over a field Φ and let Σ be an extension field of Φ. Prove that $(\bigcap \mathfrak{R}_\alpha)_\Sigma = \bigcap \mathfrak{R}_{\alpha\Sigma}$ holds in $\mathfrak{R}_\Sigma$.

9. Two-sided ideals in the ring of linear transformations. As in the finite case we define the *rank* $\rho(A)$ of a linear transformation A as the dimensionality of the image space $\mathfrak{R}A$. Similarly the *nullity* $\nu(A)$ is the dimensionality of the *null space* $\mathfrak{R}_A$ of vectors z such that $zA = 0$. If $\mathfrak{R}_1$ is a complement of $\mathfrak{R}_A$ so that $\mathfrak{R} = \mathfrak{R}_1 \oplus \mathfrak{R}_A$, then A is a 1–1 mapping of $\mathfrak{R}_1$ onto $\mathfrak{R}_1A = \mathfrak{R}A$. Hence dim $\mathfrak{R}A$ = dim $\mathfrak{R}_1$, and this implies, as before, that

$\rho(A) + \nu(A) = \dim \mathfrak{R}$. Also we can easily prove the following important relations:

1. $\rho(A + B) \leq \rho(A) + \rho(B)$.
2. $\rho(AB) \leq \min (\rho(A), \rho(B))$.

These formulas allow us to define certain proper two-sided ideals in the ring $\mathfrak{L}$ of linear transformations of an infinite dimensional vector space. Let e be any infinite cardinal such that $e \leq b = \dim \mathfrak{R}$ and let $\mathfrak{L}_e$ be the set of linear transformations A_e such that

$$\rho(A_e) < e.$$

Since $\rho(-A_e) = \rho(A_e)$, $\mathfrak{L}_e$ contains the negative of every linear transformation in this set. Also since the sum of two cardinals, one of which is infinite, is the larger of the two cardinals we see, using **1.**, that $\mathfrak{L}_e$ is closed under addition. Finally **2.** shows that $\mathfrak{L}_e$ is closed under multiplication by arbitrary elements of $\mathfrak{L}$. Hence $\mathfrak{L}_e$ is a two-sided ideal.

If e and e' are two infinite cardinals $\leq b$ and $e < e'$, then $\mathfrak{L}_e \subset \mathfrak{L}_{e'}$; for evidently $\mathfrak{L}_e \subseteq \mathfrak{L}_{e'}$. Moreover, there exist linear transformations of any given rank $\leq b$; for we can obtain such transformations by choosing a basis (e_i) of $\mathfrak{R}$ and a subset (e_k) which has the given cardinal number. Then the linear transformation such that $e_k \to e_k$ and $e_i \to 0$ for $i \neq k$ has the required rank. Now the linear transformations of rank e are in $\mathfrak{L}_{e'}$ but not in $\mathfrak{L}_e$. Hence $\mathfrak{L}_e \neq \mathfrak{L}_{e'}$ and $\mathfrak{L}_e \subset \mathfrak{L}_{e'}$. Thus we see that the correspondence $e \to \mathfrak{L}_e$ is 1–1.

The main result on two-sided ideals of $\mathfrak{L}$ is that the ideals $\mathfrak{L}_e$ are the only proper two-sided ideals in $\mathfrak{L}$. For the proof of this fact we require the following

Lemma. *If A and $B \in \mathfrak{L}$ and $\rho(B) \leq \rho(A)$, then there exist P and Q in $\mathfrak{L}$ such that $B = PAQ$.*

Proof. We write $\mathfrak{R} = \mathfrak{R}_1 \oplus \mathfrak{N}_A$ where $\mathfrak{N}_A$ is the null space of A and similarly $\mathfrak{R} = \mathfrak{R}_2 \oplus \mathfrak{N}_B$. Let (x_k) be a basis for $\mathfrak{R}_1$ and (y_λ) be a basis for $\mathfrak{R}_2$. Since $\dim \mathfrak{R}_1 = \dim \mathfrak{R}A \geq \dim \mathfrak{R}B = \dim \mathfrak{R}_2$, we can set up a 1–1 correspondence $y_\lambda \to x_\lambda$ of (y_λ) into (x_k). Let P be the linear transformation which maps y_λ into x_λ

and maps $\mathfrak{N}_B$ into 0. Since the vectors $x_\lambda A$ form a linearly independent set, there exists a linear transformation Q such that $(x_\lambda A)Q = y_\lambda B$. Then

$$y_\lambda PAQ = x_\lambda AQ = y_\lambda B,$$
$$zPAQ = 0 \quad \text{if} \quad z \in \mathfrak{N}_B.$$

Hence $PAQ = B$ as required.

Suppose now $\mathfrak{B}$ is any proper two-sided ideal in $\mathfrak{L}$. Let e be the smallest infinite cardinal number such that $e > \rho(B)$ for all $B \in \mathfrak{B}$. If $\mathfrak{B}$ contains a linear transformation B of rank $b =$ dim $\mathfrak{R}$ and A is any linear transformation of $\mathfrak{R}$, then rank $A \leq$ rank B. Hence by the preceding lemma, $A = PBQ \in \mathfrak{B}$. Thus $\mathfrak{B} = \mathfrak{L}$ contrary to hypothesis. Thus $b > \rho(B)$ for every $B \in \mathfrak{B}$ and therefore $e \leq b$. Since $e \leq b$, the definitions show that $\mathfrak{B} \subseteq \mathfrak{L}_e$. On the other hand, let C be any element of $\mathfrak{L}_e$. Then $\rho(C) < e$ and if $\rho(C)$ is infinite there must exist a $B \in \mathfrak{B}$ such that $\rho(B) \geq \rho(C)$. By the lemma this implies that $C \in \mathfrak{B}$. If $\rho(C)$ is finite, we can argue in the same way that C is in $\mathfrak{B}$ unless $\rho(C) > \rho(B)$, for every B. At any rate we can conclude that $\mathfrak{B}$ contains every linear transformation of rank one. We shall now show that any transformation of finite rank is a sum of transformations of rank 1. To prove this we write $\mathfrak{R}C = [y_1, y_2, \cdots, y_m]$ where the y_i are linearly independent. Then for any x

$$xC = \phi_1(x)y_1 + \phi_2(x)y_2 + \cdots + \phi_m(x)y_m.$$

It is clear that the $\phi_i(x)$ are linear functions and that the mapping C_i such that $xC_i = \phi_i(x)y_i$ is a linear transformation of rank one. Evidently $C = C_1 + C_2 + \cdots + C_m$ and, since we know that the $C_i \in \mathfrak{B}$, we see that $C \in \mathfrak{B}$. Thus every C in $\mathfrak{L}_e$ is in $\mathfrak{B}$ and $\mathfrak{B} = \mathfrak{L}_e$. This proves

Theorem 5. *Let $\mathfrak{L}$ be the ring of linear transformations in an infinite dimensional vector space $\mathfrak{R}$. For each infinite cardinal $e \leq$ dim $\mathfrak{R}$ define $\mathfrak{L}_e$ to be the totality of linear transformations of rank $< e$. Then $\mathfrak{L}_e$ is a proper two-sided ideal in $\mathfrak{L}$ and any proper two-sided ideal coincides with one of the $\mathfrak{L}_e$.*

We have seen that, if e and e' are two infinite cardinals $\leq b$ and $e < e'$, then $\mathfrak{L}_e \subset \mathfrak{L}_{e'}$. Theorem 5 therefore shows that the correspondence $e \to \mathfrak{L}_e$ is a lattice isomorphism of the set of infinite cardinals $< b$ on the set of proper ideals in $\mathfrak{L}$. In particular we see that $\mathfrak{L}$ has at least one proper two-sided ideal, $\mathfrak{L}_e$ where e is aleph null. This ideal consists of the transformations of finite rank and in the sequel we denote it as $\mathfrak{F}$. Clearly $\mathfrak{F}$ is contained in every two-sided ideal $\neq 0$ of $\mathfrak{L}$.

10. Dense rings of linear transformations. Some of the results on the ring of linear transformations of a finite dimensional vector space can be extended to arbitrary dense rings of linear transformations. We recall that a set $\mathfrak{A}$ of linear transformations is dense in the finite topology if and only if for any two ordered finite sets of vectors $(x_1, x_2, \cdots, x_m)$, $(y_1, y_2, \cdots, y_m)$ such that the x's are linearly independent, there exists an $A \varepsilon \mathfrak{A}$ such that $x_iA = y_i$, $i = 1, 2, \cdots, m$.

We note first that any dense ring $\mathfrak{A}$ of linear transformations is an *irreducible set of endomorphisms*. By this we mean that the only subgroups of the additive group $\mathfrak{R}$ which are mapped into themselves by $\mathfrak{A}$ are $\mathfrak{R}$ and 0. Thus, let $\mathfrak{S}$ be such a subgroup. If $\mathfrak{S} \neq 0$, it contains a vector $x \neq 0$. Then if y is any vector in $\mathfrak{R}$, we can find an $A \varepsilon \mathfrak{A}$ such that $xA = y$. Hence $y \varepsilon \mathfrak{S}$. Since y is arbitrary this shows that $\mathfrak{S} = \mathfrak{R}$.

Another noteworthy result on dense rings is that the only endomorphisms of the group $\mathfrak{R}$ which commute with every $A \varepsilon \mathfrak{A}$ are the scalar multiplications. This is an extension of Theorem 2 of the preceding chapter. The proof of the earlier result carries over word for word to the present situation.

It appears to be very difficult to construct and to classify all dense rings of linear transformations. However, a good deal can be done for an important subclass of the class of dense rings, namely, those which contain non-zero transformations of finite rank. In the remainder of this section and the next two sections we shall be concerned with the theory of rings of this type.

We give first a method for constructing such rings which will turn out to be completely general. We begin with a total subspace $\mathfrak{R}'$ of the conjugate space $\mathfrak{R}^*$ and we let $\mathfrak{L}(\mathfrak{R}' \mid \mathfrak{R})$ denote the totality of linear transformations A whose transposes A^* map $\mathfrak{R}'$ into itself. We recall that A^* is the linear transformation $f \to fA^*$ where $fA^*(x) = f(xA)$. Now let $\phi_1, \phi_2, \cdots, \phi_m$ be arbitrary elements of $\mathfrak{R}'$, $u_1, u_2, \cdots, u_m$ arbitrary vectors in $\mathfrak{R}$ and let $F = \sum_1^m \phi_i \times u_i$. We recall that this means that F is the linear transformation $x \to \sum_1^m \phi_i(x)u_i$. Thus F is of finite rank. Moreover, if f is any linear function, then $fF^*(x) = \Sigma\phi_i(x)f(u_i)$. Thus fF^* is a linear combination of $\phi_1, \phi_2, \cdots, \phi_m$ and hence $F^* \varepsilon \mathfrak{L}(\mathfrak{R}' \mid \mathfrak{R})$. This shows that $\mathfrak{L}(\mathfrak{R}' \mid \mathfrak{R})$ contains non-zero transformations of finite rank. Let $\mathfrak{F}(\mathfrak{R}' \mid \mathfrak{R}) = \mathfrak{F} \cap \mathfrak{L}(\mathfrak{R}' \mid \mathfrak{R})$ the totality of finite rank transformations contained in $\mathfrak{L}(\mathfrak{R}' \mid \mathfrak{R})$.

We prove next that $\mathfrak{F}(\mathfrak{R}' \mid \mathfrak{R})$ is dense in $\mathfrak{L}$. Thus let $(y_1, y_2, \cdots, y_m)$ be linearly independent and let $(u_1, u_2, \cdots, u_m)$ be arbitrary in $\mathfrak{R}$. Since $\mathfrak{R}'$ is total, it contains linear functions ϕ_i such that $\phi_i(y_j) = \delta_{ij}$, $i, j = 1, 2, \cdots, m$. Then the linear transformation $F = \Sigma\phi_i \times u_i \,\varepsilon\, \mathfrak{F}(\mathfrak{R}' \mid \mathfrak{R})$ and $y_iF = u_i$, $i = 1, \cdots, m$, as required.

It is clear from the definition of $\mathfrak{L}(\mathfrak{R}' \mid \mathfrak{R})$ that this set is a subring of $\mathfrak{L} = \mathfrak{L}(\mathfrak{R}, \mathfrak{R})$ ($= \mathfrak{L}(\mathfrak{R}^* \mid \mathfrak{R})$). Likewise $\mathfrak{F}(\mathfrak{R}' \mid \mathfrak{R})$ is a subring of $\mathfrak{L}$. Now let $\mathfrak{A}$ be any subring of $\mathfrak{L}(\mathfrak{R}' \mid \mathfrak{R})$ which contains $\mathfrak{F}(\mathfrak{R}' \mid \mathfrak{R})$. Then since $\mathfrak{F}(\mathfrak{R}' \mid \mathfrak{R})$ is dense, $\mathfrak{A}$ is dense. Also obviously $\mathfrak{A}$ contains non-zero transformations of finite rank.

Conversely, let $\mathfrak{A}$ be any dense ring of linear transformations such that $\mathfrak{A} \cap \mathfrak{F} \neq 0$. Let $\mathfrak{R}'$ be the subspace of $\mathfrak{R}^*$ defined by

$$\mathfrak{R}' = \sum_{F \varepsilon \mathfrak{A} \cap \mathfrak{F}} \mathfrak{R}^* F^*, \tag{7}$$

that is, $\mathfrak{R}'$ is the smallest subspace of $\mathfrak{R}^*$ containing the spaces $\mathfrak{R}^*F^*$ where $F \,\varepsilon\, \mathfrak{A} \cap \mathfrak{F}$. Then we shall show that $\mathfrak{R}'$ is total and that $\mathfrak{F}(\mathfrak{R}' \mid \mathfrak{R}) \subseteq \mathfrak{A} \subseteq \mathfrak{L}(\mathfrak{R}' \mid \mathfrak{R})$. The totality of $\mathfrak{R}'$ will follow from the following lemma which is of interest in itself.

Lemma 1. *If $\mathfrak{A}$ is an irreducible ring of endomorphisms, any non-zero two-sided ideal $\mathfrak{B}$ of $\mathfrak{A}$ is irreducible.*

Proof. If x is a non-zero element of the group $\mathfrak{R}$ in which $\mathfrak{A}$ acts, then the set $x\mathfrak{B}$ of elements xB, B in $\mathfrak{B}$, is a subgroup invariant relative to $\mathfrak{A}$. Hence either $x\mathfrak{B} = \mathfrak{R}$ or $x\mathfrak{B} = 0$. If $x\mathfrak{B} = 0$, then the collection $\mathfrak{S}$ of x's with this property is not 0. But $\mathfrak{S}$ is a subgroup invariant relative to $\mathfrak{A}$ also. Hence $\mathfrak{S} = \mathfrak{R}$. Thus $\mathfrak{R}\mathfrak{B} = 0$ contrary to $\mathfrak{B} \neq 0$. We have therefore proved that $x\mathfrak{B} = \mathfrak{R}$ holds for every non-zero x. The irreducibility of $\mathfrak{R}$ relative to $\mathfrak{B}$ is an immediate consequence of this result.

We can now prove that the space $\mathfrak{R}'$ defined by (7) is total. We know that $\mathfrak{F}$ is a two-sided ideal in $\mathfrak{L}$. Hence $\mathfrak{F} \cap \mathfrak{A}$ is a non-zero two-sided ideal in $\mathfrak{A}$ so that by Lemma 1 $\mathfrak{F} \cap \mathfrak{A}$ is irreducible. If $x \neq 0$ is in $\mathfrak{R}$, then we can find an $F \varepsilon \mathfrak{F} \cap \mathfrak{A}$ such that $xF \neq 0$. Also we can find a linear function f such that $f(xF) \neq 0$. Then $fF^*(x) \neq 0$ and $fF^* \varepsilon \mathfrak{R}'$. Hence $\mathfrak{R}'$ is total.

Now let $A \varepsilon \mathfrak{A}$. Then we have the following relations

$$\mathfrak{R}'A^* = \Sigma\mathfrak{R}^*F^*A^* = \Sigma\mathfrak{R}^*(AF)^* \subseteq \mathfrak{R}',$$

since $\mathfrak{F} \cap \mathfrak{A}$ is an ideal in $\mathfrak{A}$. Hence we have proved that $\mathfrak{A} \subseteq \mathfrak{L}(\mathfrak{R}' \mid \mathfrak{R})$. In order to prove the other inclusion relation $\mathfrak{F}(\mathfrak{R}' \mid \mathfrak{R}) \subseteq \mathfrak{A}$ we require the following

Lemma 2. *Let F be a transformation of finite rank and let $\mathfrak{R}F = [u_1, u_2, \cdots, u_m]$ where the u_i are linearly independent. Then F can be written in one and only one way as $\sum_1^m \phi_i \times u_i$ where the $\phi_i \varepsilon \mathfrak{R}^*$. Moreover, the ϕ_i are linearly independent and $\mathfrak{R}'F^* = [\phi_1, \phi_2, \cdots, \phi_m]$ holds for every total subspace $\mathfrak{R}'$ of $\mathfrak{R}^*$.*

Proof. We have seen before (p. 258) that F has the form $\sum_1^m \phi_i \times u_i$. The uniqueness has also been noted previously in our discussion of direct products (p. 201 and p. 256). Thus the first statement of the lemma holds. Now let $f \varepsilon \mathfrak{R}^*$. Then $fF^* = \Sigma\phi_i f(u_i)$; hence $\mathfrak{R}^*F^* \subseteq [\phi_1, \phi_2, \cdots, \phi_m]$. On the other hand, if $\mathfrak{R}'$ is dense, then we can find g_i in $\mathfrak{R}'$ such that $g_i(u_j) = \delta_{ij}$. Then $g_iF^* = \phi_i$ so that $[\phi_1, \phi_2, \cdots, \phi_m] \subseteq \mathfrak{R}'F^*$. This completes the proof.

Now let $F \varepsilon \mathfrak{F}(\mathfrak{R}' \mid \mathfrak{R})$ and write $F = \Sigma\phi_i \times u_i$ where the u_i form a basis for $\mathfrak{R}F$. Then $\mathfrak{R}'F^* = [\phi_1, \phi_2, \cdots, \phi_m]$ so that the $\phi_i \varepsilon \mathfrak{R}'$. Thus all the $\phi_i \varepsilon \mathfrak{R}^*F_1^* + \mathfrak{R}^*F_2^* + \cdots + \mathfrak{R}^*F_n^*$ for suitable $F_j \varepsilon \mathfrak{A} \cap \mathfrak{F}$. We apply the preceding lemma to the F_j and write $F_j = \sum_k \psi_{jk} \times v_{jk}$ where the v_{jk} form a basis for $\mathfrak{R}F_j$ and the ψ_{jk} form a basis for $\mathfrak{R}^*F_j^*$. Then $\phi_i = \sum_{j,k} \psi_{jk}\mu_{jk,i}$ for suitable $\mu_{jk,i}$ in Δ. Since $\mathfrak{A}$ is dense and the v_{jk} for fixed j are linearly independent, we can find $A_{ji} \varepsilon \mathfrak{A}$ such that

$$v_{jk}A_{ji} = \mu_{jk,i}u_i \tag{8}$$

for $k = 1, 2, \cdots, i = 1, 2, \cdots, m$. Now the transformation $\sum_{j,i} F_j A_{ji} \varepsilon \mathfrak{F} \cap \mathfrak{A}$ and

$$\begin{aligned}\sum_{j,i} F_j A_{ji} &= \sum_{j,i} \left(\sum_k \psi_{jk} \times v_{jk}\right) A_{ji} = \sum_{i,j,k} \psi_{jk} \times v_{jk}A_{ji} \\ &= \sum_{i,j,k} \psi_{jk} \times \mu_{jk,i}u_i = \sum_i \left(\sum_{j,k} \psi_{jk}\mu_{jk,i}\right) \times u_i \\ &= \Sigma\phi_i \times u_i = F.\end{aligned}$$

Thus $F \varepsilon \mathfrak{F} \cap \mathfrak{A}$. This completes the proof of the following structure theorem.

Theorem 6. *Let $\mathfrak{R}'$ be a total subspace of $\mathfrak{R}^*$ and let $\mathfrak{L}(\mathfrak{R}' \mid \mathfrak{R})$ be the totality of linear transformations A whose transposes map $\mathfrak{R}'$ into itself. Then any subring $\mathfrak{A}$ of $\mathfrak{L}(\mathfrak{R}' \mid \mathfrak{R})$ which contains $\mathfrak{F}(\mathfrak{R}' \mid \mathfrak{R}) = \mathfrak{F} \cap \mathfrak{L}(\mathfrak{R}' \mid \mathfrak{R})$ is a dense ring of linear transformations containing non-zero transformations of finite rank. Conversely, any dense ring of linear transformations which contains non-zero finite rank transformations can be obtained in this way.*

Our arguments actually establish somewhat more than we have stated in the theorem. Thus we have the formula (7) for the total subspace $\mathfrak{R}'$ determined by the given ring $\mathfrak{A}$. Moreover, it is easy to see from our discussion that $\mathfrak{F}(\mathfrak{R}' \mid \mathfrak{R}) = \mathfrak{R}' \times \mathfrak{R}$ the totality of mappings $\Sigma\phi_i \times u_i$, $\phi_i \varepsilon \mathfrak{R}'$, $u_i \varepsilon \mathfrak{R}$. If we begin with a ring between $\mathfrak{L}(\mathfrak{R}' \mid \mathfrak{R})$ and $\mathfrak{F}(\mathfrak{R}' \mid \mathfrak{R})$, then the space determined by (7) is $\mathfrak{R}'$ itself. We leave the verification of these statements to the reader.

The main theorem can also be formulated in a more symmetric fashion in terms of dual spaces. For this purpose let $\mathfrak{R}$ and $\mathfrak{R}'$ be dual relative to the bilinear form g. Then we have the natural equivalence R of $\mathfrak{R}'$ onto a total subspace $\mathfrak{S}^*$ of $\mathfrak{R}^*$: If $y' \varepsilon \mathfrak{R}'$, then $y'R$ is the linear function $g_{y'}(x) = g(y', x)$ (cf. p. 137). If $A \varepsilon \mathfrak{L}(\mathfrak{S}^* \mid \mathfrak{R})$, then $A' = RA^*R^{-1}$ is a linear transformation of $\mathfrak{R}'$ into itself. We shall call A' *the transpose of A relative to g*. As in the finite dimensional case, A' satisfies the condition

$$g(xA, y') = g(x, y'A') \tag{9}$$

for all $x \varepsilon \mathfrak{R}$ and all $y' \varepsilon \mathfrak{R}'$. Conversely, let A be any linear transformation in $\mathfrak{R}$ for which there exists a linear transformation A' in $\mathfrak{R}'$ such that (9) holds. Then $g_{y'}A^* = g_{y'A'} \varepsilon \mathfrak{S}^*$ so that $A \varepsilon \mathfrak{L}(\mathfrak{S}^* \mid \mathfrak{R})$. It follows that A' is the transpose of A relative to g. It is now natural to denote $\mathfrak{L}(\mathfrak{S}^* \mid \mathfrak{R})$ also as $\mathfrak{L}(\mathfrak{R}' \mid \mathfrak{R})$. The transformations of finite rank in this ring have the form $\Sigma y_i' \times x_i$ where, as usual, this denotes the mapping $x \rightarrow \Sigma g(x, y_i')x_i$. The transpose in $\mathfrak{R}'$ of $\Sigma y_i' \times x_i$ is $\Sigma y_i' \times' x_i$ where $y'(\Sigma y_i' \times x_i) = \Sigma y_i' g(x_i, y')$. Hence the sets $\mathfrak{F}(\mathfrak{R}' \mid \mathfrak{R}) = \mathfrak{F} \cap \mathfrak{L}(\mathfrak{R}' \mid \mathfrak{R})$ and $\mathfrak{F}(\mathfrak{R} \mid \mathfrak{R}')$ (defined in the same way) correspond in the mapping $A \rightarrow A'$.

The dual space formulation of Theorem 6 can now be given.

Theorem 6′. *Let $\mathfrak{R}$ and $\mathfrak{R}'$ be dual vector spaces and let $\mathfrak{L}(\mathfrak{R}' \mid \mathfrak{R})$ denote the totality of linear transformations in $\mathfrak{R}$ which have transposes in $\mathfrak{R}'$. Then any subring of $\mathfrak{L}(\mathfrak{R}' \mid \mathfrak{R})$ which contains $\mathfrak{F}(\mathfrak{R}' \mid \mathfrak{R}) = \mathfrak{F} \cap \mathfrak{L}(\mathfrak{R}' \mid \mathfrak{R})$ is a dense ring of linear transformations in $\mathfrak{R}$ containing non-zero transformations of finite rank. Conversely, any dense ring of linear transformations which contains non-zero finite rank transformations can be obtained in this way.*

It is clear from this formulation that the set $\mathfrak{A}'$ of transposes A' in $\mathfrak{R}'$ is a dense ring in $\mathfrak{R}'$ containing transformations of finite rank. The mapping $A \rightarrow A'$ is an anti-isomorphism of $\mathfrak{A}$ onto $\mathfrak{A}'$.

EXERCISES

1. Let $\mathfrak{R}$ have a denumerable basis over Δ and let $\mathfrak{A}$ be the totality of linear transformations whose matrices relative to this basis have the form

$$\begin{bmatrix} A & 0 \\ 0 & 0 \end{bmatrix}$$

where A is a finite matrix. Show that $\mathfrak{A}$ is a dense ring.

2. Let $\mathfrak{R}$ be as in 1. and let $\mathfrak{R}'$ be the total subspace $[\phi_1, \phi_2, \cdots]$ where $\phi_i(e_j) = \delta_{ij}$ for (e_j) a basis. Show that $A \varepsilon \mathfrak{L}(\mathfrak{R}' \mid \mathfrak{R})$ if and only if its matrix relative to (e_i) is both row finite and column finite.

3. Let U and V be the linear transformations in $\mathfrak{R}$ over Φ whose matrices are

$$\begin{bmatrix} 0 & 1 & & & \\ 0 & 0 & 1 & & \\ & & 0 & 1 & \\ & & & \ddots & \ddots \end{bmatrix}, \quad \begin{bmatrix} 0 & & & & \\ 1 & 0 & & & \\ 0 & 1 & 0 & & \\ & & 1 & \ddots & \\ & & & & \ddots \end{bmatrix}.$$

Show that the algebra $\Phi[U, V]$ generated by U and V is dense and contains transformations of finite rank.

4. Verify that, if the vectors (e_i) and (ϕ_i) are complementary $(\phi_i(e_k) = \delta_{ik})$, then the linear transformations $e_{ik} = \phi_i \times e_k$ satisfy the multiplication table for matrix units.

5. (Litoff) Prove that any finite subset of $\mathfrak{F}(\mathfrak{R}' \mid \mathfrak{R})$ can be imbedded in a subring which is isomorphic to a finite matrix ring Δ_n.

6. Prove that any right ideal of $\mathfrak{F}(\mathfrak{R}' \mid \mathfrak{R}) = \mathfrak{R}' \times \mathfrak{R}$ has the form $\mathfrak{S}' \times \mathfrak{R}$ where $\mathfrak{S}'$ is a subspace of $\mathfrak{R}'$ and that any left ideal has the form $\mathfrak{R}' \times \mathfrak{S}$, $\mathfrak{S}$ a subspace of $\mathfrak{R}$. (Cf. § 3–4 of Chapter VIII.)

7. Prove that $\mathfrak{F}(\mathfrak{R}' \mid \mathfrak{R})$ is simple.

8. Prove that A has a transpose in $\mathfrak{R}'$ if and only if A is a continuous mapping of $\mathfrak{R}$, endowed with the $\mathfrak{R}'$-topology, into itself.

11. Isomorphism theorems. We take up now the question of isomorphism of dense rings of linear transformations containing non-zero transformations of finite rank. For this purpose we shall use the structure theorem in its original form, namely, that $\mathfrak{A}$ can be sandwiched in as $\mathfrak{L}(\mathfrak{R}' \mid \mathfrak{R}) \supseteq \mathfrak{A} \supseteq \mathfrak{F}(\mathfrak{R}' \mid \mathfrak{R})$ where $\mathfrak{R}'$ is a total subspace of $\mathfrak{R}^*$. We have seen (Ex. 6) that every right ideal of $\mathfrak{F}(\mathfrak{R}' \mid \mathfrak{R})$ has the form $\mathfrak{S}' \times \mathfrak{R}$ where $\mathfrak{S}'$ is a subspace of $\mathfrak{R}'$. Now if A is any linear transformation in $\mathfrak{R}$, then $(\Sigma\psi_i \times x_i)A = \Sigma\psi_i \times x_iA$. Hence $\mathfrak{S}' \times \mathfrak{R}$ is in reality a right ideal in $\mathfrak{L}$ and *a fortiori* in $\mathfrak{A}$. If $\mathfrak{S}' = [\psi]$ is a one-dimensional subspace of $\mathfrak{R}'$, then it is clear that $\mathfrak{S}' \times \mathfrak{R}$ is a minimal right ideal of $\mathfrak{F}(\mathfrak{R}' \mid \mathfrak{R})$. Since it is clear that any right ideal of $\mathfrak{A}$ contained in $\mathfrak{F}(\mathfrak{R}' \mid \mathfrak{R})$ is a right ideal of $\mathfrak{F}(\mathfrak{R}' \mid \mathfrak{R})$, our remark shows that $[\psi] \times \mathfrak{R}$ is a minimal right ideal of $\mathfrak{A}$.

We can now list the following facts about $\mathfrak{A}$: 1) $\mathfrak{A}$ is an irreducible ring of endormophisms in $\mathfrak{R}$, 2) the set of endomorphisms which commute with every A in $\mathfrak{A}$ is the set of scalar multiplications, and 3) $\mathfrak{A}$ possesses minimal right ideals. These results enable us to carry over the discussion in § 5 of Chapter VIII on the isomorphism of rings of linear transformations of finite dimensional vector spaces.

We sketch the argument that we used before. Let $\mathfrak{J}$ be a minimal right ideal in $\mathfrak{A}$ and let x be a vector such that $x\mathfrak{J} \neq 0$. Then $x\mathfrak{J}$ is a subgroup of $\mathfrak{R}$ mapped into itself by $\mathfrak{A}$. Hence $x\mathfrak{J} = \mathfrak{R}$. We can conclude as before that the mapping χ: $B \to xB$ is an operator isomorphism of $\mathfrak{J}$ onto $\mathfrak{R}$.

We now consider two isomorphic rings $\mathfrak{A}_i$, $i = 1, 2$, where $\mathfrak{L}(\mathfrak{R}_i' \mid \mathfrak{R}_i) \supseteq \mathfrak{A}_i \supseteq \mathfrak{F}(\mathfrak{R}_i' \mid \mathfrak{R}_i)$ and $\mathfrak{R}_i$ is a vector space over Δ_i, $\mathfrak{R}_i'$ a total subspace of linear functions in $\mathfrak{R}_i$. Let $A_1 \rightarrow A_1\phi$ be an isomorphism of $\mathfrak{A}_1$ onto $\mathfrak{A}_2$. Choose a minimal right ideal $\mathfrak{J}_1$ in $\mathfrak{A}_1$ and let $\mathfrak{J}_2 = \mathfrak{J}_1\phi$. Let ψ_i be an operator isomorphism of $\mathfrak{J}_i$ into $\mathfrak{R}_i$ determined as above. Then by the argument of § 5, Chapter VIII, the mapping $U = \psi_1^{-1}\phi\psi_2$ is a 1–1 semi-linear transformation of $\mathfrak{R}_1$ onto $\mathfrak{R}_2$ and

$$A_1\phi = U^{-1}A_1U \tag{10}$$

holds for all A_1 in $\mathfrak{A}_1$. In particular we see as before that the base division rings are isomorphic and $\mathfrak{R}_1$ and $\mathfrak{R}_2$ have the same dimensionality.

In the finite case this is all one needs to say. In the infinite case, however, there is an important additional remark which should be made. This concerns the transpose of the semi-linear transformation U. We proceed to define this concept for any semi-linear transformation.

Let S be any semi-linear transformation of $\mathfrak{R}_1$ into $\mathfrak{R}_2$ and let s be the isomorphism of Δ_1 onto Δ_2 associated with S. Then $(\alpha x_1)S = \alpha^s(x_1S)$ for any x_1 in $\mathfrak{R}_1$. Now let $f(x_2)$ be any linear function in $\mathfrak{R}_2$ and set

$$g(x_1) = f(x_1S)^{s^{-1}}.$$

Then it is clear that g is a linear function on $\mathfrak{R}_1$. Also it can be verified directly that the mapping S^*: $f \rightarrow g$ is a semi-linear transformation of $\mathfrak{R}_2^*$ into $\mathfrak{R}_1^*$ with associated isomorphism s^{-1}. We call S^* *the transpose* of S. As for linear transformations

$$(S_1S_2)^* = S_2^*S_1^*. \tag{11}$$

This implies that, if S has the inverse S^{-1} (a semi-linear transformation of $\mathfrak{R}_2$ onto $\mathfrak{R}_1$ with isomorphism s^{-1}) so that $SS^{-1} = 1 = S^{-1}S$, then $S^*(S^{-1})^* = 1 = (S^{-1})^*S^*$. Hence S^* also has an inverse. This is equivalent to saying that, if S is a 1–1 mapping of $\mathfrak{R}_1$ onto $\mathfrak{R}_2$, then S^* is 1–1 of $\mathfrak{R}_2^*$ onto $\mathfrak{R}_1^*$. We observe that F_1 is a transformation of finite rank in $\mathfrak{R}_1$ if and only if $U^{-1}F_1U$ is of finite rank in $\mathfrak{R}_2$. It follows that ϕ maps $\mathfrak{F}(\mathfrak{R}_1' \mid \mathfrak{R}_1)$ onto $\mathfrak{F}(\mathfrak{R}_2' \mid \mathfrak{R}_2)$. If we refer to formula (7) for the $\mathfrak{R}_i'$, we can verify that U^* maps $\mathfrak{R}_2'$ onto $\mathfrak{R}_1'$. Thus,

$$\mathfrak{R}_2'U^* = \Sigma\mathfrak{R}_2^*F_2^*U^* = \Sigma\mathfrak{R}_2^*U^*F_1^*$$
$$= \Sigma\mathfrak{R}_1^*F_1^* = \mathfrak{R}_1'.$$

We can therefore state the following isomorphism theorem:

Theorem 7. *Let* $\mathfrak{R}_i$, $i = 1, 2$, *be a vector space over* Δ_i *and let* $\mathfrak{R}_i'$ *be a total subspace of linear functions on* $\mathfrak{R}$. *Suppose that* $\mathfrak{A}_i$ *is a subring of* $\mathfrak{L}(\mathfrak{R}_i' \mid \mathfrak{R}_i)$ *containing* $\mathfrak{F}(\mathfrak{R}_i' \mid \mathfrak{R}_i)$ *and let* ϕ *be an isomorphism of* $\mathfrak{A}_1$ *onto* $\mathfrak{A}_2$. *Then there exists a* 1–1 *semi-linear transformation* U *of* $\mathfrak{R}_1$ *onto* $\mathfrak{R}_2$, *whose transpose maps* $\mathfrak{R}_2'$ *onto* $\mathfrak{R}_1'$, *such that* $A_1\phi = U^{-1}A_1U$ *holds for all* A_1 *in* $\mathfrak{A}_1$.

This theorem has a number of interesting consequences. We give one of these here, a generalization of Ex. 5, p. 237.

Corollary 1. *Let* $\mathfrak{R}$ *be a vector space over a field* Φ *and let* $\mathfrak{R}'$ *be a total subspace of* $\mathfrak{R}^*$. *Then every automorphism of* $\mathfrak{L}(\mathfrak{R}' \mid \mathfrak{R})$ *which leaves the elements of the center fixed is inner.*

Proof. Evidently $\mathfrak{L}(\mathfrak{R}' \mid \mathfrak{R})$ contains the set Φ_l of scalar multiplications. Also since the only endomorphisms which commute with all the elements of $\mathfrak{L}(\mathfrak{R}' \mid \mathfrak{R})$ are the scalar multiplications, the set Φ_l is the center of $\mathfrak{L}(\mathfrak{R}' \mid \mathfrak{R})$. If $A \to A\phi$ is an automorphism in $\mathfrak{L}(\mathfrak{R}' \mid \mathfrak{R})$, then there is a semi-linear transformation U such that $A\phi = U^{-1}AU$. If u is the associated automorphism in Φ_l, then $U^{-1}\alpha_lU = (\alpha^u)_l$. Assume now that $\alpha_l\phi = \alpha_l$ for all α. Then $U^{-1}\alpha_lU = \alpha_l\phi = \alpha_l$. Hence $\alpha_l = (\alpha^u)_l$ and so $u = 1$. Thus U is linear. Since U^* maps $\mathfrak{R}'$ into itself, $U \in \mathfrak{L}(\mathfrak{R}' \mid \mathfrak{R})$ by definition. Thus ϕ is inner.

This result includes, of course, the following special case:

Corollary 2. *Every automorphism of the ring* $\mathfrak{L}$ *of linear transformations of a vector space over a field which leaves the elements of the center fixed is inner.*

EXERCISE

1. Let Δ be any division ring with the property that the only automorphisms in Δ leaving the elements of the center fixed are inner. Prove that, if $\mathfrak{R}$ is a vector space over Δ, then every automorphism of a ring $\mathfrak{L}(\mathfrak{R}' \mid \mathfrak{R})$ which leaves the elements of the center fixed is inner.

12. Anti-automorphisms and scalar products. We consider now the problem of finding conditions that a dense ring of linear transformations containing non-zero transformations of finite rank possess an anti-automorphism. It is convenient to assume here the second formulation of the structure theorem, that is, that $\mathfrak{A}$ is given as a subring of $\mathfrak{L}(\mathfrak{R}' \mid \mathfrak{R})$ containing $\mathfrak{F}(\mathfrak{R}' \mid \mathfrak{R})$ where $\mathfrak{R}$ and $\mathfrak{R}'$ are dual relative to a bilinear form g. We introduce the division ring Δ' anti-isomorphic to Δ and let $\alpha \rightarrow \alpha^t$ be a fixed anti-isomorphism of Δ onto Δ'. Then we can regard $\mathfrak{R}$ as a right vector space over Δ' if we take $x\alpha^t = \alpha x$. Similarly, $\mathfrak{R}'$ is a left vector space over Δ' if $\alpha^t x' = x'\alpha$.

Now suppose that $A \rightarrow A\Psi$ is an anti-automorphism in $\mathfrak{A}$. If A' denotes the transpose of A relative to g, then $A \rightarrow A'$ is an anti-isomorphism of $\mathfrak{A}$ onto a ring $\mathfrak{A}'$ and $\mathfrak{L}(\mathfrak{R} \mid \mathfrak{R}') \supseteq \mathfrak{A}' \supseteq \mathfrak{F}(\mathfrak{R} \mid \mathfrak{R}')$. It follows that the mapping $A' \rightarrow A\Psi$ is an isomorphism of $\mathfrak{A}'$ onto $\mathfrak{A}$. Hence by the isomorphism theorem there exists a semi-linear transformation V of the left vector space $\mathfrak{R}'$ (over Δ') onto the left vector space $\mathfrak{R}$ such that

$$A\Psi = V^{-1}A'V \tag{12}$$

holds for all $A \varepsilon \mathfrak{A}$. The semi-linear transformation V can be used to define a scalar product in $\mathfrak{R}$; for we can put

$$h(x, y) = g(x, yV^{-1}). \tag{13}$$

Then it is clear that

$$h(x_1 + x_2, y) = h(x_1, y) + h(x_2, y)$$
$$h(x, y_1 + y_2) = h(x, y_1) + h(x, y_2)$$
$$h(\alpha x, y) = \alpha h(x, y).$$

Moreover,

$$\begin{aligned} h(x, \alpha y) &= g(x, (\alpha y)V^{-1}) = g(x, \alpha^{v^{-1}}(yV^{-1})) \\ &= g(x, (yV^{-1})\alpha^{v^{-1}t^{-1}}) = g(x, yV^{-1})\alpha^{v^{-1}t^{-1}} \\ &= h(x, y)\alpha^{v^{-1}t^{-1}} \end{aligned}$$

where v is the isomorphism of Δ' onto Δ associated with V. Since v^{-1} is an isomorphism of Δ onto Δ' and t^{-1} is an anti-isomorphism of Δ' onto Δ, $v^{-1}t^{-1}$ is an anti-automorphism in Δ. We write

$\bar{\alpha} = \alpha^{v^{-1}t^{-1}}$; hence $h(x, \alpha y) = h(x, y)\bar{\alpha}$ and h is a scalar product in $\mathfrak{R}$ relative to $\alpha \rightarrow \bar{\alpha}$.

It is easy to see that the non-degeneracy of g and the properties of V imply the non-degeneracy of h. Hence if we regard $\mathfrak{R}$ as a right vector space over Δ by taking $x\bar{\alpha} = \alpha x$, then $\mathfrak{R}$ is dual with itself relative to h. We prove next that $\mathfrak{A}$ is a subring of $\mathfrak{L}(\mathfrak{R} \mid \mathfrak{R})$ containing $\mathfrak{F}(\mathfrak{R} \mid \mathfrak{R})$. Thus if $A \in \mathfrak{A}$, then

$$\begin{aligned} h(xA, y) &= g(xA, yV^{-1}) = g(x, yV^{-1}A') = g(x, yV^{-1}A'VV^{-1}) \\ &= g(x, yA\Psi V^{-1}) = h(x, yA\Psi). \end{aligned}$$

Thus $A \in \mathfrak{L}(\mathfrak{R} \mid \mathfrak{R})$ and its transpose relative to h is the image $A\Psi$ under the given anti-isomorphism. On the other hand, let $F \in \mathfrak{F}(\mathfrak{R} \mid \mathfrak{R})$. Then

$$\begin{aligned} g(xF, y') &= h(xF, y'V) = h(x, y'VF\Psi) = h(x, y'VF\Psi V^{-1}V) \\ &= g(x, y'VF\Psi V^{-1}); \end{aligned}$$

hence F has a transpose relative to g and $F \in \mathfrak{A}$. We have therefore proved that $\mathfrak{L}(\mathfrak{R} \mid \mathfrak{R}) \supseteq \mathfrak{A} \supseteq \mathfrak{F}(\mathfrak{R} \mid \mathfrak{R})$ and that $A \rightarrow A\Psi$ is the transpose mapping relative to h.

We have not yet fully exploited the fact that $A\Psi \in \mathfrak{A}$. We shall show next that this condition implies that $h(x, y)$ is a *weakly hermitian* scalar product in the sense that there exists a 1–1 semi-linear transformation Q of $\mathfrak{R}$ onto itself such that

$$h(x, y) = \overline{h(y, xQ)} \tag{14}$$

holds for all $x, y \in \mathfrak{R}$. Consider the mapping $y_1 \times x_1$, that is, $x \rightarrow h(x, y_1)x_1$. We know that this belongs to $\mathfrak{F}(\mathfrak{R} \mid \mathfrak{R})$ and that its transpose is $y \rightarrow y_1 h(x_1, y) = h(x_1, y)^a y_1$ where $\alpha \rightarrow \alpha^a$ is the inverse of $\alpha \rightarrow \bar{\alpha}$. Since the transpose mapping sends $\mathfrak{A}$ into itself, $y \rightarrow h(x_1, y)^a y_1$ coincides with a linear transformation $\Sigma u_i \times v_i$. A simple argument shows that in reality our transformation has the form $z_1 \times y_1$. It follows that $h(x_1, y)^a = h(y, z_1)$ holds for all y. If we take the bar of both sides, we obtain $h(x_1, y) = \overline{h(y, z_1)}$. Thus for each x there exists a z such that $h(x, y) = \overline{h(y, z)}$ holds for all y. Now the non-degeneracy of h implies that z is uniquely determined by x; hence $x \rightarrow z$ is a mapping

Q of $\mathfrak{R}$ into itself. Clearly Q is an endomorphism. Moreover,

$$h(\alpha x, y) = \alpha h(x, y) = \overline{\alpha h(y, xQ)} = \overline{h(y, xQ)\alpha^a}$$
$$= \overline{h(y, \alpha^{a^2}(xQ))};$$

hence $(\alpha x)Q = \alpha^{a^2}(xQ)$ and Q is semi-linear with associated automorphism a^2. The non-degeneracy of h implies that Q is 1–1. Finally, since every element of $\mathfrak{F}(\mathfrak{R} \mid \mathfrak{R})$ is a transpose, Q is a mapping onto $\mathfrak{R}$. This proves that h is weakly hermitian.

Conversely, assume that h is a non-degenerate weakly hermitian scalar product in $\mathfrak{R}$ over Δ. Let $A \varepsilon \mathfrak{L}(\mathfrak{R} \mid \mathfrak{R})$ and let A' now denote its transpose relative to h. Then,

$$\overline{h(xA', y)} = h(yQ^{-1}, xA') = h(yQ^{-1}A, x)$$
$$= \overline{h(x, yQ^{-1}AQ)}.$$

Hence $h(xA', y) = h(x, yQ^{-1}AQ)$. This shows that $A' \varepsilon \mathfrak{L}(\mathfrak{R} \mid \mathfrak{R})$ and that $A'' = Q^{-1}AQ$. Hence $A \rightarrow A'$ is an anti-automorphism of $\mathfrak{L}(\mathfrak{R} \mid \mathfrak{R})$. We can summarize our results as follows:

Theorem 8. *Let $\mathfrak{A}$ be a dense ring of linear transformations in $\mathfrak{R}$ over Δ containing non-zero transformations of finite rank. Assume that $\mathfrak{A}$ possesses an anti-automorphism $A \rightarrow A\Psi$. Then Δ has an anti-automorphism $\alpha \rightarrow \bar{\alpha}$ and there exists a non-degenerate weakly hermitian scalar product in $\mathfrak{R}$ such that $\mathfrak{A}$ is included between $\mathfrak{L}(\mathfrak{R} \mid \mathfrak{R})$ and $\mathfrak{F}(\mathfrak{R} \mid \mathfrak{R})$ and such that $A \rightarrow A\Psi$ coincides with the transpose mapping relative to h. Conversely, if $\mathfrak{R}$ has a non-degenerate weakly hermitian scalar product, then the transpose mapping in $\mathfrak{L}(\mathfrak{R} \mid \mathfrak{R})$ is an anti-automorphism.*

We impose next the condition that $A \rightarrow A'$ $(= A\Psi)$ is involutorial, that is, that $A'' = A$ holds for all $A \varepsilon \mathfrak{A}$. By the relation $A'' = Q^{-1}AQ$ derived above, our condition is equivalent to $Q^{-1}AQ = A$ for all $A \varepsilon \mathfrak{A}$. The latter holds if and only if $Q = \mu_l$ a scalar multiplication. Thus $A \rightarrow A'$ is involutorial if and only if

$$h(x, y) = \nu\overline{h(y, x)} \tag{15}$$

for a fixed $\nu(= \bar{\mu})$ and all x, y. We shall now show that we can replace h by a suitable multiple $s(x, y) = h(x, y)\tau$ which is either hermitian or skew hermitian. We note first that iteration of (15) gives

$$h(x, y) = \nu \overline{\overline{h(x, y)}} \bar{\nu}. \tag{16}$$

If we choose x, y so that $h(x, y) = 1$, this gives $\bar{\nu} = \nu^{-1}$. If $\nu = -1$, h is skew hermitian and there is nothing to prove. If $\nu \neq -1$, then we set $\tau = (\nu + 1)^{-1}$ and we verify that

$$\tau^{-1}\bar{\tau} = (\nu + 1)(\bar{\nu} + 1)^{-1} = (\nu + 1)(\nu^{-1} + 1)^{-1} = \nu.$$

Then if $s(x, y) = h(x, y)\tau$, we can verify that s is a scalar product whose anti-automorphism is $\alpha \to \alpha^* \equiv \tau^{-1}\bar{\alpha}\tau$. Also

$$\begin{aligned} s(y, x)^* &= \tau^{-1}\overline{s(y, x)}\tau = \tau^{-1}\bar{\tau}\overline{h(y, x)}\tau \\ &= \nu\overline{h(y, x)}\tau = h(x, y)\tau = s(x, y); \end{aligned}$$

hence s is hermitian. We have therefore proved the following

Theorem 9. *Let $\mathfrak{A}$ and Ψ be as in Theorem 8, and assume that $\Psi^2 = 1$. Then Δ has an involutorial anti-automorphism and there exists a non-degenerate hermitian or skew hermitian scalar product s in $\mathfrak{R}$ such that $A \to A\Psi$ coincides with the transpose mapping relative to h. Conversely, if $\mathfrak{R}$ has a non-degenerate hermitian or skew hermitian scalar product, then the transpose mapping in $\mathfrak{L}(\mathfrak{R} \mid \mathfrak{R})$ is an anti-automorphism.*

We remark also that in view of Ex. 4, page 151, we can suppose that our scalar product is either hermitian or skew symmetric. The latter possibility can hold only if $\Delta = \Phi$ a field.

EXERCISES

1. Let h be a non-degenerate hermitian or skew hermitian scalar product in $\mathfrak{R}$. Prove that, if $\mathfrak{S}$ is a finite dimensional non-isotropic subspace of $\mathfrak{R}$, then $\mathfrak{R} = \mathfrak{S} \oplus \mathfrak{S}'$ where $\mathfrak{S}'$ is the orthogonal complement of $\mathfrak{S}$. ($\mathfrak{S}$ is *non-isotropic* if $\mathfrak{S} \cap \mathfrak{S}' = 0$, see page 151.)

2. (Rickart) A linear transformation U is h *unitary* if $h(xU, yU) = h(x, y)$ holds for all x, y in $\mathfrak{R}$. A unitary transformation I such that $I^2 = 1$ is called an *involution*. Assume the characteristic of Δ is $\neq 2$ and prove that, if I is an involution, then there exists a decomposition $\mathfrak{R} = \mathfrak{R}_+ \oplus \mathfrak{R}_-$ where $\mathfrak{R}_+$ and $\mathfrak{R}_-$ are non-isotropic and orthogonal and $xI = x$ for x in $\mathfrak{R}_+$ and $xI = -x$ for x in $\mathfrak{R}_-$.

13. Schur's lemma. A general density theorem. The range of applications of the results which we obtained for dense rings of linear transformations can be considerably broadened. We are going to show that these results apply to arbitrary irreducible rings of endomorphisms; for we shall prove that the two concepts —dense ring of linear transformations and irreducible ring of endomorphisms—are fully equivalent. We have seen that every dense ring of linear transformations is an irreducible set of endomorphisms. It remains to prove the converse.

Thus suppose that $\mathfrak{A}$ is an irreducible ring of endomorphisms of a commutative group $\mathfrak{M}$. Our first step is to introduce a division ring Δ relative to which $\mathfrak{M}$ is a vector space and $\mathfrak{A}$ is a set of linear transformations. This step can be taken because of the following fundamental lemma.

Schur's lemma. *If $\mathfrak{A}$ is an irreducible ring of endomorphisms in a commutative group $\mathfrak{M}$, then the ring $\mathfrak{B}$ of endomorphisms that commute with every $A \in \mathfrak{A}$ is a division ring.*

Proof. Let B be any non-zero element in $\mathfrak{B}$. The image group $\mathfrak{M}B$ is invariant relative to $\mathfrak{A}$. This is immediate from the commutativity of B with the elements of $\mathfrak{A}$. Since $\mathfrak{M}B \neq 0$, the irreducibility of $\mathfrak{A}$ implies that $\mathfrak{M}B = \mathfrak{M}$. Next let $\mathfrak{N}$ be the kernel of the endomorphism B. Again we can verify that $\mathfrak{N}$ is an $\mathfrak{A}$-subgroup. Also $\mathfrak{N} \neq \mathfrak{M}$ since $B \neq 0$. Hence $\mathfrak{N} = 0$. This means that B is 1–1. Thus we see that B is an automorphism of $\mathfrak{M}$ (onto itself). The inverse mapping B^{-1} is also an endomorphism. Clearly B^{-1} commutes with every A in $\mathfrak{A}$. Hence $B^{-1} \in \mathfrak{B}$. We have therefore proved that every $B \neq 0$ of $\mathfrak{B}$ has an inverse in $\mathfrak{B}$. Thus $\mathfrak{B}$ is a division ring.

Since the ring of endomorphisms $\mathfrak{B}$ is a division ring containing the identity endomorphism, the group $\mathfrak{M}$ together with $\mathfrak{B}$ constitutes a right vector space. Here, of course, the scalar product xB, x in $\mathfrak{M}$, B in $\mathfrak{B}$ is simply the image of x under B. In conformity with our consistent emphasis on left vector spaces we shall now regard $\mathfrak{M}$ as a left vector space. We let Δ denote a division ring anti-isomorphic to $\mathfrak{B}$. Then if $\beta \rightarrow B$ is a definite anti-isomorphism of Δ onto $\mathfrak{B}$, the product $\beta x = xB$ turns $\mathfrak{M}$ into a left vector space over Δ. Evidently the elements of $\mathfrak{A}$ are linear transformations in $\mathfrak{R}$ over Δ (or $\mathfrak{R}$ over $\mathfrak{B}$). Also we know that the only endomorphisms that commute with every $A \varepsilon \mathfrak{A}$ are the scalar multiplications.

From now on we assume that $\mathfrak{A} \neq 0$. Then if $\mathfrak{N}$ denotes the set of z such that $zA = 0$ for all A, $\mathfrak{N} \neq \mathfrak{R}$. But $\mathfrak{N}$ is a subgroup invariant relative to $\mathfrak{A}$. Hence $\mathfrak{N} = 0$. This result means that, if x is any vector $\neq 0$, then there exists an $A \varepsilon \mathfrak{A}$ such that $xA \neq 0$. Moreover, it is clear that the set $x\mathfrak{A}$ of images xA of the fixed vector x is an $\mathfrak{A}$-subgroup. Again by the irreducibility of $\mathfrak{A}$ we conclude that $x\mathfrak{A} = \mathfrak{R}$. Thus if $x \neq 0$ and y is any vector, then there exists an A in $\mathfrak{A}$ such that $xA = y$. We have therefore proved that A is 1-fold transitive in the sense of the following definition.

A set $\mathfrak{A}$ of linear transformations in $\mathfrak{R}$ is *k-fold transitive* if, given any two ordered sets of $l \leq k$ vectors $(x_1, x_2, \cdots, x_l)$, $(y_1, y_2, \cdots, y_l)$ such that the x's are linearly independent, there exists an $A \varepsilon \mathfrak{A}$ such that $x_iA = y_i$ for $i = 1, 2, \cdots l$.

We continue our analysis of irreducible sets of endomorphisms by proving next that $\mathfrak{A}$ is two-fold transitive. Here we shall make use of the fact that the scalar multiplications are the only endomorphisms which commute with all the A in $\mathfrak{A}$. As a preliminary to the proof we note that a ring of linear transformations is k-fold transitive if 1) $\mathfrak{A}$ is 1-fold transitive, and 2) if $(x_1, x_2, \cdots, x_l)$ are $l \leq k$ linearly independent vectors, then for any $i = 1, 2, \cdots, l$, there exists a linear transformation E_i in $\mathfrak{A}$ such that

$$x_jE_i = 0 \quad \text{for} \quad j \neq i \quad \text{and} \quad x_iE_i \neq 0.$$

For if E_i, $i = 1, 2, \cdots, l$, exists, then we can find a B_i in $\mathfrak{A}$ such that $x_iE_iB_i = y_i$. Then $A = \Sigma E_iB_i$ has the required properties $x_iA = y_i$, $i = 1, 2, \cdots, l$. Now take $l = 2$ and suppose on the contrary that there is no E in $\mathfrak{A}$ such that $x_1E = 0$ but $x_2E \neq 0$. Then if B is any element of $\mathfrak{A}$ such that $x_1B = 0$ also $x_2B = 0$. This fact implies that the correspondence $x_1A \rightarrow x_2A$, A varying in $\mathfrak{A}$ is single-valued. For if $x_1A = x_1A'$, A and A' in $\mathfrak{A}$, then $x_1B = 0$ for $B = A - A'$. Hence $0 = x_2B = x_2(A - A')$ and $x_2A = x_2A'$. Now we know that the set of images $x_1\mathfrak{A}$ is the whole space $\mathfrak{R}$. Also it is clear that our mapping is a homomorphism. Hence it is an endomorphism of $\mathfrak{R}$. If C is any linear transformation in $\mathfrak{A}$, then

$$(x_1A)C = x_1AC \to x_2AC = (x_2A)C$$

and this shows that the mapping $x_1A \to x_2A$ commutes with C. It follows that this mapping is a scalar multiplication, that is, there exists a $\beta \varepsilon \Delta$ such that $x_2A = \beta(x_1A)$ holds for all A. Thus $(x_2 - \beta x_1)A = 0$ for all A. Hence $x_2 = \beta_1 x_1$ and this contradicts the linear independence of x_1 and x_2.

Our final step is to show that $\mathfrak{A}$ is dense in $\mathfrak{L}$ or, what is the same thing, $\mathfrak{A}$ is k-fold transitive for all k. We shall, in fact, prove somewhat more, namely, we shall show that *any two-fold transitive ring of linear transformations is dense*. Suppose $\mathfrak{A}$ has this property and assume that we know already that $\mathfrak{A}$ is k-fold transitive for a particular k. Then the result will follow by induction if we can show that, if $x_1, x_2, \cdots, x_{k+1}$ are linearly independent vectors, then there exists a transformation F in $\mathfrak{A}$ such that $x_iF = 0$ for $i \leq k$ but $x_{k+1}F \neq 0$. By the induction assumption we know that there exist E_j in $\mathfrak{A}$ such that

$$x_iE_j = \delta_{ij}x_i, \quad i, j = 1, 2, \cdots, k.$$

We set $E = \sum_1^k E_i$ and we consider first the case in which $x_{k+1}E \neq x_{k+1}$. Then $x_{k+1}E - x_{k+1} \neq 0$ and there exists an A in $\mathfrak{A}$ such that $(x_{k+1}E - x_{k+1})A \neq 0$. Then if $F = EA - A$

$$x_{k+1}F = x_{k+1}(EA - A) = (x_{k+1}E - x_{k+1})A \neq 0.$$

On the other hand, $x_iE = x_i$ for $i \leq k$; hence $x_iEA = x_iA$ and $x_iF = 0$. Suppose next that $x_{k+1}E = x_{k+1}$. Then we assert that there is an $i \leq k$ such that $x_{k+1}E_i$, x_i are linearly independent; for, otherwise, $x_{k+1}E_i = \beta_i x_i$ and

$$x_{k+1} = x_{k+1}E = \Sigma x_{k+1}E_i = \Sigma \beta_i x_i$$

contrary to the linear independence of $x_1, x_2, \cdots, x_{k+1}$. Now let $x_{k+1}E_i$ and x_i be linearly independent for a particular i. Then since $\mathfrak{A}$ is two-fold transitive, there exists a $B \varepsilon \mathfrak{A}$ such that $x_{k+1}E_iB \neq 0$ but $x_iB = 0$. If we set $F = E_iB$, we find that $x_jF = x_jE_iB = 0$ for $j \neq i$ and $\leq k$ and $x_iF = 0$ but $x_{k+1}F \neq 0$. This proves our assertion and completes the proof of the following

General density theorem. *Let $\mathfrak{A}$ be any irreducible ring of endomorphisms $\neq 0$ in a commutative group $\mathfrak{R}$ and regard $\mathfrak{R}$ as a left vector space over a division ring Δ anti-isomorphic to the division ring of endomorphisms which commute with every A in $\mathfrak{A}$. Then $\mathfrak{A}$ is a dense ring of linear transformations in $\mathfrak{R}$ over Δ.*

14. Irreducible algebras of linear transformations. The theorems of the preceding section can also be applied to irreducible algebras of linear transformations, and in this form they give some results which are fundamental in the theory of group representations. We proceed to derive these results. Thus we begin with a vector space $\mathfrak{R}$ over a field Φ and with an algebra $\mathfrak{A}$ of linear transformations in $\mathfrak{R}$ over Φ. The assumption that $\mathfrak{A}$ is an algebra means that $\mathfrak{A}$ is closed under the compositions of addition and multiplication and under multiplication by elements of Φ (or Φ_l).

Assume now that $\mathfrak{A}$ is irreducible as a set of linear transformations (cf. § 1, Chapter IV). Thus we are assuming that the only subspaces of $\mathfrak{R}$ which are invariant relative to $\mathfrak{A}$ are $\mathfrak{R}$ and 0 or, equivalently, that the set $(\mathfrak{A}, \Phi_l)$ is an irreducible set of endomorphisms. We shall now show that, if $\mathfrak{A} \neq 0$, then $\mathfrak{A}$ itself is irreducible as a set of endomorphisms. To prove this let x be any vector $\neq 0$ in $\mathfrak{R}$. Consider the set $x\mathfrak{A}$ of vectors xA. Since $\mathfrak{A}$ is an algebra, $x\mathfrak{A}$ is a subspace. Also it is clear that $x\mathfrak{A}$ is $\mathfrak{A}$-invariant. Hence, either $x\mathfrak{A} = \mathfrak{R}$ or $x\mathfrak{A} = 0$. If the second alternative holds, then the set $\mathfrak{N}$ of vectors z such that $z\mathfrak{A} = 0$ contains non-zero vectors. Clearly $\mathfrak{N}$ is a subspace and $\mathfrak{N}$ is $\mathfrak{A}$ invariant also. Hence $\mathfrak{N} = \mathfrak{R}$ and $\mathfrak{A} = 0$ contrary to assumption. We therefore have $x\mathfrak{A} = \mathfrak{R}$ for any non-zero x and this implies directly that $\mathfrak{A}$ is an irreducible set of endomorphisms.

We can now apply the results of the preceding section. For this purpose we consider the ring $\mathfrak{B}$ of endomorphisms which commute with every $A \varepsilon \mathfrak{A}$. Evidently $\mathfrak{B} \supseteq \Phi_l$. We observe next that the elements of $\mathfrak{B}$ are linear transformations. Thus let $\alpha_l \varepsilon \Phi_l$, $B \varepsilon \mathfrak{B}$. Then for any $A \varepsilon \mathfrak{A}$ we have

$$\begin{aligned} 0 &= B(\alpha_l A) - (\alpha_l A)B = (B\alpha_l)A - \alpha_l(AB) \\ &= (B\alpha_l)A - \alpha_l(BA) = (B\alpha_l)A - (\alpha_l B)A \\ &= (B\alpha_l - \alpha_l B)A. \end{aligned}$$

Now, if $B\alpha_l - \alpha_l B \neq 0$, then we can find a vector x such that $y = x(B\alpha_l - \alpha_l B) \neq 0$. Then $yA = 0$ for all A, and this contradicts the irreducibility of $\mathfrak{A}$. Thus $B\alpha_l = \alpha_l B$ for every α_l, and B is a linear transformation. We have therefore proved that $\mathfrak{B}$ is also the totality of linear transformations which commute with the elements of $\mathfrak{A}$. Clearly $\mathfrak{B}$ is an algebra of linear transformations. By Schur's lemma $\mathfrak{B}$ is a division algebra.

We now follow the procedure of the preceding section and introduce the division algebra Δ anti-isomorphic to $\mathfrak{B}$. Since $\mathfrak{B}$ contains Φ_l we can suppose that $\Delta \supseteq \Phi$. Also it is easy to see that we can regard $\mathfrak{R}$ as a left vector space over Δ in such a way that the scalar multiplication by the elements of the subset Φ is the original scalar multiplication. The main density theorem now states that $\mathfrak{A}$ is a dense set of linear transformations of $\mathfrak{R}$ over Δ.

We shall now specialize our results to obtain some classical theorems on algebras of linear transformations. We assume that Φ is algebraically closed and that $\mathfrak{R}$ is finite dimensional over Φ. Let $\mathfrak{A}$ be an irreducible algebra of linear transformations in $\mathfrak{R}$ over Φ. Let B be a linear transformation which commutes with every element of $\mathfrak{A}$. Then B is an endomorphism that commutes with every element of the irreducible set $(\mathfrak{A}, \Phi_l)$. Hence by Schur's lemma either B is 0 or B is non-singular. Now let ρ be a root of the characteristic polynomial of B. Then $C = B - \rho_l$ commutes with every $A \varepsilon \mathfrak{A}$. But det $C = 0$ so that C is singular. Hence $C = 0$ and $B = \rho_l$. We have therefore proved that the only linear transformations which commute with every $A \varepsilon \mathfrak{A}$ are the scalar multiplications. The same result can also be established for irreducible sets of linear transformations. Thus if Ω is such a set, then the enveloping algebra $\mathfrak{A}$ (cf. § 6 of Chapter IV) of Ω is an irreducible algebra of linear transformations. Moreover, if B is a linear transformation which commutes with every $A \varepsilon \Omega$, then B commutes with every $A \varepsilon \mathfrak{A}$. Hence we can state the following theorem which is one of the most useful special cases of Schur's lemma.

Theorem 10. *Let Ω be an irreducible set of linear transformations in a finite dimensional vector space $\mathfrak{R}$ over an algebraically closed field. Then the only linear transformations which commute with every $A \varepsilon \Omega$ are the scalar multiplications.*

An immediate consequence of this result and of the density theorem is

Burnside's theorem. *If $\mathfrak{A}$ is an irreducible algebra $\neq 0$ of linear transformations in a finite dimensional vector space over an algebraically closed field, then $\mathfrak{A} = \mathfrak{L}$ the complete algebra of linear transformations.*

Proof. By Theorem 10 the division algebra of linear transformations commuting with the $A \varepsilon \mathfrak{A}$ is Φ_l. Hence by the density theorem, $\mathfrak{A}$ is a dense set of linear transformations of $\mathfrak{R}$ over Φ. Since $\mathfrak{R}$ has a finite basis over Φ, this implies that $\mathfrak{A} = \mathfrak{L}$.

INDEX

Graduate Texts in Mathematics

continued from page ii

93 DUBROVIN/FOMENKO/NOVIKOV. Modern Geometry — Methods and Applications Vol. I.
94 WARNER. Foundations of Differentiable Manifolds and Lie Groups.
95 SHIRYAYEV. Probability, Statistics, and Random Processes.
96 ZEIDLER. Nonlinear Functional Analysis and Its Applications I: Fixed Points Theorem.